Understanding Complex Systems

Future scientific and technological developments in many fields will necessarily depend upon coming to grips with complex systems. Such systems are complex in both their composition – typically many different kinds of components interacting simultaneously and nonlinearly with each other and their environments on multiple levels – and in the rich diversity of behavior of which they are capable. The Springer Series in Understanding Complex Systems series (UCS) promotes new strategies and paradigms for understanding and realizing applications of complex systems research in a wide variety of fields and endeavors. UCS is explicitly transdisciplinary. It has three main goals: First, to elaborate the concepts, methods and tools of complex systems at all levels of description and in all scientific fields, especially newly emerging areas within the life, social, behavioral, economic, neuro- and cognitive sciences (and derivatives thereof); second, to encourage novel applications of these ideas in various fields of engineering and computation such as robotics, nano-technology and informatics; third, to provide a single forum within which commonalities and differences in the workings of complex systems may be discerned, hence leading to deeper insight and understanding. UCS will publish monographs, lecture notes and selected edited contributions aimed at communicating new findings to a large multidisciplinary audience.

More information about this series at http://www.springer.com/series/5394

Hassan Qudrat-Ullah
Editor

Innovative Solutions for Sustainable Supply Chains

Springer

Editor
Hassan Qudrat-Ullah
KFUPM
College of Industrial Management
Dhahran, Saudi Arabia

ISSN 1860-0832 ISSN 1860-0840 (electronic)
Understanding Complex Systems
ISBN 978-3-030-06833-2 ISBN 978-3-319-94322-0 (eBook)
https://doi.org/10.1007/978-3-319-94322-0

This Springer imprint is published by the registered company Springer Nature Switzerland AG.
The registered company address is: Gewerbestrasse 11, 6330 Cham, Switzerland.

To my grandson
Abdurahman Hassan
-Hassan Qudrat-Ullah-

Preface

The purpose of this edited volume is to provide the reader with state-of-the-art perspectives and model-based applications to better understand the dynamics of sustainable supply chains of the twenty-first century. This collection of innovative contributions by the scholars of the field presents a paradigmatic shift in thinking about sustainable supply chain complexity for the challenges of the twenty-first century. The book is less about abstract possibilities and esoteric techniques and more about understanding the dynamics of complex systems as they pertain to sustainable supply chains. Through various perspectives such as agent-based modeling, system dynamics modeling, and different methodologies, we believe we have improved understanding in a wide range of supply chain settings at critical encounters with complexity. We are hopeful that this collection will provide a synergy that can leverage a new level of understanding in order to theorize and empirically explore the emergent and process nature of complexity across levels of analysis that foster insights into the dynamics of supply chain systems. Furthermore, we hope to encourage further theoretical and empirical research that perhaps may be interwoven by the works presented herein and in advancing new methodological perspectives to include multilevel and cross-level analysis to better understand the dynamics of sustainability-focused supply chains.

In the quest for innovative solutions for the management of sustainable supply chains, many challenges lie ahead. Specifically, as we move toward displacing traditional thinking, founded in "bottom-line"-focused decisions, to one where plural logics of "triple bottom" coexist under conditions of uncertainty and ambiguity, the need for systematic and integrated solutions for sustainable supply chains becomes pronounced. Our aim here has been to focus our attention on the whole (supply chain system) and its interrelationships. It is our hope that this book will stimulate a new way of thinking as a proclamation of a new era of resource constraints and renewed focus on "integrative" solutions.

Content Overview

The integrating theme of this book is principle-based perspectives and modeling-based solutions to deal with dynamics of sustainable supply chains. The book contains four parts. Part I presents the introduction and preview of *Innovative Solutions for Sustainable Supply Chains*. Part II of this book, "The Unique Perspectives for the Sustainable Supply Chains," examines the theoretical approaches and perspectives that uniquely support decision-making in sustainable supply chains including "Determinants of ITF R&D Technology Commercialization in Logistics and Supply Chain Industries: R&D Technologist Perspective"; "A Meta-analysis of Sustainable Supplier Selection Approaches"; "Green Supply Chain Management, Green Innovations, and Green Practices"; "The Road Towards a Circular Economy: The Role of Modular Product Designs in Supply Chains"; "Policy Design for Sustainable Supply Chain Through Trainings"; and "The Application of System Dynamics for a Sustainable Procurement Operation." Part III of this book, "The Innovative Solutions for Industry-Specific Supply Chains," provides empirical evidence for the applications of various modeling and simulation approaches including system dynamics, agent-based modeling, and econometrics. These innovative modeling-based solutions include "Alternative Energy Policy for Mitigating the Asynchrony of the Wind-Power Industry's Supply Chain in Brazil"; "Resilience as Basis for Sustainability: Shortages in Production Supply Chains for Essential Consumer Goods"; "Sustainable Food Supply Chain Management: An Integrated Framework and Practical Perspectives"; "Designing a Sustainable Tourism Supply Chain: A Case Study from Asia"; and "Risk Assessment of the Timber Supply Chain in Southern Ontario Using Agent-Based Simulation."

Finally, Part IV of this book provides an overview of key findings and future research directions on thematic areas: (i) "The Unique Perspectives for the Sustainable Supply Chain" and (ii) "The Innovative Solutions for Industry-Specific Supply Chains."

Dhahran, Saudi Arabia Hassan Qudrat-Ullah

Acknowledgments

I would like to thank everyone, mentioned herein or not, for their continued support in helping to bring this book to completion. Most importantly, I am sincerely grateful to the contributing authors of this book. Their support of this endeavor enabled us to platform the collective lessons presented in the book. I also acknowledge the work and knowledge of the members of our review panel. My appreciation also goes to all the people at Springer, USA, especially Christopher and HoYing with whom I corresponded for their advice and facilitation in the production of this book. I would like to thank Taub Jeffrey and the production team from Springer, for their help in the final production of this book. I am grateful to my family, Tahira Qudrat, Anam Qudrat, Ali H. Qudrat, Umer K. Qudrat, and Umael H. Qudrat for their encouragement and support throughout this endeavor. Special thanks and appreciation goes to my mother, Fazeelat Begum, and my mother-in-law, Saira Bano, whose prayers and wishes provide unique strengths to me for such taxing tasks.

Finally, the editor would like to acknowledge the financial support provided by the Deanship of Scientific Research (DSR) at King Fahd University of Petroleum and Minerals (KFUPM) for funding this work through project No. BW171002.

Hassan Qudrat-Ullah
Dammam, Saudi Arabia
April, 2018

Contents

Contributors

Ali Asgary York University, Toronto, Canada

Tashfeen M. Azhar School of Business and Economics, University of Management and Technology, Lahore, Pakistan

Sherif Barrad Massachusetts Institute of Technology, Cambridge, MA, USA

Kong Bieng Chuah Department of Systems Engineering and Engineering Management, City University of Hong Kong, Kowloon, Hong Kong

Umar Burki School of Business, University of South-Eastern Norway, Kongsberg, Norway

Siu Cheung Ho Department of Systems Engineering and Engineering Management, City University of Hong Kong, Kowloon Tong, Hong Kong

Federico Cosenz Department of European Studies and International Integration (DEMS), Università Degli Studi di Palermo, Palermo, Italy

Lize Duminy Bern University of Applied Sciences, School of Engineering, Biel, Switzerland

Isaac Dyner Faculty of Natural Science and Engineering, Universidad Jorge Tadeo Lozano, Bogotá, Colombia

Stéphane Gagnon Université du Québec en Outaouais, Gatineau, QC, Canada

Stefan N. Groesser Bern University of Applied Sciences, School of Engineering, Biel, Switzerland

A. D. Nuwan Gunarathne University of Sri Jayewardenepura, Nugegoda, Sri Lanka

Griffith University, Gold Coast, Australia

Milton M. Herrera Universidad Jorge Tadeo Lozano, Bogotá, Colombia

Department of European Studies and International Integration (DEMS), Università Degli Studi di Palermo, Palermo, Italy

Marcin Lewandowski Ecostrat Inc, Toronto, Canada

Deemantha G. Navaratne University of Sri Jayewardenepura, Nugegoda, Sri Lanka

Thomas Nowak WU (Vienna University of Economics and Business), Institute for Production Management, Vienna, Austria

Amanda E. Pakianathan University of Sri Jayewardenepura, Nugegoda, Sri Lanka

Yasasi T. Perera N. University of Sri Jayewardenepura, Nugegoda, Sri Lanka

Hassan Qudrat-Ullah College of Industrial Management, King Fahd University of Petroleum Engineering and Minerals, Dhahra, Saudi Arabia

Vivek Shrestha College of Management, Mahidol University, Bangkok, Thailand

Fuminori Toyasaki York University, School of Administrative Studies, Toronto, ON, Canada

Felix Tuczek Institute for Transport and Logistics Management, WU (Vienna University of Economics and Business), Vienna, Austria

Raul Valverde Concordia University, Montreal, QC, Canada

Tina Wakolbinger Institute for Transport and Logistics Management, WU (Vienna University of Economics and Business), Vienna, Austria

Winai Wongsurawat College of Management, Mahidol University, Bangkok, Thailand

Ijaz Yusuf School of Business and Economics, University of Management and Technology, Lahore, Pakistan

Part I
Introduction: An Overview of Sustainable Supply Chains

Chapter 1
Innovative Solutions for Sustainable Supply Chains: An Introduction

Hassan Qudrat-Ullah

1.1 Introduction

Managing twenty-first-century supply chains in a sustainable manner, a multibillion dollar industry, has become an ever-increasing challenge. Given the increasing significance of socioeconomic and environmental aspects, the management of sustainable supply chains has become a complex, dynamic task. Multiple and often conflicting objectives of stakeholders including suppliers, manufacturers and service providers, and retailers add to the complex nature of decisions that modern-day managers of supply chains face. With unprecedented technological developments and innovations at hand, sustainability can be maximized for all the activities of a supply chain including service concept, product design, material sourcing and procurement, manufacturing processes, delivery of the final product, and end-of-life management of the product. Here are two topical issues pertaining to the development and growth of sustainable supply chains across the domains (as shown in Fig. 1.1):

(i) How do various perspectives (e.g., green supply chains, design modularity, R&D, investments, and education and training) provide actionable insights and decisions pertaining to the design, development, and growth of sustainable supply chains?
(ii) How do various modeling approaches (e.g., system dynamics, econometric, and agent-based modeling) provide innovative solutions to the issues pertaining to the design, development, and growth of sustainable supply chains?

H. Qudrat-Ullah (✉)
College of Industrial Management, King Fahd University of Petroleum Engineering and Minerals, Dhahra, Saudi Arabia
e-mail: hassan.qudratullah@kfupm.edu.sa

H. Qudrat-Ullah (ed.), *Innovative Solutions for Sustainable Supply Chains*, Understanding Complex Systems, https://doi.org/10.1007/978-3-319-94322-0_1

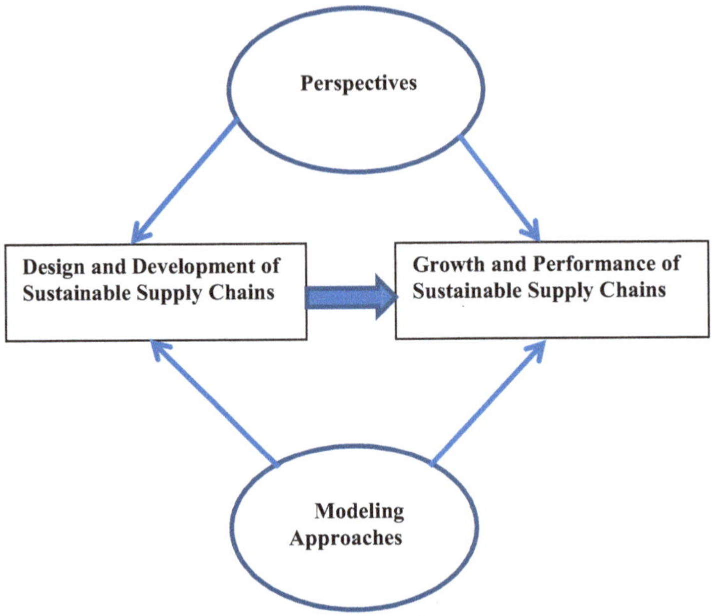

Fig. 1.1 Perspectives and modeling approaches for sustainable supply chains

Consequently, the sustainable supply chain issues and problems require a systematic and integrated approach. Modeling and simulation in general and system dynamics and agent-based modeling in particular have the capabilities to deal with the complexity of sustainable supply chain-related problems. Therefore, the primary aim of this book is to present the latest decision-making perspectives, tools, techniques, and insightful and innovative solutions that decision-makers can utilize to overcome the challenges that their sustainable supply chains face.

In an attempt to provide some unique perspectives and innovative solutions for sustainable supply chains in the twenty-first century, we issued the call for contributions in this volume. Specifically, we sought help from the system dynamics and agent-based modeling community. Consequently, several different examples of perspectives and modeling approaches, with a common unifying goal of "improving the performance of sustainable supply chains" are provided in this volume.

1.1.1 Methodology

In our call for contributions on "The Innovative Solutions for Sustainable Supply Chains," we went through various email lists of professional bodies. We also posted the call for chapters on message boards of a few international conferences on the related topics. Personal invitations were sent to target authors as well. We received 19 "two-to-three-page" abstracts as the expression of interest. Based on the initial screening by our review panel, the authors of 12 chapters were invited to submit the complete chapter. All 12 chapters received from the contributors went through a double-blind review process. The reports from the independent reviewers were sent to the authors to address the issues and incorporate the suggestions made by the reviewers. Only 11 chapters made it to final stage of acceptance. The final versions of chapters have been edited and included in this volume.

1.1.2 Research Categories

We applied a thematic approach to classify our contributions for this book. The chapters thus complied are classified into four categories following the structure of the book. The first category, the current one, presents the introduction and preview of "Innovative Solutions for Sustainable Supply Chains." The second category examines the unique perspectives for enhancing the performance of the sustainable supply chains including *Determinants of ITF R&D Technology Commercialization in Logistics and Supply Chain Industries: R&D Technologist Perspective*; *A Meta-analysis of Sustainable Supplier Selection Approaches*; *Green Supply Chain Management, Green Innovations, and Green Practices*; *The Road Towards a Circular Economy: The Role of Modular Product Designs in Supply Chains*; *Policy Design for Sustainable Supply Chain Through Trainings*; and *The Application of System Dynamics for a Sustainable Procurement Operation.*

Next, the third category showcases the state-of-the-art innovative solutions for industry-specific supply chains including *Alternative Energy Policy for Mitigating the Asynchrony of the Wind-Power Industry's Supply Chain in Brazil*, *Resilience as Basis for Sustainability: Shortages in Production Supply Chainsfor Essential Consumer Goods*, *Sustainable Food Supply Chain Management: An Integrated Framework and Practical Perspectives*, *Designing a Sustainable Tourism Supply Chain: A Case Study from Asia*, and *Risk Assessment of the Timber Supply Chain in Southern Ontario Using Agent-Based Simulation.* Finally, the last category discusses *conclusions and future research directions* in sustainable supply chains.

1.2 The Unique Perspectives for the Sustainable Supply Chains

Here we present six unique perspectives for enhancing the performance of twenty-first-century supply chains. These perspectives shed light on important actionable strategies and decisions pertaining to sustainable supply chains. For instance, why "supplier selection" is so critical, why be "green," what "training" of personnel means, does a "modular design" matter, and what is the role of "procurement operations" and "R&D" in improving the overall performance of a sustainable supply chain.

1.2.1 R&D Technologists' Perspective

When it comes to the unique perspectives for the sustainable supply chains, R&D technologists' perspective presents some interesting insights. In Chap. 2, by S.C.Ho and K.B.Chuah, the risks and challenges faced by companies in Hong Kong's logistics and supply chain industries are discussed (Ho & Chuah, 2018). Authors define successful commercialization and adoption of research and development (R&D) project deliverables as the whole R&D project completed, commercialized, and adopted in the industry. In their chapter, they examine the R&D technologists' perspective, the determinant factors of Innovation and Technology Fund (ITF) R&D technology commercialization and adoption in Hong Kong's logistics and supply chain industries. R&D technologists face various challenges of developing technology that matches industry needs. They investigate how R&D technologists think their efforts can align with industry needs during the development process. They consider this study as the first attempt of a multi-perspective study to bridge the gaps from idea generation, R&D and subsequent technology implementation, and adoption by industry. In this study, qualitative and quantitative data collected from R&D technologists has been carried out using a proposed market-driven project management model, market-science-technology-application-market (MSTAM); the results will shed light on latter study from other perspectives.

1.2.2 A Meta-analysis-Based Perspective

A meta-analysis-based perspective provides some critical pathways for future researchers in the domain of sustainable supply chains. In this context, in Chap. 3, by Felix Tuczek and Tina Wakolbinger, the authors post that supplier selection is a key factor in determining the sustainability performance of supply chains. In this chapter, they conduct a meta-analysis of literature on supplier selection approaches considering all three dimensions of sustainability. To close the gap in current research, they enhance it with a review of recent papers. They present a classification of methods and criteria applied to sustainable supplier selection. In terms of methods, they see the use of combined methods as a trend. They argue that combined methods allow for the integration of subjective judgment of decision-makers with quantifiable parameters. Specifically, environmental and social criteria demand methods to deal with qualitative as well as quantitative factors are suitable. Analyzing mathematical approaches applied to specific industries, they identified that literature is lagging in considering environmental issues in supplier selection approaches in the automotive industry, an example of a high-polluting industry. They also found that social criteria are hardly integrated into selection approaches in the apparel industry, an example in labor-intensive industries. Thus, they suggest that future research should focus on industry-related criteria tailored for the respective purposes. Again, future researchers on sustainable supply chains will certainly cherish this chapter.

1.2.3 Green Perspective on Supply Chain Management

Do green innovations and green practices have an impact on sustainable supply chains? Umar Burki, in Chap. 4, attempts to address this topical question (Burki, 2018). In this chapter, his objective is to provide a broad understanding of the role and impact of green innovations and green practices in green supply chains. The author advances various assertions including (i) green supply chains are an extension of ordinary supply chains, (ii) green innovations and green practices function as key antecedents in transforming an ordinary supply chain into a green supply chain, and (iii) effective green supply chains apply a coordinated triple bottom-line (TBL) approach to simultaneously realize environmental, economic, and social goals in supply chain operations. He posits that external and internal factors play an instrumental role in the adoption of green innovations and green practices in supply chain operations. He identifies some of these factors as changing consumer demand, government regulations and institutional pressures, technological innovations, and stakeholders' attitude toward long-term environment sustainability. Based on a comprehensive review of green supply chain management literature, he finds that (i) when businesses integrate green innovations and green practices in their supply chains, they achieve encouraging triple bottom-line outcomes and (ii) transparency and truthful compliance to green innovations and practices in supply chain operations are some of the biggest challenges in green supply chains. This research provides some actionable insights for the decision-makers in sustainable supply chains.

1.2.4 Product Design Modularity Perspective on Sustainable Supply Chains

How does product design modularity impact supply chains is an interesting question that the authors of Chap. 5, Thomas Nowak, Fuminori Toyasaki, and Tina Wakolbinger, investigate (Nowak et al., 2018). According to these authors, product modularity has become a well-established concept for new product design that leads to accelerated product development and increases a company's abilities for mass customization. Recently, product modularity also starts receiving attention as a promising mechanism for circular economy because it is expected to mitigate negative product-related environmental effects. While effects of product modularity are well investigated in forward supply chains, little attention has been paid to implications of modular designs for reverse logistics operations. This study explores the links between anticipatory and reactive supply chain strategies, consumer preferences, as well as a company's optimal product design decision by considering reverse logistics operations. These relationships are explored by comparing two optimization problems: one for a company following a reactive and one for a company following an anticipatory strategy. Based on the numerical analysis of the mathematical models, the paper provides recommendations for policy-makers trying to encourage manufacturers to pursue the path toward a circular economy. The recommendations highlight the important role that consumer behavior plays and also show possible trade-offs between different sustainability goals. This chapter provides some general recommendations for policy-makers to encourage suppliers and producers to adapt sustainability practices.

1.2.5 Training Perspective on Sustainable Supply Chains

Both soft and hard skills training play a critical role in productivity and prosperity of a business firm albeit a sustainable supply chain firm. In Chap. 6, Ijaz Yosuf and Tafsheen M. Azhar provide a systematic analysis of the role of training in enhancing the performance of supply chains (Yosuf & Azhar, 2018). They begin their chapter with two assertions: (i) trainings imparted to the company employees are prerequisites for organizational transformation, and (ii) impact of the training appears in the form of changed behavior and attitude of the employees that contribute significantly for the enhancement of the supply chain score of the focal firm. In this chapter the authors discuss:

> The types of trainings generally categorized in soft skills and hard skills. Training need analysis is best proven method utilized to identify the competency gaps of current employees. Soft skills trainings and hard skills trainings are designed for capacity building in order to reduce the gap and raising the employee productivity towards the sustainable supply chain management. Soft skills trainings not only change the attitude and behavior of the employee but as well enhance the motivational level of the employees that ultimately contribute in terms of better product quality and waste reduction. Hard skills trainings improve the technical capabilities of the workers. Reduced waste percentage, improved process settings, declining cost of quality, mistake proofing in product design and enhanced productivity are the contributing factors for sustained supply chain performance. Training need analysis is most appropriate method in the case company for assessing the competency gap. Training budget is allocated accordingly to reduce the competency gap.

According to the authors, the major objective of this chapter is to design the plausible policies for enhanced supply chain performance conducting experimentation with the simulated system dynamics model; what type of training is required more and how significantly this training impacts the supply chain score for enhanced supply chain performance are the research questions being explored; experimentation with the model unveils the underlying symptoms and keeps on playing with the model to make the system better behaved; and trainings usually considered as expenditure can be a valuable asset if its effectiveness improves the supply chain performance. They contribute with a system dynamics simulated model which the users can apply to design the policy streams for improved supply chain performance.

1.2.6 System Dynamics Perspective on a Procurement Function Within the Supply Chain

In the final chapter of this section, Chap. 7, the authors, Sherif Barrad, Raul Valverde, and Stéphane Gagnon, present a system dynamics perspective on an important yet often neglected functional area of any sustainable supply chain, the procurement function (Barrad et al., 2018). First they introduce and describe the system dynamics approach as (i) system dynamics is an approach to modeling complex systems using feedback loops to explain relationships between variables and to reflect their nonlinear interdependencies through time, along with their underlying driving forces (Sterman, 2000), and (ii) systems are graphically represented by a set of active nodes, with qualitative and quantitative attributes, along with passive nodes, modeled as flows and stocks acting as buffers between active nodes. In the context of this chapter, they use the system dynamics approach to model the operations of a procurement function within the supply chain. In this chapter, they propose a system dynamics interpretation of procurement drivers and a link to the operational and strategic levels of decision-making.

1.3 The Innovative Solutions for Industry-Specific Supply Chains

After the reader of this book has learned about six unique perspectives unified with a common goal, improving the performance of sustainable supply chain, here in this section, we present five state-of-the-art solutions, rich with actionable insights, for industry-specific supply chains including the (i) wind-power industry, (ii) essential consumer goods industry, (ii) food industry, (iv) tourism industry, and (v) timber industry. Insights and lessons learned from these industry-specific supply chains, however, are applicable equally well for other industries and jurisdictions.

1.3.1 Wind-Power Industry's Supply Chain in Brazil

This chapter deals with the asynchrony of the wind-power industry supply chain in Brazil. The authors of this Chap. 8, Milton M. Herrera, Isaac Dyner, and Federico Cosenz, present a systematic and model-based analysis of the climate-induced dynamics of wind power in Brazil (Herrera et al., 2018). Here is how they describe the issue:

> High dependency on hydroelectricity has revealed drawbacks in the security of power supplies as a consequence of the climate variability in South America. Under these conditions, Brazil is starting to consider alternative renewable sources for energy production, seeking to avoid periods of scarcity, while also promoting clean technologies in its electricity market. Since 2004, wind power has shown a significant rise in terms of installed capacity in this

> country. Despite increases in wind power units, Brazil suffers from delays in setting up its transmission infrastructure, which affects the performance of the wind-power supply chain.

In this chapter, they present a simulation model that helps assess the long-term effects of an alternative sustainable energy policy, which may contribute to overcoming the asynchrony between renewables generation policy and the insufficiency of transmission infrastructure. Using lessons learned from simulations, they conclude that the transmission industry in Brazil requires appropriate investment incentives for just-in-time synchrony with the expansion of the wind industry.

1.3.2 Essential Consumer Goods Industry Supply Chain in Europe

The author, Stefan Grossor, of this Chap. 9, draws on multiple perspectives to develop key indicators of a sustainable supply chain in this industry (Grossor, 2018). He portrays the issue as follows:

> Some consumer products, termed essential consumer goods, are crucial to sustaining health or even life. A shortage in supply of essential consumer goods can have tangible negative impacts on society. This study applies this topic to the case of inexpensive, generic, injectable oncological medication shortages in Europe. Cancer patient outcomes including survival rates, as well as treatment costs are significantly influenced by oncological medicines shortages. Even though the problem is well documented and universally acknowledged, a lack of data has deterred any quantitative solution-oriented studies. However, a structural model can provide reliable insight in cases where data is unavailable or unreliable through relying on structural validation.

In this study, he then proposes the first causal model showing the underlying structure of the European inexpensive, generic, injectable oncological medications supply chain. He identifies the most common causes of supply shortages and develops a quantitative supply chain model with the ability to simulate causes of identified shortages and proposes the key performance indicators to evaluate the sustainability of the supply chains in question from several perspectives. He concludes with two future research issues: "(i) our study calls for a quantitative comparison and robust sensitivity analysis of all primary and secondary causes of medicines shortages using the proposed model, and (ii) both existing and new policy recommendations regarding oncological medicines shortages in Europe should be studied quantitatively."

1.3.3 Designing a Sustainable Tourism Supply Chain

The tourism industry encompasses a web of supply chains when multiple actors play their role to the success of various new ventures especially in Asia. Relying on a real case study, Wongsurawat et al., in Chap. 10, present an in-depth analysis of a start-up firm in the context of sustainability (Wongsurawat & Shrestha, 2018. In their own words, this is the summary of their work in this chapter:

> This chapter presents an in-depth look into a startup with the innovative idea of providing authentic travel experience by matching tourists with locals who are willing to share their native skills or traditional knowledge. With numerous offices offering services in cities across South and Southeast Asia, the online social enterprise has proven to be financially sustainable. Using the experience of this social venture as a case study, this chapter will analyze the challenges involved in redesigning a new service supply chain that creates novel value for customers, and at the same time shifts a greater share of benefits generated from tourism to previously marginalized players.

1.3.4 Sustainable Management of Global Food Supply Chain

An integrated framework for sustainable food supply chain management is described in Chap. 11, "Sustainable Food Supply Chain Management: An Integrated Framework and Practical Perspectives" by Gunarathne, A D Nuwan, Navaratne, Deemantha G, Pakianathan, Amanda E, and Perera N Yasasi T. They posit and provide some an interesting conclusion by saying, "With the uncontrollable growth of world population, in some parts of the world millions of people are at the risk of hunger while in certain countries there is a significant waste of food. This imbalance in the global food supply chain increasingly demands sustainable management to "end hunger", as envisaged in the sustainable development goals. Considering the importance of food supply chains, the purpose of this chapter is to provide an overview of sustainable food supply chain management and to propose an integrated framework. With this objective, the chapter provides an overview of sustainable development and the importance of food supply chain management to address the global food crisis (Gunarathne et al., 2018). It also provides an overview of the theoretical background of sustainable food supply chains. The chapter also presents an integrated framework for sustainable food supply chain management by extending prior work on the subject. In every section, carefully selected mini cases are provided to bring in more insights into the points/arguments mentioned therein."

1.3.5 Risk Assessment of the Timber Supply Chain in Southern Ontario

In the final chapter of this section, Chap. 12, "Risk Assessment of the Timber Supply Chain in Southern Ontario Using Agent-Based Simulation," Marcin Lewandowsk and Ali Asgary demonstrate the utility of agent-based modeling approach in analyzing the supply-demand dynamics in the timber industry (Lewandowsk & Asgary, 2018). According to the authors:

> The bioenergy sector has been experiencing significant growth in the last two decades. That said, the industry faces many challenges, mainly focused around the understanding of feedstock supply risk. Developers and investors cannot properly price risk without raw material supply chain risk understanding, making the development of the bioenergy industry slower than it would otherwise be. Currently biofuel, or wood pellet, production in Ontario requires wood chips supplied by existing sawmills. The supply of wood chips in turn depends on the supply of timber. A model was developed here simulating the timber supply chain in Southern Ontario. The objective of the simulation was to show the applicability of computer simulation methods in determining the most resilient areas from a perspective of a developer looking to build a new biofuel plant. The simulation presented here, developed in Any Logic 7.3.5, is considered a base simulation. That is, it can be improved upon to simulate different disturbances, or add/change experiment assumptions. The simulation is therefore a first version of a useful tool that has a potential to improve the understanding of risk among biofuel developers and investors.

Although this chapter applies an agent-based modeling to better understand the dynamics of supply chains in the timber industry, the utility of such models is limited but can be applied to analyze the complex demand-supply dynamics present in most of the supply chains of the twenty-first century.

1.4 Concluding Remarks

At the outset of this project, we set the objective of this edited volume as "to present the latest decision making tools, techniques, and insightful and innovative solutions that decision makers can utilize to overcome the challenges that their sustainable supply chains face." Both eminent and distinguished as well emerging scholars of the supply chain domain responded to our call for contributions with their unique solution-oriented research. We are successful in showcasing, here in this book, 11 chapters covering a range of perspectives and model-based solutions unified by a common goal: to improve the performance of sustainable supply chains.

We have six leading contributions presenting unique perspectives including *R&D technologist's perspective, a meta-analysis-based perspective, green's perspective on supply chain management, product design modularity perspective on sustainable supply chains, trainings' perspective on sustainable supply chains, and system dynamics's perspective on a procurement function within the supply chain.* These perspectives provide insights and solid reasons as to what should be done and why it should be done to build and sustain integrated supply chains for the twenty-first century.

With regard to the theme of the innovative solutions for the sustainable supply chains, we have five state-of-the-art applications of system dynamics, econometric, and agent-based models: *wind-power industry's supply chain in Brazil, essential consumer goods industry's supply chain in Europe, sustainable management of global food supply chain, designing a sustainable tourism supply chain*, and *risk assessment of the timber supply chain in Southern Ontario*. These applications address the critical question of "how to do" by demonstrating the design, development, and application of model-based solutions aimed at improving the performance of sustainable supply chains.

It is worth noting that although the model-based contributions in this volume have been applied to only five specific industry-related issues, the insights presented and the model structure developed are generic enough to be applied to the supply chains of other domains e.g., healthcare, energy, aviation, and retail industry. Likewise, several chapters in this book have presented "future research opportunities" for the researchers in the domain of sustainable supply chains to avail.

References

Barrad, S., Valverde, R., & Gagnon, S. (2018). The application of system dynamics for a sustainable procurement operation. In H. Qudrat-Ullah (Ed.), *Innovative solutions for sustainable supply chains*. New York: Springer, USA (in press).

Burki, U. (2018). Green Supply Chain Management, Green Innovations, and Green Practices. In H. Qudrat-Ullah (Ed.), *Innovative Solutions for Sustainable Supply Chains*. New York: Springer, USA (in press).Tuczek, F., and Wakolbinger, T. (2018). A Meta-Analysis of Sustainable Supplier Selection Approaches. In: Qudrat-Ullah, H. (ed.). Innovative solutions for sustainable supply chains. Springer, USA: New York (in press).

Gunarathne, A., Nuwan, D., Navaratne, D. G., Pakianathan, A. E., Perera, N., & Yasasi, T. (2018). Sustainable food supply chain management: An integrated framework and practical perspectives. In H. Qudrat-Ullah (Ed.), *Innovative solutions for sustainable supply chains*. New York: Springer, USA (in press).

Herrera, M., Dyner, I., & Cosenz, F. (2018). The wind-power industry's supply chain in Brazil. In H. Qudrat-Ullah (Ed.), *Innovative solutions for sustainable supply chains*. New York: Springer, USA (in press).

Ho, S. C., & Chuah, K. B. (2018). Determinants of ITF R&D Technology Commercialization in logistics and supply chain industries: R&D technologist perspective. In H. Qudrat-Ullah (Ed.), *Innovative solutions for sustainable supply chains*. New York: Springer, USA (in press).

Lewandowsk, M., & Ali Asgary, A. (2018). Risk assessment of the timber supply chain in southern Ontario using agent-based simulation. In H. Qudrat-Ullah (Ed.), *Innovative solutions for sustainable supply chains*. New York: Springer, USA (in press).

Nowak, T., Toyasaki, T., & Wakolbinger, T. (2018). The road towards a circular economy: The role of modular product designs in supply chains. In H. Qudrat-Ullah (Ed.), *Innovative solutions for sustainable supply chains*. New York: Springer, USA (in press).

Stefan Grossor, S. (2018). Resilience as basis for sustainability: Shortages in production supply chains for essential consumer goods. In H. Qudrat-Ullah (Ed.), *Innovative solutions for sustainable supply chains*. New York: Springer, USA (in press).

Sterman, J. D. (2000). *Business dynamics: Systems thinking and modeling for a complex world*. New York: McGraw-Hill.

Wongsurawat, W., & Shrestha, V. (2018). Designing a sustainable tourism supply chain-a case study from Asia. In H. Qudrat-Ullah (Ed.), *Innovative solutions for sustainable supply chains*. New York: Springer, USA (in press).

Yosuf, I., & Azhar, T. (2018). Policy Design for Sustainable Supply Chains through trainings. In H. Qudrat-Ullah (Ed.), *Innovative solutions for sustainable supply chains*. New York: Springer, USA (in press).

Part II
The Unique Perspectives for the Sustainable Supply Chains

Chapter 2
Determinants of ITF R&D Technology Commercialization in Logistics and Supply Chain Industries: R&D Technologist Perspective

Siu Cheung Ho and Kong Bieng Chuah

Abbreviations

AGV	Autonomous-guided vehicles
A-M	Application to market
API	Application programming interface
CCTV	Closed-circuit television
EMI	Early manufacturing involvement
ESI	Early supplier involvement
IOT	Internet of Things
IP	Intellectual property
IT	Innovation and technology
ITB	Information Technology Bureau
ITF	Innovation and technology fund
KPI	Key performance indicator
LBS	Location-based service
M-S	Market to science
MSTAM	Market, science, technology, application, and market
NPD	New product development
OM	Organization manager
PC	Project coordinator
PKI	Public key infrastructure
PM	Project management
PRD	Pearl River Delta
R&D	Research and development

S. C. Ho (✉) · K. B. Chuah
Department of Systems Engineering and Engineering Management,
City University of Hong Kong, Kowloon, Hong Kong
e-mail: scho25-c@my.cityu.edu.hk; mebchuah@cityu.edu.hk

H. Qudrat-Ullah (ed.), *Innovative Solutions for Sustainable Supply Chains*,
Understanding Complex Systems, https://doi.org/10.1007/978-3-319-94322-0_2

RFID	Radio-frequency identification
RO	Research questions
ROI	Return on investment
RT	Requirements traceability
SD	Standard deviation
SDC	Sensor data cloud
SI	System integrator
SME	Small- and medium-sized enterprises
S-T	Science to technology
STAM	Science, technology, application, and market
T-A	Technology to application
VR	Virtual reality

2.1 Introductions

According to Research Report (2011), in Hong Kong, for at least the past 10 years, more than 98% of enterprises in Hong Kong are small- and medium-sized enterprises (SMEs). They may be small individually, but together they employ 48% of the total working population over 1.3 M employees, and their combined business values was nearly 57% that of total business values in Hong Kong in 2009. SME may not have enough resources to support and enable research and development (R&D) technology in their company. They may have some concerns on the value of R&D Technology such as cost, implementation time, extra manpower, actual value, understanding of staff, etc. This study investigates how R&D Technologists think their efforts can align with the industry needs during the development process. It also examines from R&D Technologist's perspective the determinant factors of innovation and technology fund (ITF) and R&D Technology commercialization and adoption in Hong Kong Logistics and Supply Chain Industries. This research aims to find out which critical issues affect R&D project deliverables commercializing and productizing to the industry. An R&D Technologist always faces challenges of developing technology that matches industry needs. This research investigates how R&D Technologists think their efforts can align with the industry needs during the development process.

This is the first attempt of the multi-perspective study to bridge the gaps from idea generation, R&D, and subsequent technology implementation and adoption by industry. This research describes background of study, ITF R&D project in Logistics and Supply Chain Industry in Hong Kong, literature review, theoretical background, and research methodology, analysis of findings, and discussion and conclusions. Qualitative and quantitative of the data collected from R&D Technologists have been carried out using a proposed market-driven project management model, MSTAM, which stands for market, science, technology, application, and market.

This study will address four distinct research questions (RQ) as follows:

RQ1: What are the main reason or major barriers why ITF's R&D results commercialization and adoption rate is not good in Hong Kong Logistics and Supply Chain Industry?
RQ2: What are the appropriate actions or elements to motivate the industry users or top management in the company to adopt or license the R&D results in their company?
RQ3: What are the key considerations for them to adopt new technology or R&D results in Hong Kong?
RQ4: How to facilitate, adopt, or integrate the R&D deliverables in the Logistics and Supply Chain Industry in Hong Kong?

2.1.1 Challenge of ITF R&D Project in Hong Kong

According to the ITF (2013), the audit selected projects for ITF, the project completed in the period form May 2008 to December 2012. The report shown that 12 projects are with post-completion evaluation, and the results have no technology breakthrough, limit adoption, and no successful commercialization. These projects cost ranging from HKD 1 million to HKD 19 million. None of the 12 projects had commercialized technology breakthrough to the industry. The report showed the figure of project cost of new projects commenced in the period of 2011–2012 to 2014–2015. The total project cost cumulative in 9 years is HKD 238.6 million for five research centers in Hong Kong. The total commercialization income in five research centers was HKD 174.9 million. The commercialization income compared with project cost of new projects ratio was 4.13%. As indicated above, industry contribution was very low. In the current situation, R&D centers were facing two key issues in the ITF R&D project, namely, R&D project development and commercialization rate was low and low project turnover rate.

2.1.2 Challenge of ITF R&D Project in Logistics and Supply Chain Industry in Hong Kong

In accordance with the HKTDC (2016) figure, the merchandise trade performance index shows that the total exports, reexports, imports, and total trade figures continue to decrease from 2014 to the first quarter of 2016. The retail sales growth, visitor arrival growth ratio, and service trade performance of total trade figure were falling shapely, due to the Hong Kong immigration control of China visitor. Hong Kong Logistics and Supply Chain companies are facing a big challenge in these years. The major economic indicators indicate that the total HK exports in 2015 are over HKD 2396.9 billion. This is a big figure for Hong Kong. Therefore, Logistics and Supply Chain Industries are still one of the main pillar industries in Hong Kong.

In this dynamic competition business environment, if the company wants to survive in this dynamic business market, they should explore and push up their core capability. R&D Technology is an effective tool to help them sustain their business and increase their core competitive advantage.

The Innovation and Technology Bureau (ITB) was formally established on 20 November 2015. The bureau is responsible for formulating holistic policies relating to innovation and technology; strengthening the coordination among the government, industry, academia, and research sectors; and expediting the development of innovation, technology, and related industries in Hong Kong. R&D project quantity will be increased in this coming few years (ITB 2016). HKSAR have revised the policy to promote smart production and attract high value-added technology industries as well as high value-added manufacturing processes that are a new direction for local industry experts (Sheila Lam 2016). According to this information, R&D project demand for Logistics and Supply Chain Industry will surge in the coming years. This study promotes the further deployment of innovative and advanced technologies to spearhead the development of "reindustrialization" in Hong Kong. All research centers, universities, industries, and research institution in Hong Kong are facing extreme challenge to manage R&D project from idea generation, R&D and subsequent technology implementation, and adoption by industry. Effective and efficiency management of R&D research project is needed. This study provides holistic understanding for managing project risk identification and navigation which is beneficial for transfer and development of the R&D project deliverables to the Logistics and Supply Chain Industry in Hong Kong for moving their business forward.

2.1.3 Chapter Content Outline

The rest of this chapter comprises of nine sections. Section 2.2 describes the existing enabling technology in Hong Kong Logistics and Supply Chain Industry. Section 2.3 states the linear and nonlinear model of innovation and the role of R&D process in innovation. Section 2.4 lists the research conceptual framework and research methodology for this research study. Section 2.5 provides the conceptual model of ITF's R&D project commercialization. Section 2.6 identifies the ITF R&D ecosystem in Hong Kong Logistics and Supply Chain Industries, key stakeholder interests, and concerns of ITF R&D projects, introduces a basic research model of R&D development, and defines a conceptual model of ITF's R&D project commercialization and critical gap analysis in different stages of the proposed model. Section 2.7 proposes the ITF R&D projects development proposition for the proposed model. Section 2.8 is data collection and analysis. Finally, Sect. 2.9 discusses the qualitative and quantitative analysis and summarizes the findings.

2.2 Enabling Technology in Hong Kong Logistics and Supply Chain Industry

Refer to LSCM (2017) and LC Paper (2015); in recent years, Hong Kong has gradually come to a consensus that it is necessary to diversify the development of our industrial structure where the innovation and technology have a vital role to play. The most popular R&D Technology in Hong Kong Logistics and Supply Chain industry are (1) Infrastructure Information Technology System, (2) Internet of things (IOT) and RFID technology, (3) location-based service (LBS) technology, (4) logistics and supply chain analytics and applications, (5) Supply Chain Security (Authentication/Logistics Security), and (6) Financial Services and e-Commerce Technology. Developing core technological competencies and enhancing the efficiency in the Logistics and Supply Chain Industry for increasing the company capability, a research institute in Hong Kong has undertaken industry-oriented research projects involving industry and universities.

Currently, ITF R&D enabling technology for Hong Kong Logistics and Supply Chain Industry is applied in different areas. Referring to LC Paper (2015, 2016, 2017), some examples of the ITF R&D project enabling research and commercial adoption in Hong Kong Logistics and Supply Chain Industry are:

1. Logistic and transport – Developed and integrated a smart IOT cross-boundary logistics for cargo truck monitoring system to enhance the custom clearance time without boundary stopover and provide end-to-end seamless transshipment.
2. Airport technology initiative – Video-based technologies are applicable vertically to airports around the world or horizontally to other closed-circuit television (CCTV)-ready facilities like shopping malls and exhibition centers with indoor, outdoor, and mixed environment to monitor allocations of resources inside premises and a virtual reality (VR) technology in real applications to perform effective logistics operations and supply chain management to attune the technology with real cases and operations in actual training and performance evaluation.
3. Construction industry – Multiple hardware components installed to the safety belt system. It senses the attached status of the safety belt, the lanyard, and the lifeline. Proper usage of safety belts depends on both the employer and the employee. When the system is available to the market, it provides one more tool for the employer and supervisors to remind workers to put on appropriate safety measures in the working site. This technology contributes to raise the construction site safety level in Hong Kong.
4. E-commerce industry – e-Cheque Apps Development Cloud aims to facilitate e-Cheque payment to enhance the public confidence in this innovative payment instrument, as well as to facilitate financial institutions, payment service providers, and business to build an application on e-Cheque and as a result promote the use of e-Cheque in Hong Kong. This enabling technology ensures an e-Cheque will be removed or destroyed beyond recovery. In e-Transactions using e-Cheque,

it is transferred among digital wallets and will be saved in payer's or payee's devices.

5. Location-based service (LBS) – Indoor GPS technologies introduce the conventional GPS positioning technologies from outdoor usage to indoor applications. The project deliverables aim to provide positioning accuracy to 2–5 m for mobile phone users in indoor environment in Hong Kong, and location analytics application programming interface (API) has been developed which allows users to create other mobile applications that require track-trace technology and movement-based recommendation.
6. Smart community service – Smart IoT infrastructure and technologies provide an easy-to-use tracking platform for better monitoring of elders during outings and sensor data cloud (SDC) system developed to track and monitor the delivery and movement of visual products since harvest to delivery.
7. Retail – Product authentication infrastructure and systems at retail point enables product integrity in the supply chain from the point of certified from manufacturing to the point of retail, it proposes a running code authentication method on radio-frequency identification (RFID) anti-counterfeit labels and a public key infrastructure (PKI) based code-issuing server. At the retail end, the authentication-enabled kiosk has been developed for end-customers to validate the product integrity;
8. Robotics application – A new robot design of autonomous-guided vehicles (AGVs) which can be fitted and operated in small and tightly packed warehouses. It has great potential for warehouses in Hong Kong and Pearl River Delta (PRD) areas as they are generally compact.

Table 2.1 Enabling technology in Hong Kong Logistics and Supply Chain Industry

Logistics and Transport	Airport	Construction	e-Commerce

➢ E-lock tech for ITFS ➢ Airfreight advance arrival info ➢ Mesh routers for ports ➢ Sensing for HK Post & museum	➢ HKIA's indoor navigation ➢ RFID baggage system ➢ Hybrid RFID reader ➢ VR Operation Training System	➢ BIM for Site Safety ➢ Rear Alarm for Vehicle ➢ RFID Safety Belt system	➢ e-cheque ➢ Internet Infrastructure Protection
Location Based Service	**Smart Community Service**	**Retail**	**Robotics**
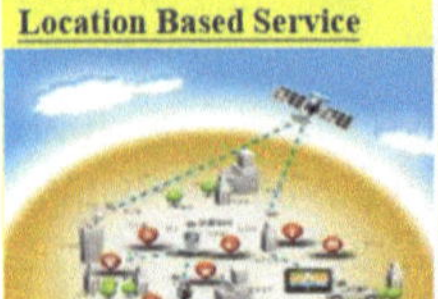			
➢ Indoor GPS ➢ Location Analytics for ➢ Exhibition ➢ Beidou GIS infrastructure	➢ Eyebank's ocular tissue ➢ monitoring ➢ Baby Tag monitoring system ➢ Various RFID techs for TWGH	➢ Product authentication ➢ Low-cost readership ➢ Pearl authentication ➢ Mobile RFID reader	➢ Autonomous Guided Vehicles ➢ Intelligent Robot Baggage Automation ➢ Mobile Robotic System

Table 2.1 summarizes some current enabling technologies for Logistics and Supply Chain Industry in Hong Kong. As can been seen, the ITF R&D enabling technologies can be applied in different environment and industries.

2.3 The Role of R&D Process in Innovation

The role of R&D process in innovation is the roadmap to determine the R&D project whether success or failure. The R&D process index like the key performance indicator (KPI) and a compass steering the R&D Technologist develop the R&D product to the market. Start from the ideas generation, idea screening, concept testing, business analysis, product development, test marketing, market entry, and evaluation. Shown in Fig. 2.1a, the work flow of linear model of new product development

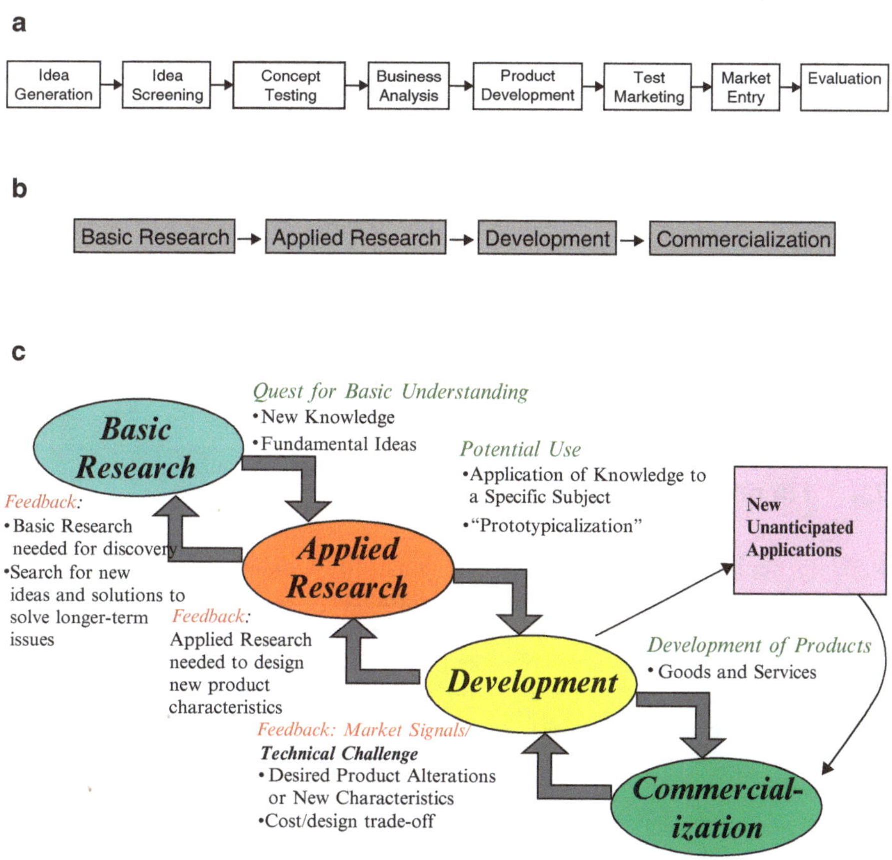

Fig. 2.1 (**a**) Linear model of new product development. (**b**) The linear model of innovation. (**c**) A nonlinear model of innovation

is a consecutive process from *Nil* to *Have*. Normally, R&D is grouped together as one process. Research is focus to be science-oriented, and development is the mechanism from science concept to the commercial product and services. The role of R&D process in innovation included market pull and science push, and it involves R&D Technologist and industry user. Collaboration and alignment between each other are needed.

According to Wessner (2005), the linear model of innovation involves four key processes from basic research to applied research, then development, and finally to commercialization (Fig. 2.1b). The innovation process starts from discovery to innovation to commercialization. In this process involves consecutive operation process and challenges from science push side to market pull side.

Currently, interactive communication and information sharing can be enhanced for R&D process in innovation that technological breakthroughs can be achieved. Referring to Fig. 2.1c, two-way interactive collaboration can communicate in different stages. This method collects feedback from each stage to emphasize the R&D commercialization opportunity. Starting from basic research stage to applied research stage, the new knowledge and fundaments ideas from the basic research quest for basic understanding of innovation to applied research. The basic research feedback and findings will be returned to the basic research for further exploration.

In the applied research stage to development stage, it involves different application of knowledge to specific subject and development for prototypicalization. From the development stage to commercialization stage, the new unanticipated applications and R&D deliverables to the market and the market feedback will be returned for product and services improvement. Finally, the information flow continuously runs in circle for R&D process. Therefore, the proactive communication channel can improve the operation of R&D processes, and these are the key success factors of innovation.

2.4 Theoretical Background

R&D Technologist and industry user are constantly looking for technology solutions that can improve company capability efficiency and sustain company's core competitiveness in this dynamic business environment. Qualitative and quantitative data collected from R&D Technologists have been carried out against a proposed project management model. The basic proposed model will be defined in Sect. 2.6. A thorough understanding of R&D Technologist perspectives can enhance the R&D projects development process and enable technology adoption for facilitating R&D deliverables commercialization to Logistics and Supply Chain in Hong Kong. Novel R&D project management mechanism solutions are required in order to speed up and create the value in the R&D process. Refer to Chapman & Hyland (2004); this study said that *Product innovation is recognized as key conditions of business*

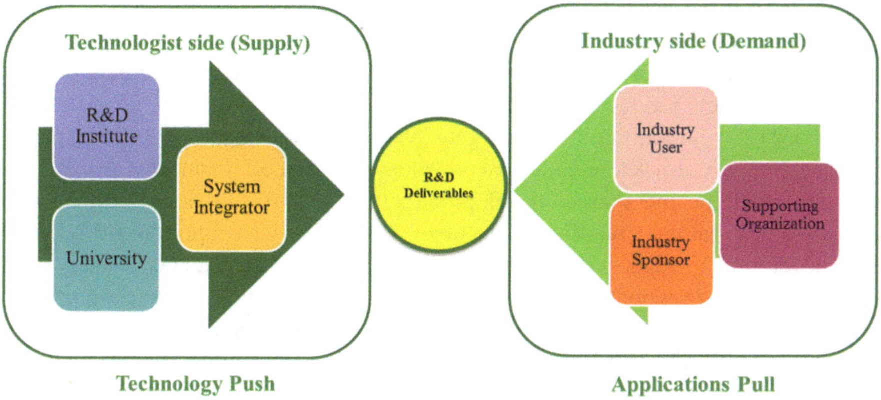

Fig. 2.2 ITF R&D ecosystem in Hong Kong Logistics and Supply Chain Industries

success. Therefore, successfully commercialized R&D project deliverables to industry can increase the competitive advantage for the industry.

For the traditional innovation process, there are two directions to review the innovation process as technology push and application pull approach. R&D development and market adoption are necessary factors seen as a receptacle of the results of the R&D activity. The R&D result needs to match and combine with pure technology push and market pull forces (Vittorio Chiesa 2001). Figure 2.2 is ITF R&D project development ecosystem in Hong Kong. It involves different stakeholders in R&D Technologist side and industry side like R&D institute, university, and system integrator (SI) in the supply side and industry user, industry sponsor, and supporting organization in the demand side. Also, Roberts (1988) identified technology push and market pull innovation processes have different elements and factors linked together. The push and pull approaches can review the recognition of technical feasibility and potential market demand. The key function is search, use, test, and response between technology side and market side for balancing the need of both sides.

2.5 Conceptual Model of ITF's R&D Project Commercialization

Cooper (1983) identifies the five success factors of new product development (NPD) which are (1) *understanding of user's need*, (2) *efficiency of development*, (3) *characteristics of management and managers*, (4) *effectiveness of communications (internal and external)*, and (5) *magnitude of marketing efforts*. A single variable cannot lead the new product success. In the new product development process, there are many variables influencing the development process. The developer should find out which critical variables lead the new product success. Cooper (1987) also states that *an innovation strategy is essential to an effective new product program.* The new product development should be focused on three areas as application, customer, and technologies. These areas need to be aligned in the development process leading to the new product success.

Cooper (2011) modified the stage-gate system and introduced pre-development stage in Fig. 2.3a to define each new development process from idea generation, scoping, build business case, development, testing and validation, and launch. Copper states that the front-end or pre-development stage is important for NDP. In each development stage, starting from gate 1 to gate 5 like as gatekeeper to examine the previous stage whether suitable to continuous to the next stage development process to ensure the new product direction and specification are correct.

Phaal et al. (2011) also provide an extension of the basic model, STAM (science, technology, application, and market), from scientific enquiry to commercial products. The authors agree the market information is a critical data for developing technology. In STAM model, it outlines the first phases as *precursor science* to second phase as *embryonic technology* and the third phases as *future application* to the final stage as *growth market.* This model points out the gap between each development stage from basic research, applied research, to experimental development (Phaal et al. 2011) (Fig. 2.3b).

2.6 Research Model Development

In most studies of innovation development, the coupling model of innovation is based on the needs of society and the marketplace and state of the art in technology and production. Rothwell and Zegveld (1985) state the innovation process involves six key elements such as idea generation, development, prototype production, manufacturing, marketing and sales, and market place. Phaal et al. (2011) also provide an extension of the basic model, STAM, from scientific enquiry to commercial products. Both authors agree the market information is a critical data for developing technology. Refer to Seo-Kyun et al. (2009); leaning and external networking is a significant factor to influence the innovation performance of the firms. To improve the technology commercialization capabilities, SME must avoid narrowly on R&D, and it should invest in the capabilities to commercialize the R&D results in the company. For the project success factros, Daniel et al. (2014) investigate show that the applicability of structural complexity metrics in early development stages and the uncertain information and potential errors to the fuzziness of the early development stage can be identified in advance and pushing up the project success. Therefore, this research study modifies STAM model to develop a new model, MSTAM (market, science, technology, application, and market) to study the market need first before starting the development process (See Fig. 2.4a).

Fareena and Lillian (2000) examine the adoption of new technologies. To broaden the adoption rate of the technology in the organizations, it should focus on individual levels. It helps to understand the key players concerned and perception of the technology and what are the characteristics of the company. For the conceptual

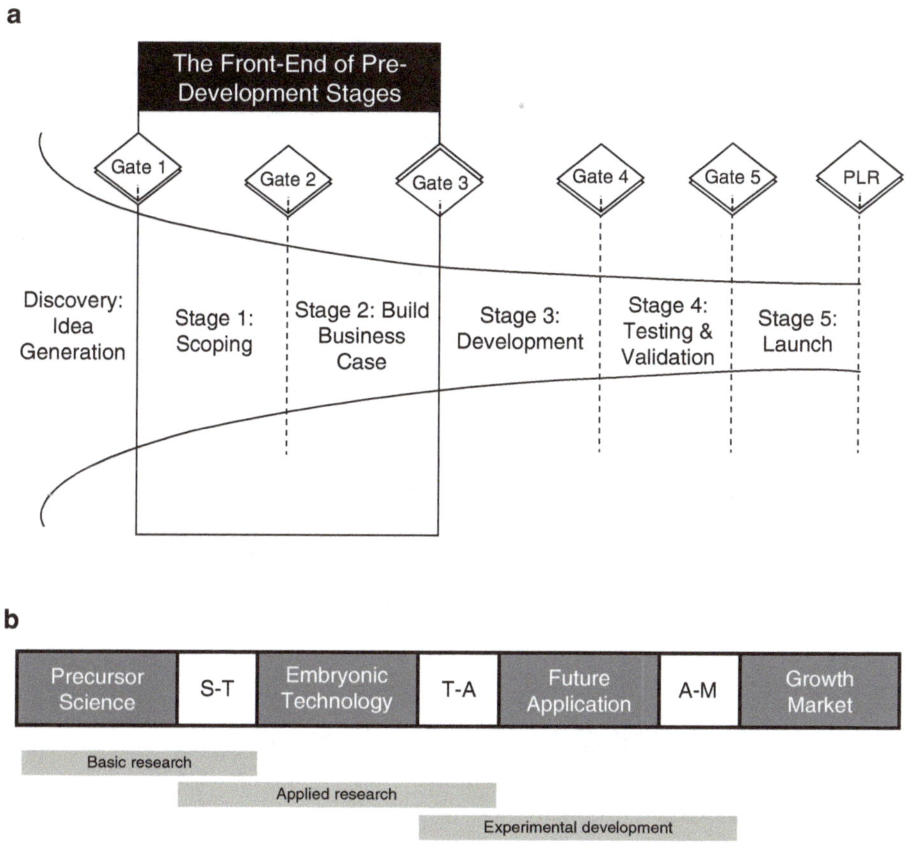

Fig. 2.3 (**a**) The pre-development stage in the stage-gate system make or break major – new-product projects (Cooper 2011). (**b**) STAM model. (Adapted from Phaal et al. (2011) and the OECD model)

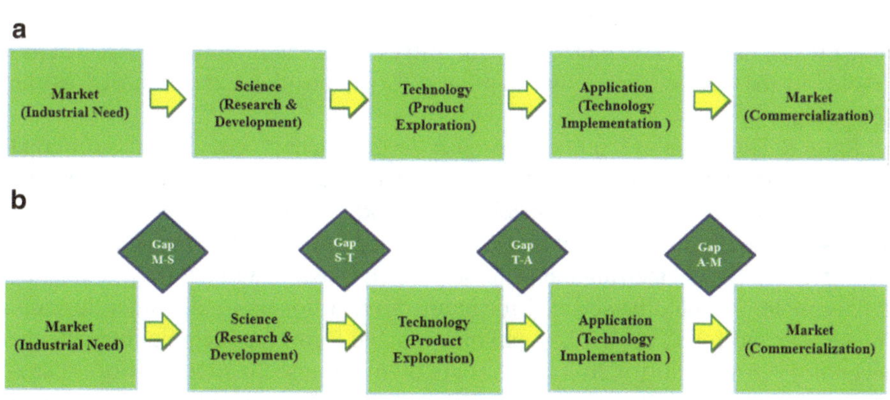

Fig. 2.4 (**a**) MSTAM model. (**b**) Critical gap analysis in MSTAM model

framework: To enhance the study by Phaal et al. (2011), whose focus is mainly on scientific enquiry to commercial products, this study focuses not only on the role of design demonstrators in scientific exploration but also on those variables related to gap analysis between each phase to identify the inhibiting factors in the development process. Yayoi Hirose (2012) demonstrates the cases to show how the knowledge transfers from the technology researcher to the users. The analysis results show that the member's engagement overwhelmingly overcame several cultural gaps between technology researchers and potential users which can be impediments to knowledge transfer and generation. This is proved that gap analysis can enhance the technology transfer process (Fig. 2.4b).

Innovation performance is innovativeness exploration which is defined that enabling technology can enhance the capacity into the new process, product, or idea in particular organization (Hult et al. 2004). Refer to the LSCM Survey (2008), the challenge of R&D project development of technology adoption in the industry. The company concerns are on three key problems as difficult to access return on investment (ROI), lack of industry and government support, and limited budget. Refer to Hult et al., (2004) and LSCM Survey (2008), results, the findings is to push up the technology adoption to the market, the first step should be identify the barriers in the technology adoption and explore which interested technology initiatives such as improvement of operational efficiency, customer services and reduce human errors etc.

2.7 ITF R&D Projects Development Proposition

The critical analysis of the effectiveness of R&D process assessment for ITF R&D project development and commercialization in Hong Kong Logistics and Supply Chain Industry, the research. The research framework is divided into two areas as R&D Technologist perspective and industry user perspective. This is the first research study focusing on the expectations of R&D Technologist's concern on the R&D project deliverables and value in the ITF R&D process such as how to shorten the R&D time, improve the R&D deliverables quality, enhance control mechanism of the R&D management, motivate the industry adopt, and promote the R&D deliverables to the market. From the literatures, 12 propositions are proposed as the following.

Refer to Uda Hashim (2002); the project development process includes different requirements and conditions need to be collected such as technical requirement, customer requirement, product specification, and so on. Bangchao Wang (2016) suggests a requirements traceability (RT) to trace the critical step of the development process for transferring the technology from academia to industry. A suitable RT technology should be selected for industry practitioners in the initial development stage and found out the key external variable factors for helping integrate the technology to the industry. Joko and Anggi (2016) propose the strategies for enhancing the system implementation processes, and it should be based on the *real needs* of the user to define which modules are suitable for the technology application. These lead to the authors' first proposition.

P1: *Collect Industry Requirement at the R&D Initial Stage*

John E. Ettlie (1995) states the early manufacturing involvement (EMI) is a significant factor to reduce the change request of new product development and emphasis on the return on investment (ROI) of the new product. Venlakaisa et al. (2009) also agree the early supplier involvement (ESI) can improve the development processes. The benefits of ESI are shorter lead time, better quality, reduced costs, increased level of commitment, and have better solutions. For the project success factors, Daniel et al. (2014) investigate the project success factors depend on how to know the uncertainty in the early development process and stated that *If the structural dependencies within a system are fully known, the complexity can be handled.* Harshada et al. (2016) express user participation in the early project development stage. The developer can easily know the user real expectation and difficulties they face in every day. For the aforementioned literature, the second proposition is:

P2: *Industry User Involvement in the R&D Project Concept Development Stage*

Joko and Anggi (2014) identify the system modules requirement for the development process in different enterprises. The foundation system module specification is listed in detail for identifying which same functional modules can be grouped and installed for the business processes. Sofija et al. (2016) found that some technical people did not participate in the engineering development activities and the project development requirements are not detailed enough for acceptance test requirements. It triggers some internal and external issue such as underestimating the complexity of requirement, client needs, unexpected results occurred, etc. Kellner et al. (2016) propose model-based development from functional structure, customer requirements, and detailed system requirement diagram to transform the information along the development process and interaction of different components of the overall system. Therefore, a basic product specification and customer requirements are project development tools to guide that the project goes to the right direction. Zhao et al. (2009) propose high levels of collaboration with suppliers in the early development stage to align the company goals with suppliers. This is an essential part for the company to consider the relationship, knowledge sharing, and difficulty with strategic suppliers in the product development process. Summarizing the findings, the third proposition is:

P3: *Basic R&D Product Specification Established in Science Stage of Research and Development*

Marcos and Luiz (1995) present a project's environment system evaluating the relationship between customer, project, and supplier. Relationship management with risk planning for internal and external stakeholder should build up the relationship with each other. The project's environment system is a blueprint of the project for the project team to manage the project smoothly. A Gatekeeper is a critical role in the R&D process, Yuji et al. (2006) propose a role of middle man and research consultant to improve the communication between researcher and customer to overcome the gap between each other for improving the communication for R&D management. Souhaib et al. (2015) study the relation between requirement engineering technique and challenge in the software industry. The findings show the challenges and techniques of the project should be linked together in a systematic way to define the critical tasks in advance for project management. Ana et al. (2016) suggest a project, program, and portfolio management concepts to address the project issues in the project development for R&D organizations. The key components are problem identification and motivation, objective definition, design and development, demonstration, evaluation, and communication. Xiaohua Sun (2016) introduces a risk management method of engineering project. The process of risk management includes five modules such as risk identification, risk estimation, risk evaluation, risk response, and monitoring and controlling the risk. This risk management technique is an effective tool for R&D project management. Based on this literature, the forth proposition is:

P4: *A Gatekeeper Identifies the Project Risk in Advance for Clarifying Administrative and Technical Issue in the R&D Project (Project Risk Management)*

Dong-dong and Ping (2009) found that supplier early involvement can provide a significant benefit in the NPD process. It examined the performance of the company in six areas as quality, cost, delivery, development potentiality, technical capabilities, and cooperation capability that can be improved. Kamil and Ali (2014) agreed that industry sponsors join in the research project; it helps the researcher to acquire the industry information and skills in the right direction to create the new technology for solving the real issue of industry. Bruno et al. (2016) proposed a requirements engineering and management process in the concept phase of NPD. The process is structured into five sub-processes: (1) requirements engineering management, (2) elicitation, (3) analysis, (4) specification, and (5) verification and validation. Different stakeholders are involved to deal with the unstable NPD environment. Franken et al. (2016) also agreed the end users involvement in the software development process in this collaboration method. The developer collects the direct feedback and expectation from users. Consequently, the gap between the developers and users has been eliminated. The fifth proposition is:

P5: *Industry User Involvement in the Development Process*

Misoo et al. (2006) stated "Various meetings and related data accompanied in the progress of R&D projects are to be managed for successful projects." The R&D projects have many information and new information need to be shared with each project stakeholder in the R&D project for exploration. According to Mamoru and Hiroshi (2016) investigation, the results for the project fail are caused by three factors, these are project management ability, technology, and people skills in project management. Therefore, simplifying the R&D process and regular project meeting are the essential tools for project management. Liliana et al. (2016) defined the evaluation procedure to improve the project management (PM) in a systematic way in large public sector projects. There are nine project management components suggested as integrated management, requirement management, contract management, time management, cost management, team management, risk management, reporting and communication management, and quality management to the PM system.

Fagno et al. (2016) suggest using systematic review project mechanism to identify the key steps and processes in the information technology (IT) project. Most barriers of the IT projects are relating to communication and human resources issue. To summarize the literature, the sixth proposition is suggested as:

P6: *Regular Project Status Update and Unification Schedule Management*

Mats and Jeflrey (2000) propose the specification-based prototyping to control the development process using process control system. The process control system has four variable models, these are monitored variables, software input, software output, and controlled variables, to transfer the system requirements and product specification to software requirements for producing a mock-up prototype for pilot's display. The specification-based prototyping is translated from high-level systems requirements to detail requirements of software to allow the customers to know easily and participate in the development process. Ana et al. (2015) present a prototype to conduct the test procedure and pretesting to validate the testing results in advance. The prototype can be obtained in the results in a short period of time. Grandjonc and Khattabi (2016) provide a project roadmap to execute a project from project start to project end. The system dynamic model divides from into four stages: (1) requirements definition, (2) preliminary design, (3) first prototype implementation, and (4) final design and implementation. In the middle of project, the project requirements have been updated for validating the first prototype performance. The literature proved the prototype and preliminary results have a great support for development of the project.

P7: *Prototype Sample and Preliminary Results* ***(Proof of Concept)***

Boulter (2004) studies the intellectual property (IP) issues for R&D management. The company has some considerations for IP application like as the cost affordable, funding concern, uniqueness, the boundaries of technical transfer, time and resources support, restrictions, and so on. L. Boulter also suggests to utilize pre- and post-contract management to protect the research project. Lenina SVB (2016) proposes a series of IP protection procedures to let the inventor know how to protect their invention. IP management is a key element to enhance the commercialization sucesss, Herlandí et al. (2016) present a process model to protect the technology via IP management to leverage technologies transfer, invent, and develop for institution transfer or licensing the technology to the potential organization. Qing and Shawn (2017) expressed that "Intellectual Property is inextricably linked to the innovative development of mass innovation spaces" and agreed the IP protection is an important element for managing the R&D project. Jian and Qing (2017) propose an integrated IP assessment system for universities for enhancing the protection and development of IP creation. The proposed system has eight categories such as IP filings, IP attainment, IP implementation, IP protection, IP management, IP environment system, and IP commercialization. According to the literature, IP protection is important to protect the developed innovation technology and pre-IP registration. Therefore, the eighth proposition is:

P8: *Pre-Intellectual Property (IP) Registration*

To enhance the commercialization success to the Logistics and Supply Chain Industry, an effective promotion for the devloped product to the industry is needed, Cheng-Mei et al. (2013) state the enterprise success factor of the business is dependent on the quick knowledge and technology application of the innovative product application in this competitive business market. The suggestion of this study are (1) enhancing the links and capacity among academia, research centers, and business sectors and (2) promoting germination activities in early R&D project development stage to enhance the performance technology industry incubation. For NPD, Robert and Jinhui (2016) identify the role of manufacturer and supplier from conceptualization stage to design stage and production stage. The analyzed findings examined innovations through supply chain partnerships. As mentioned in P6, Mats and Jeflrey (2000) stated that the specification-based prototyping is translated from high-level systems requirements to detail requirements of software. The prototyping and results allow the customers to know easily the developed product in the developmental stage.

Requirements traceability (RT) is a key indicator for industry selection and evaluation. The researcher it can easy to trace which RT technologies transfer whether maturity for industrial practitioners (Bangchao Wang 2016).

P9: *Promote the Prototype and Results to the Industry*

Konstantinos et al. (2008) propose a virtual and augmented reality approach to support collaborative product design evaluation, demonstration, and customization. The major collaboration functions are (1) user/project management, (2) message/chat, (3) project versioning, (4) calendar/scheduler, and (5) file sharing/browser. The preliminary product design and specification can be shared among the developer, collaborator, and user. The user can see the product design of prototype in advance and comment during the development process to let all parties more interactive. Kawtar et al. (2015) worked closely with industry vendor to develop and test the prototype product. This research encourages industrial user to join in the development-specific application and develop product integration in their company for collecting the actual feedback of the industrial user. To ensure the developed product quality fulfill the user's need, trial test is needed. Gilbert et al. (2016) performed the trial in two organizations to test the developed software systems whether can overcome the industrial issue not only functional requirements, safe and reliable. Industry validation is needed in the product development process to examine the usefulness, value, usability, and results oriented from the initial assessment by the industrial user. Industry validation is the critical process in the R&D project. To examine and evaluate the industry's need, trial implementation in the industry premises is proposed.

P10: *Trial Implementation in Industry Premises*

Hartley et al. (1997) propose the supplier's involvement in the product development to provide the technical support for increasing their contribution in the overall product development process. The suppliers provide a special insight of product specification to match closely to the actual market need. Suleiman et al. (2002) differentiate the commercialization factors of disruptive and sustaining technologies on variety of issues of market focus in three areas such as size of firm commercializing the technology, technology sources, and risk of involvement with new technology. The external support from the external research sources can build up the firm's capabilities. Cheng-Mei et al. (2013) analyzed the innovation commercialization development fostering issue in Taiwan from 2007 to 2010. The reason of commercialization development is not active and less market-oriented in the early stage of R&D. The suggestion is to enhance the communication and connection and increase the business sector adoption capacity of new technology in the company among research institutions, universities, and business sector. Norhadilah and Azmawani (2014) reviewed 182 research articles to identify the success factor of innovation commercialization performance. Those factors are defined as organizational resources, knowledge management, strategic orientation, organizational support, contextual factor, and lastly collaboration. The results are the contextual and organization support; these are key factors of innovation commercialization. Christof and Carlos (2016) propose a role of requirements engineering to address the development process, culture, and technology dimensions for helping the company technology integration success. The requirements of engineering have been defined in eight modules as stakeholders, value, requirements, process, architecture, validation, integration, and continuity. Summarized the literature finding, the 11th proposition is:

P11: *Technical Support and Finalize Product Specification*

Amadi-Echendu and Rasetlola (2011) suggest two approaches for technology commercialization; these are linear model with feedback interaction and functional approach for emphasizing continuous linkages with networked agents. These approaches are started from technology to application, product or service, and market. The results of the finding are functional approach better than linear model. Michael and Stephen (2013) propose a system dynamic model to define a causal loop diagram for identifying the adoption rate of potential adopters and population. The system dynamic model divides from potential adopters, semi-adopters, and former adopters. In the adoption process, the main entities for IT adoption in the complex environment are promoters, inhibitors, and imitators. In-depth understanding for this system dynamic model of the adoption processes is needed for all stakeholders. Florian and Orestis (2014) identify the critical success factors for strategic management of radical technological innovation focus in four modules, and these are (1) target market, (2) organization, (3) technology, and (4) technology commercialization process. Partnership and strategic alliances are proposed to diffuse the technology to the market. Therefore, the 12th proposition is proposed as:

P12: *Promote the Developed Technologies by various Marketing Effort*

2.8 Data Collection and Analysis

This research explores the development and application of new ITF R&D project monitoring mechanism for enhancing the R&D project deliverables commercializing and productizing the results to the industry. The market focus in this research study includes quantitative and qualitative data analysis. Through the survey methodology to collect the data from the R&D Technologist, it defines the target survey group, questionnaire development, design of qualitative and quantitative research of R&D Technologist, and industry people working in Hong Kong Logistics and Supply Chain Industry. The suggested integrated approach enables the decision-makers to use a five-point Likert scale as 1 = strongly disagree, 2 = somewhat disagree, 3 = neutral, 4 = somewhat agree, and 5 = strongly agree. This method illustrates the R&D Technologist perspective analysis and finds out which critical factor(s) enhance the R&D project operation for solving the gap between both parties.

From January to June 2017, the author discussed during the period from January to June 2017, the author discussed with some R&D Technologists in the R&D institute, university, system integrator (SI) company, and industry partner to conduct a qualitative and quantitative survey in Hong Kong with 73 R&D Technologists, who are key developer and top management from these organizations. The survey questionnaire is an open-ended question with assumption. This study aims to develop a new project management model, MSTAM, to have a good grasp of factors influencing R&D project commercialization through a multi-perspective methodology. This is the first attempt of the multi-perspective study to bridge the gaps from idea generation, R&D and subsequent technology implementation and adoption by industry. It proposes an evaluation framework for conducting commercialization strategy to integrate the R&D Technology deliverables embedded in Logistic and Supply Chain Industry. For critical analysis of the effectiveness of R&D process assessment for ITF R&D project development and commercialization in Hong Kong Logistics and Supply Chain Industry. The research framework focuses on R&D Technologist perspective, the expectation of R&D Technologist, industry concern on the R&D project deliverables, and value. This research aims to identify the objectives for launching the efficiency monitoring mechanism for enhancing the R&D project effective and efficient to commercialize the R& D project results and adopt in the industry.

Through the questionnaire to collect the R&D Technologists comment and experiential insightful in the R&D project, the author summarized the feedback and key concern of the R&D Technologists to investigate which critical factors are the key issues in the R&D development process and what factors can facilitate the R&D results commercializing and productizing to the industry effectively. This survey is divided into two categories, the first is investigating the determinants of ITF's R&D projects results commercialization and adoption for Hong Kong Logistics and Supply Chain Industry, and the second is focusing on what critical factors influence the R&D project results commercialization and adoption to the Hong Kong Logistics and Supply Chain Company and how well the industry user is willing to use this R&D results in the corporate.

2.8.1 *Qualitative Analysis*

Semi-structured and in-depth interviews were conducted through face-to-face interview, telephone conversation, and online survey. In the R&D development ecosystem Refer to Fig. 2.2, the R&D Technologists are in R&D institute, university, and SI companies (industrial partner). In the first question, there are different views of the importance of the R&D development process that may accelerate the R&D process time. Summarized is the key comment for the stakeholders as follows.

Some R&D Institute respondents commented that profit sharing and intellectual property (IP) rights are the key concern of the stakeholders. An agreement and simple procedure are the main factors that motivate the user to adopt the certain technology.

> In the context of HK, profit sharing and IP protection are usually the interests and concerns of the stakeholders. If the users have a demand on a certain technology, an agreement could be more easily achieved.

Some respondents from the university commented that due to the multiple stakeholders involved in the R&D projects and their interest for the applicability of the new technology, control mechanism is needed to balance each stakeholder.

> Although multiple stakeholders are involved, they have one common interest: the applicability of the new technology in the industry.

Some respondents from SI companies and industrial partner commented that they are concern on the conflict of interest, time, and development cost. The value of the technology is a common consideration of the top management of the company.

> The only thing I can think of is conflict of interests in terms of priority, time of development and costs increment to be incurred.

Secondly, the findings disclosed many respondents responded that most companies in Hong Kong are SME. The company does not have additional resources to use those new technologies. Also, the market trend and industrial model are changing quickly. The existing model and process mechanism cannot fulfil the requirements of commercialization.

> The quality of deliverables, in terms of usability and novelty, is one of the main issues. Most of ITC's projects stay and end in the prototype stage. To sell such deliverables, novelty and IP are typically the main strength. But this is a relatively weak part of most projects which tend to focus on downstream development. High valued R&D is usually in IP creation like that a breakthrough idea would be of high value even if a full system is not included. However, the focus is on downstream projects. In general, the ITF funding is not sufficient to cover the resources needed to build a production grade system.

R&D shall be generated from market needs and not from R&D or academic side. Seek for technical support from overseas institute or consultant of necessary.

It looks to me that the objective of ITF's R&D, instead of supporting the industries, is to subsidize the university technical teams so they can survive. I believe that ITF's R&D has down-graded the university whereby university should be focused on theoretical inventions instead of integrations and replications which should be the business of industrial companies. In the nutshells, R&Ds undertaken in governmental bodies used to be bureaucratic which is too such expensive and time consuming. Niche market and relevant industrial model changes quick overnight and that the existing LSCM model cannot meet the requirements of commercialization.

Thirdly, most respondents concerns the cost, time, and infrastructure setup; it involved the resources investment and allocation.

Focus on midstream IP creation and take a longer timeline for potential projects, rather than, stress on immediate cost recovery.

Correct the mind-set to encourage error and room for improvement. Put the return of monetary investment at a lower significance. Involve professionals from other industries. This is the age for cross-industry collaboration.

Establishment of less expansive infrastructural ecosystem for the industries so they can survive in this local base.

Lastly, the respondents commented that R&D timeframe, development cost, R&D results quality, and financial and technical support to the industry are the main concerns for the key players to consider in adopting the technology in their company.

The key considerations for industry users to adopt new technology or R&D results in Hong Kong may include the amount of future revenue due to adopting these results and the difficulties to implement these results in their customized business situation.

Maybe it's time to cultivate industry players to have a mind-set of making use of ITF R&D resources as a R&D branch/unit to their companies. The R&D model changes from the old (resource-impossible) in-house R&D unit to the new in-house ideas/concept/needs plus outsourced R&D works to university. It's also a way to train up local talents for industries.

Led by the Government. Government departments should not set high barriers for R&D organizations or start-ups to supply new technologies for usage in government projects. KPIs should be given for government departments to adopt new technologies. When R&D solutions are proven in government projects, industry users will have confidence to use the solutions in commercial projects.

2.8.2 Quantitative Analysis

The second part of this survey through the qualitative responses from the R&D Technologists uses the quantitative assessment to evaluate the ratings of importance for these determinant factors of influencing ITF R&D project deliverables commercializing and productizing factors of time, control, quality, and motivation. The mean score, standard deviation, and determinant factors are listed in Table 2.2. The mean score range is from a minimum of 3.41 to a maximum of 4.42 and the standard deviation minimum range is around 0.28–0.34. To conclude, there are 25 items over 4 mean score and 14 items whose mean score rating is below 4. These are 64.1% and 35.9%, respectively.

1. Time factors – The highest mean score factor is 4.37 for the item of the project coordinator (PC) collecting industry need. The lowest mean score factor is 3.89 for the item of contract and agreement conditions need to be signed within 1 month after the project approval.
2. Quality factors – The highest mean score factor is 4.42 for the item of PC collecting the industry need/requirement in the proposal stage. The lowest mean score factor is 3.81 for the item of SI which joins in the final R&D process.
3. Control factors – The highest mean score is 4.37 for the item of PC collecting the industry need in the proposal stage. The lowest score factor is 3.53 for the item of patent needs to be applied in advance or before the project end.
4. Motivating factors – The highest mean score is 4.34 for the item of PC providing the technical support to system integrator to transfer the R&D Technology to the industry. The lowest score factor is 3.41 for the item of R&D results, and project deliverables pricing needs to be fixed in advance or before the project end.

These factors of each area are reallocated into different gap analysis. The mean score, standard deviation, and determinant factors are listed in Table 2.3. For questions 5.3 and 5.8, the question description is the same and explained to all respondents, and this was duplicated. Therefore, the data would be counted for the first impression of the responds; the data for question 5.8 has been ignored.

For the gap 1 analysis from market to science (M-S), the highest mean score factor is 4.42 for the PC collecting industry need and requirement in the proposal stage for the quality factor.

In gap 2, science to technology (S-T), the highest mean score is 4.33 for the factor of control of OM simplifies the technical and administrative inquiry from government to project team to eliminate the misunderstanding between government and project team.

For technology to application (T-A) gap 3, the highest mean score is 4.16 in control factor of OM which simplifies the technical and administrative inquiry from government to project team to eliminate the misunderstanding between government and project team.

For the gap 4 analysis from application to market (A-M), the highest mean score is 4.34 in motivating factors of PC who provides the technical support to system integrator to transfer the R&D Technology to the industry. Also, the lowest mean score item obtained in this survey is 3.41. The R&D results and project deliverables pricing need to be fixed in advance or before the project end.

Table 2.2 Determinant factor of commercialization of ITF R&D project deliverables in R&D Technologist perspective

Question no	Assumption factor of ITF R&D deliverables commercialization	Mean score	SD
Time factors of ITF R&D project			
2.1	Project coordinator (PC) collects the industry need/requirement in the proposal stage	4.37	0.30
2.2	Industry sponsors support the R&D project in the R&D proposal stage to speed up the approval process	4.08	0.32
2.3	Industry sponsors join in the R&D project vetting meeting to response the inquiry for increasing the funding award opportunity	4.12	0.30
2.4	Organization manager (OM) reviews the R&D proposal and suggests what information needs to be provided or amended to comply with the funding guideline to the project team	4.18	0.31
2.5	OM simplifies the technical and administrative inquiry from government to project team to eliminate the misunderstanding between government and project team	4.33	0.31
2.6	OM reviews the contract and agreement to communicate with each stakeholder to speed up the whole process schedule	4.22	0.32
2.7	OM monitors and inspects the whole R&D development deliverables and process in each milestone	3.90	0.32
2.8	Contract and agreement conditions need to be signed within 1 month after the project approval	3.89	0.33
2.9	Benefit and IP right sharing needs to be identified before the funding vetting and approval stage	3.93	0.34
Quality factors of ITF R&D project			
3.1	PC collects the industry need/requirement in the proposal stage	4.42	0.29
3.2	PC specifies the R&D project specification and outcome in detail in the proposal	4.15	0.33
3.3	Industry sponsors provide the technical or financial support to project team in the R&D development process	4.11	0.33
3.4	Industry sponsors join in the R&D development process and provide the user feedback to project team	4.08	0.33
3.5	OM reviews the R&D proposal and suggests what information needs to be provided or amended to comply with the funding guideline to project team	4.03	0.34
3.6	OM monitors and inspects the whole R&D development deliverables and process in each milestone	3.86	0.29
3.7	OM inspects the R&D final result before the final progress review meeting	3.95	0.33
3.8	System integrator (SI) joins in the final R&D process	3.81	0.34
3.9	Prototype samples are provided in the midterm project progress for industry user pretesting	3.86	0.33
Control factors of ITF R&D project			
4.1	PC collects the industry need in the proposal stage	4.37	0.30
4.2	PC specifies the R&D project specification and outcome in detail in the R&D proposal	4.15	0.32

(continued)

Table 2.2 (continued)

Question no	Assumption factor of ITF R&D deliverables commercialization	Mean score	SD
4.3	Industry sponsors join in the R&D project vetting meeting to response the inquiry for increasing the funding award opportunity	4.03	0.33
4.4	OM reviews the R&D proposal and suggests what information needs to be provided or amended to comply with the funding guideline to project team	4.06	0.33
4.5	OM simplifies the technical and administrative inquiry from government to project team to eliminate the misunderstanding between government and project team	4.16	0.32
4.6	OM reviews the contract and agreement to communicate with each stakeholder to enhance the whole process	4.00	0.33
4.7	Patent needs to be applied in advance or before the project end	3.53	0.28
4.8	Benefit and IP right sharing needs to be identified before the handing vetting and approval stage	3.86	0.34
4.9	Contract and agreement condition needs to be signed within 1 month after the project approval	3.81	0.34
Motivating factors of ITF R&D project			
5.1	PC provides the technical support to system integrator to transfer the R&D Technology to the industry	4.34	0.30
5.2	PC specifies the R&D project specification and outcome in detail in the R&D project	4.01	0.33
5.3	R&D results and project deliverables pricing need to be fixed in advance or before the project end	3.41	0.32
5.4	Prototype samples are provided in the midterm project progress for promotion	3.78	0.33
5.5	Promote final R&D results/product in the big event or exhibition before launching the product to the market	4.13	0.33
5.6	Top management's commitment to steer the R&D results adoption in the company	4.27	0.32
5.7	Training support to operation staff from project team and system integrator	4.12	0.33
5.8	R&D results and project deliverables pricing need to be fixed in advance or before the project end	3.55	0.32
5.9	Benefit and IP right sharing needs to be identified before the funding vetting and approval stage	3.78	0.32
5.10	Government provides the funding support to industry user to adopt the R&D results	4.30	0.29
5.11	R&D research institute and university provide assistance to industry user to adopt the R&D results	4.15	0.32
5.12	Government, R&D research institute and university assist the industry to open up new market	4.10	0.32

*Five-point scale: 1 = strongly disagree, 2 = somewhat disagree, 3 = neutral, 4 = somewhat agree, 5 = strongly agree

Table 2.3 Gap analysis for ITF R&D project deliverables in R&D Technologist perspective

Question no	Assumption factor of ITF R&D deliverables commercialization	Factors	Mean score	SD
Gap analysis from market to science (M-S)				
2.1	Project coordinator (PC) collects the industry need/ requirement in the proposal stage	Time	4.37	0.30
2.2	Industry sponsors support the R&D project in the R&D proposal stage to speed up the approval process	Time	4.08	0.32
2.3	Industry sponsors join in the R&D project vetting meeting to respond the inquiry for increasing the funding award opportunity	Time	4.12	0.30
3.1	PC collects the industry need/requirement in the proposal stage	Quality	4.42	0.29
4.1	PC collects the industry need in the proposal stage	Control	4.37	0.30
4.3	Industry sponsors join in the R&D project vetting meeting to response the inquiry for increasing the funding award opportunity	Control	4.03	0.33
4.6	OM reviews the contract and agreement to communicate with each stakeholder to enhance the whole process	Control	4.00	0.33
4.8	Benefit and IP right sharing needs to be identified before the funding vetting and approval stage	Control	3.86	0.34
4.9	Contract and agreement condition need to be signed within 1 month after the project approval	Control	3.81	0.34
5.9	Benefit and IP right sharing needs to be identified before the funding vetting and approval stage	Motivating	3.78	0.32
Gap analysis from science to technology (S-T)				
2.4	Organization manager (OM) reviews the R&D proposal and suggests what information needs to be provided or amended to comply with the funding guideline to the project team	Time	4.18	0.31
2.5	OM simplifies the technical and administrative inquiry from government to project team to eliminate the misunderstanding between government and project team	Control	4.33	0.31
2.6	OM reviews the contract and agreement to communicate with each stakeholder to speed up the whole process schedule	Time	4.22	0.32
2.8	Contract and agreement conditions need to be signed within 1 month after the project approval	Time	3.89	0.33
2.9	Benefit and IP right sharing needs to be identified before the funding vetting and approval stage	Time	3.93	0.34
3.2	PC specifies the R&D project specification and outcome in detail in the proposal	Quality	4.15	0.33
3.5	OM reviews the R&D proposal and suggests what information needs to be provided or amended to comply with the funding guideline to project team	Quality	4.03	0.34
4.2	PC specifies the R&D project specification and outcome in detail in the R&D proposal	Control	4.15	0.32

(continued)

Table 2.3 (continued)

Question no	Assumption factor of ITF R&D deliverables commercialization	Factors	Mean score	SD
4.4	OM reviews the R&D proposal and suggests what information needs to be provided or amended to comply with the funding guideline to project team	Control	4.06	0.33
5.2	PC specifies the R&D project specification and outcome in detail in the R&D project	Motivating	4.01	0.33
Gap analysis from technology to application (T-A)				
2.7	OM monitors and inspects the whole R&D development deliverables and process in each milestone	Time	3.90	0.32
3.3	Industry sponsors provide the technical or financial support to project team in the R&D development process	Quality	4.11	0.33
3.4	Industry sponsors join in the R&D development process and provide the user feedback to project team	Quality	4.08	0.33
3.9	Prototype samples are provided in the midterm project progress for Industry user pretesting	Quality	3.86	0.33
4.5	OM simplifies the technical and administrative inquiry from government to project team to eliminate the misunderstanding between government and project team	Control	4.16	0.32
5.4	Prototype samples are provided in the midterm project progress for promotion	Motivating	3.78	0.33
5.7	Training support to operation staff from project team and system integrator	Motivating	4.12	0.33
5.11	R&D research institute and university provide assistance to industry user to adopt the R&D results	Motivating	4.15	0.32
Gap analysis from application to market (A-M)				
3.6	OM monitors and inspects the whole R&D development deliverables and process in each milestone	Quality	3.86	0.29
3.7	OM inspects the R&D final result before the final progress review meeting	Quality	3.95	0.33
3.8	System integrator (SI) joins in the final R&D process	Quality	3.81	0.34
4.7	Patent needs to be applied in advance or before the project end	Control	3.53	0.28
5.1	PC provides the technical support to system integrator to transfer the R&D Technology to the industry	Motivating	4.34	0.30
5.3	R&D results and project deliverables pricing need to be fixed in advance or before the project end	Motivating	3.41	0.32
5.5	Promote final R&D results/product in the big event or exhibition before launching the product to the market	Motivating	4.13	0.33
5.6	Top management's commitment to steer the R&D results adoption in the company	Motivating	4.27	0.32
5.10	Government provides the funding support to industry user to adopt the R&D results	Motivating	4.30	0.29
5.12	Government, R&D research institute and university assist the industry open up new market	Motivating	4.10	0.32

*Five-point scale: 1 = strongly disagree, 2 = somewhat disagree, 3 = neutral, 4 = somewhat agree, 5 = strongly agree

2.9 Discussion and Conclusions

The aim of R&D is to improve the current technologies offered by an organization or to develop innovation that strengthens the organization's position in the marketplace. This research investigates ITF R&D project deliverables commercialization issue to the market through past literature to define the key problem in the development process and analyze the critical factors which caused among each stage of R&D phases.

Qualitative and quantitative analysis of data collected from R&D Technologists has been carried out using a proposed market-driven project management model, MSTAM. This is the first attempt of the multi-perspective study to bridge the gaps from idea generation, R&D, and subsequent technology implementation and adoption by industry. The traditional ITF R&D project operation mechanism has many monitoring procedures and operation steps to examine the R&D project progress. This traditional ITF R&D project operation mechanism may be ignored the R&D deliverable whether suitable apply to the industry in a reasonable time, need and fulfil the expectation of industry and R&D Technologist. The key issue affecting the R&D technology commercialization, it should be complicated and long project time schedule. This study contributes a novel market-driven project management model, MSTAM, to define the barrier and gap in each development process.

This is an initial study for ITF R&D project and the data only collected 73 R&D Technologist comments in the technologist side. The current R&D Technologists are still working on ITF R&D project of around 166 people in 2016. The response rating is around 44%. This survey is concentrated in the four key factors, time, quality, control, and motivation, to collect the data from the R&D Technologists to analyze the main concern and expectation for them in ITF R&D project development process. According to the research results, there are three key factors in the development process; these are:

1. PC collects the industry need/requirement in the proposal stage; this is the key element of time factor, quality factor, control factor, and critical gap from market to science in the ITF R&D process.
2. PC provides the technical support to system integrator to transfer the R&D Technology to the industry; this is the key element of motivating factor and critical gap from application to market in the ITF R&D process.
3. OM simplifies the technical and administrative inquiry from government to project team to eliminate the misunderstanding between government and project team; this is the key element in the gap of S-T and T-A in the ITF R&D process.

Referring to the captioned findings, industry need and R&D Technologist-applied research may need to be aligned for increasing the commercialization opportunity to the market. The further study will be collected for the industry point of view and a real case study or pilot implementation needs to be tested to prove the proposed model whether suitable in ITF R&D project. The author believes the survey results are a good indicator for all of us to explore how to enhance the ITF R&D project commercialization rate to the market. The analytical framework resulting from this research is proved that the proposition and MSTAM are the correct direction for assisting future research of ITF R&D project commercialization.

Acknowledgments The authors would like to take this opportunity to thank the industrial advisors, Dr. Jack So, Dr. Aldar Chan, Dr. CC Cheung, and Dr. KL Fan for providing their valuable comment for the questionnaire. The authors also would like to express their sincere thanks and warmest appreciation to those R&D Technologists for sharing their valuable thoughts and incredible insights.

Appendix: R&D Technologist Survey Questionnaire

CityU

College of Science and Engineering

Department of Systems Engineering and Engineering Management

Dear Sir/Madam,

Re: Survey on Management of Commercialization of ITF R&D Projects Deliverables

The City University of Hong Kong is conducting a research regarding what critical factors affect ITF R&D project commercializing and productizing the results to the Logistics and Supply Chain industry in Hong Kong. You are cordially invited to complete this questionnaire based on your experiences. You should be able to answer all questions in 10–15 min.

This questionnaire is part of an R&D survey for an **EngD research study** to identify the critical element and determinants factor influencing ITF's R&D results commercialization and adoption in Hong Kong Logistics and Supply Chain Company.

Click here to participate in the online survey: https://zh.surveymonkey.com/r/6SH76PJ

All specific responses will be treated as **strictly confidential** and used for academic research purposes in aggregated forum only. The research report will not release any individual company information and data. The survey result developed by this study will be feedback to the research institute for continuous improvement. Therefore your participation will contribute significantly to the research institute.

You are encouraged to invite your qualified colleagues to participate in this survey.

We sincerely hope that you could participate in this important research and believe that its findings will be useful for you and your company. For more information, please feel free to contact me at 9230–5672 or scho25-c@my.cityu.edu.hk. Your reply on or before 10 February 2017 will be highly appreciated.

Best regards,
Dee Ho
For further information about this survey, please contact:
Ho Siu Cheung
Engineering Doctorate Candidate
Systems Engineering and Engineering Management
City University of Hong Kong
(Tel: 852 9230-5672, Email: scho25-c@my.cityu.edu.hk)
Supervisor:
Dr. Chuah, Kong Bieng
Associate Professor
EngD Deputy Programme Leader & EngD Thesis Coordinator
Department of Systems Engineering and Engineering Management
City University of Hong Kong

(Tel: 852 3442-8437, Email: mebchuah@cityu.edu.hk)

Name:	
Organization:	
Job Title:	
Date:	

A. What critical factors influence the R&D project results commercialization and adoption to the Hong Kong Logistics and Supply Chain Company, and how well the industry user is willing to use this R&D results in corporate?

This study aims to find out the determinant factors of Innovation and Technology Fund (ITF)'s research and development (R&D) project results commercialization and adoption in Hong Kong Logistics and Supply Chain Industry and which critical issue affects technology transfer to the industry. R&D Technologist always faces various challenges to explore and develop new R&D project to the Logistics and Supply Chain Industry effectively and match with the industry need.

This section examines the possible factors that may influence the R&D results adoption in Hong Kong Logistics and Supply Chain Industry. Based on you experiences (or what you believe), please tick the box below each statement that most accurately reflects the extent to which you agree or disagree with following statement.

Remark:

1. Key player in the R&D ecosystem are R&D research institute, university, and system integrator (SI).
2. Organization manager (OM) is a role of research institution manager handling the whole R&D process.

3. Project coordinator (PC) is a key technologist/developer to develop and oversee overall project executions and steer project directions.
4. Project manager (PM) is a key project team member to assist the project coordinator in managing the project scheme throughout the whole development period.
5. Intellectual property (IP) refers to creations of the intellect for which a monopoly is assigned to designated owners by law.
6. System integrator (SI) is a role to transfer R&D projects results into the company.

1. How would you agree on the importance of the following tasks in the R&D development process that **may accelerate the R&D processing time**?

		Strongly disagree	Somewhat disagree	Neutral	Somewhat agree	Strongly agree
1.1	Project coordinator (PC) collects the industry need/requirement in the proposal stage	☐	☐	☐	☐	☐
1.2	Industry sponsors support the R&D project in the R&D proposal stage to speed up the approval process	☐	☐	☐	☐	☐
1.3	Industry sponsors join in the R&D project vetting meeting to respond the inquiry for increasing the funding award opportunity	☐	☐	☐	☐	☐
1.4	Organization manager (OM) reviews the R&D proposal and suggests what information needs to be provided or amended to comply with the funding guideline to the project team	☐	☐	☐	☐	☐
1.5	OM simplifies the technical and administrative inquiry from government to project team to eliminate the misunderstanding between government and project team	☐	☐	☐	☐	☐
1.6	OM reviews the contract and agreement to communicate with each stakeholder to speed up the whole process schedule	☐	☐	☐	☐	☐
1.7	OM monitors and inspects the whole R&D development deliverables and process in each milestone	☐	☐	☐	☐	☐
1.8	Contract and agreement conditions need to be signed with in 1 month after the project approval	☐	☐	☐	☐	☐
1.9	Benefit and IP right sharing need to be identified before the funding vetting and approval stage	☐	☐	☐	☐	☐

1.10 Others

2. How would you agree with the significance of the following personnel involvement in the R&D development process that **may increase the value or positively impact** of the R&D project results

		Strongly disagree	Somewhat disagree	Neutral	Somewhat agree	Strongly agree
2.1	PC collects the industry need/requirement in the proposal stage	☐	☐	☐	☐	☐
2.2	PC specifies the R&D project specification and outcome in detail in the proposal	☐	☐	☐	☐	☐
2.3	Industry sponsors provide the technical or financial support to project team in the R&D development process	☐	☐	☐	☐	☐
2.4	Industry sponsors join in the R&D development process and provide the user feedback to project team	☐	☐	☐	☐	☐
2.5	OM reviews the R&D proposal and suggests what information needs to be provided or amended to comply with the funding guideline to project team	☐	☐	☐	☐	☐
2.6	OM monitors and inspects the whole R&D development deliverables and process in each milestone	☐	☐	☐	☐	☐
2.7	OM inspects the R&D final result before the final progress review meeting	☐	☐	☐	☐	☐
2.8	System integrator (SI) joins in the final R&D process	☐	☐	☐	☐	☐
2.9	Prototype samples are provided in the mid-term project progress for Industry user pre-testing	☐	☐	☐	☐	☐

2.10 Others

3. How would you agree on the key monitoring mechanism or control process of the following that **may eliminate the project risk and solve the communication issue** in the R&D process?

		Strongly disagree	Somewhat disagree	Neutral	Somewhat agree	Strongly agree
3.1	PC collects the industry need in the proposal stage	☐	☐	☐	☐	☐
3.2	PC specifies the R&D project specification and outcome in detail inthe R&D proposal	☐	☐	☐	☐	☐
3.3	Industry sponsors join in the R&D project vetting meeting to response the inquiry for increasing the funding award opportunity	☐	☐	☐	☐	☐
3.4	OM reviews the R&D proposal and suggest what information needs to be provided or amended to comply with the funding guideline to project team	☐	☐	☐	☐	☐
3.5	OM simplifies the technical and administrative inquiry from government to project team to eliminate the misunderstanding between government and project team	☐	☐	☐	☐	☐
3.6	OM reviews the contract and agreement to communicate with each stakeholder to enhance the whole process.	☐	☐	☐	☐	☐
3.7	Patent needs to be applied in advance or before the project end	☐	☐	☐	☐	☐
3.8	Benefit and IP right sharing needs to be identified before the funding vetting and approval stage	☐	☐	☐	☐	☐
3.9	Contract and agreement condition needs to be signed with in 1 month after the project approval	☐	☐	☐	☐	☐

3.10 Others

4. How would you agree on the following promotion activities in the R&D development process that **motivate or help** industry users to adopt or license the R&D results in their company?

		Strongly disagree	Somewhat disagree	neutral	Somewhat agree	Strongly agree
4.1	PC provides the technical support to system integrator to transfer the R&D technology to the industry	☐	☐	☐	☐	☐
4.2	PC specifies the R&D project specification and outcome in detail in the R&D project	☐	☐	☐	☐	☐
4.3	R&D results and project deliverables pricing need to be fixed in advance or before the project end	☐	☐	☐	☐	☐
4.4	Prototype samples are provided in the mid-term project progress for promotion	☐	☐	☐	☐	☐
4.5	Promote final R&D results/product in the big event or exhibition before launching the product to the market	☐	☐	☐	☐	☐
4.6	Top management's commitment to steer the R&D results adoption in the company	☐	☐	☐	☐	☐
4.7	Training support to operation staff from project team and system integrator	☐	☐	☐	☐	☐
4.8	R&D results and project deliverables pricing need to be fixed in advance or before the project end	☐	☐	☐	☐	☐
4.9	Benefit and IP right sharing needs to be identified before the funding vetting and approval stage	☐	☐	☐	☐	☐
4.10	Government provides the funding support to industry user to adopt the R&D results	☐	☐	☐	☐	☐
4.11	R&D Centre and University provide assistance to industry user to adopt the R&D results	☐	☐	☐	☐	☐
4.12	Government, R&D Centre and University assist the industry open up new market	☐	☐	☐	☐	☐

4.13 Others

B. What are the determinants of ITF's R&D projects results commercialization and adoption for Hong Kong Logistics and Supply Chain Industry? (Brief reply is good enough.)

1. In general, R&D project development process involved multiple developers and users with different interests. What do you think are their interests and concern, and to what extent do you think collaboration is possible among these key developers and users?

2. In your opinion, what do you think is the main reason(s) or major barrier(s) why ITF's R&D results commercialization and adoption rate is not good in Hong Kong Logistics and Supply Chain Industry?

3. What do you think would be the solution to the problems identified in Q (2)?

4. Base on your experiences, what are the appropriate action(s) or element(s) to motivate the industry users or top management in the company to adopt or license

the R&D results in their company, or what are the key considerations for them to adopt new technology or R&D results in Hong Kong?

References

Amadi-Echendu, J. E., & Rasetlola, R. T. (2011). *Technology commercialization factors, frameworks and models*. IEEE Conference Publications, pp. 144–148.

Ana Ferraz, José Machado, & Vítor Carvalho (2015). *Prototype for determination of pre-transfusion tests based on image processing techniques*. IEEE Conference Publications, pp. 1–6.

Ana Lima, Gabriela Fernandes, & Ricardo J. Machado (2016). *Project and program management implications in the portfolio management of IT projects in applied R&D organizations*. IEEE Conference Publications, pp. 224–229.

Bangchao Wang (2016). *Requirements traceability technologies selection for industry*. IEEE Conference Publications, pp. 450–455.

Boulter, L. (2004). *Intellectual property issues for R&D management of global-industry academic links*. IEEE Conference Publications, pp. 634–645.

Bruno M. Arnaut, Denise B. Ferrari, & Marcelo Lopes de Oliveira e Souza (2016). *A requirements engineering and management process in concept phase of complex systems*. IEEE Conference Publications, pp. 1–6.

Chapman, R., & Hyland, P. (2004). Complexity and learning behaviors in product innovation. *Technovation, 24*, 553–561.

Cheng-Mei Tung, Ta-Yu Tseng, & Wan-Chang Yen (2013). *Fostering innovation commercialization at research institute and university: Strategy and policy implications*. IEEE Conference Publications, pp. 928–934.

Christof Ebert, & Carlos Henrique C. Duarte (2016). *Requirements engineering for the digital transformation: Industry panel*. IEEE Conference Publications, pp. 4–5.

Cooper, R. G. (1983). A process model for industrial new product development. *IEEE Transactions on Engineering Management, EM-30*(1), 2–11.

Cooper, R. G. (1987). Defining the new product strategy. *IEEE Transactions on Engineering Management, EM-34*(3), 184–193.

Cooper, R. G. (2011). *Winning at new products: Creating value through innovation* (4th ed.p. 194). New York: Basic Books.

Daniel Kasperek, Stefanie Fink, Sebastian Maisenbacher, Wolfgang Bauer, & Maik Maurer (2014). *Assessing the informative value of complexity metrics within design structure matrices in early development phases of complex systems*. IEEE Conference Publications, pp. 351–356.

Dong-dong Gao, & Ping, L. V. (2009). *A sorting-based management model to support early supplier involvement in new product development*. IEEE Conference Publications, pp. 496–501.

Fagno Fonseca, Patrick Letouzé, Rubens Pompeu, Leandro Garcia, Selma Regina, & George França (2016). *Barriers in information technology project management*. IEEE Conference Publications, pp. 370–375.

Fareena Sultan, & Lillian Chan (2000). The adoption of new technology: The case of object-oriented computing in software companies. *IEEE Transactions on Engineering Management, 47*(1), 106–206.

Florian Wohlfeil, & Orestis Terzidis (2014). *Critical success factors for the strategic management of radical technological innovation*. IEEE Conference Publications, pp. 1–9.

Franken, S., Kolvenbach, S., & Gräther, W. (2016). *Involving end users into collaborative software development: The showcase of CloudTeams*. IEEE Conference Publications, pp. 821–826.

Gilbert Regan, Derek Flood, & Fergal McCaffery (2016). *Research findings from an industrial trial of a traceability assessment and implementation framework*. IEEE Conference Publications, pp. 91–95.

Grandjonc Thomas, & Khattabi Yahya (2016). *Benchmarking of the CAE-MPIC interface cards using ARINC 661 standard: An example of a university-industry collaboration*. IEEE Conference Publications, pp. 1–8.

Harshada Patel, Sue Cobb, Madeline Hallewell, Mirabelle D'Cruz, Richard Eastgate, Lorenzo Picinali, & Stefano Tamascelli (2016). *User involvement in design and application of virtual reality gamification to facilitate the use of hearing aids*. IEEE Conference Publications, pp. 77–81.

Hartley, J. L., Meredith, J. R., McCutcheon, D., & Kamath, E. R. (1997). Suppliers' contributions to product development: An exploratory study. *IEEE Transactions on Engineering Management, 44*(3), 258–267.

Herlandí de Souza Andrade, Ligia M. Soto Urbina, Andrea de Oliveira N. Follador, & Roberto da Cunha Follador (2016). *Processes proposal for the intellectual property protection management in a technology licensing office from a Brazilian Scientific and Technological Institution*. IEEE Conference Publications, pp. 1672–1680.

HKTDC (2016). *Economic and trade information on Hong Kong*. Retrieved from: http://hong-kong-economy-research.hktdc.com/business-news/article/Market-Environment/Economic-and-Trade-Information-on-Hong-Kong/etihk/en/1/1X000000/1X09OVUL.htm. 20 May 2016.

Hult, G. T. M., Hurley, R. F., & Knight, G. A. (2004). Innovativeness: Its antecedents and impact on business performance. *Industrial Marketing Management, 33*, 429–438.

ITB (2016). *Innovation and Technology Bureau established*. Retrieved from: http://www.hketosydney.gov.hk/eng/press/2015/1120b.htm. 17 May 2016.

ITF (2013). *Chapter 9: Innovation and technology fund*. Retrieved from: http://www.aud.gov.hk/pdf_e/e61ch09.pdf. 17 May 2016.

Jian Lin, & Qing Lin (2017). *Analysis of the key factors of intellectual property management at art institutions*. IEEE Conference Publications, pp. 206–208.

John E. Ettlie (1995). *Early manufacturing involvement in new product development*. IEEE Conference Publications, pp. 104–109.

Joko Siswanto, & Anggi Maulida (2014). *ERP module requirements for micro, small and medium enterprise fashion industry in Bandung*. IEEE Conference Publications, pp. 183–188.

Joko Siswanto, & Anggi Maulida (2016). *Validated ERP modules requirement for micro, small and medium enterprise fashion industry*. IEEE Conference Publications, pp. 1–6.

Kamil Yavuz Kapusuz, & Ali Kara (2014). *An industry sponsored undergraduate research (UR) experience: Preliminary study on fulfillment of program outcomes and industry requirements*. IEEE Conference Publications, pp. 306–311.

Kawtar Belrhiti Alaoui, Zakaria Naimi, Ahmed Benlarabi, & Abdelkader Outzourhit (2015). *Pilot line preindustrial reactor installation for applied research in vacuum deposition techniques for the preparation and characterization of photovoltaic cells*. IEEE Conference Publications, pp. 1–3.

Kellner, A., Ahrens, M., Friedl, M., Hehenberger, P., Weingartner, L., Zeman, K., Kernschmidt, K., Feldmann, S., & Vogel-Heuser, B. (2016). *Challenges in integrating requirements in model based development processes in the machinery and plant building industry.* IEEE Conference Publications, pp. 1–6.

Konstantinos Smparounis, Dimitris Mavrikios, Menelaos Pappas, Vagelis Xanthakis, Giovanni Paolo Viganò, & Katharina Pentenrieder (2008). *A virtual and augmented reality approach to collaborative product design and demonstration.* IEEE Conference Publications, pp. 1–8.

LC Paper (2015). *Legislative Council Panel on Commerce and Industry – Progress Report of Research & Development (R&D) Centres for 2014–15 cum Comprehensive Review on R&D Centres for 2012–12 to 2014–15.* Retrieved from: http://www.legco.gov.hk/yr14-15/english/panels/ci/papers/ci20150616cb1-971-3-e.pdf. 20 May 2017.

LC Paper (2016). *Legislative Council Panel on Commerce and Industry – Progress Report of Research & Development (R&D) Centres for 2015–16.* Retrieved from: http://www.legco.gov.hk/yr15-16/english/panels/ci/papers/ci20160621cb1-1045-5-e.pdf. 20 May 2017.

LC Paper (2017). *Legislative Council Panel on Commerce and Industry – Progress Report on Research & Development Centres for 2016–17.* LC Paper No. CB (1)1144/16–17(03). Retrieved from: http://www.legco.gov.hk/yr16-17/english/panels/ci/papers/ci20170620cb1-1144-3-e.pdf. 14 July 2017.

Lenina, S. V. B. (2016). *Intellectual property protection: A step to realize make in India.* IEEE Conference Publications, pp. 2089–2092.

Liliana Guzmán, Silke Steinbach, Philipp Diebold, Thomas Zehler, Klaus Schneider, & Matthias Habbe (2016). *Evaluating the benefits of systematic project management in large public sector projects.* IEEE Conference Publications, pp. 32–35.

LSCM (2017). *R&D focus.* Retrieved from: http://www.lscm.hk/eng/channel.php?channel=rd-focus. 22 Feb 2017.

LSCM Survey (2008). *Hong Kong logistics and supply chain industry survey.* Retrieved from: http://www.haffa.com.hk/files/2_LSCM%20MI%20Project%20Survey%20Result.pdf. 18 May 2016.

Mamoru Kiyota, & Hiroshi Kubo (2016). *A training method proposal for R&D project managers.* IEEE Conference Publications, pp. 2088–2098.

Marcos Lopez Rego, & Luiz Carlos de SB Carvalho (1995). *Managing R&D projects: A systemic approach.* IEEE Conference Publications, pp. 281–286.

Mats P.E. Heimdahl, & Jeflrey M. Thompson (2000). *Specification based prototyping of control systems.* IEEE Conference Publications, vol. 1, pp. 1D3/1–1D3/8.

Michael Mutingi, & Stephen Matope (2013). *Dynamics of information technology adoption in a complex environment.* IEEE Conference Publications, pp. 1466–1471.

Misoo Kwon, Hoyeon Ryu, Gunhee Kim, & Sungdo Ha (2006). *Design of OWL ontology in R&D project management meeting.* IEEE Conference Publications, pp. 417–423.

Norhadilah Abdul Hamid, & Azmawani Abd. Rahman (2014). *A systematic literature review on the success factor of innovation commercialization performance.* IEEE Conference Publications, pp. 199–204.

Phaal, R., O'Sullivan, E., Routley, M., Ford, S., & Probert, D. (2011). A framework for mapping industrial emergence. *Technological Forecasting and Social Change, 78*(2), 217–230.

Qing Lin, & Shawn Wong (2017). *A study of intellectual property protection for mass innovation spaces.* IEEE Conference Publications, pp. 973–975.

Research Report (2011). *The road to sustainable development of Hong Kong SMEs – the present and future.* Retrieved from: https://www.smefund.tid.gov.hk/english/sdf/deliverables/Sustainabledev_SMEs_research_report_eng.pdf. 19 May 2016.

Robert Chiang, I., & Jinhui Wu, S. (2016). Supplier involvement and contract design during new product development. *IEEE Transactions on Engineering Management, 63*(2), 248–258.

Roberts, E. B. (1988). Managing invention and innovation. *Research Technology Management, 31*(1), 13–19.

Rothwell, R., & Zegveld, W. (1985). *Reindustrialization and technology.* Harlow: Longman.

Seo-Kyun Kim, Bong-Gyou Lee, & Kyoung-Seok Oh (2009). *The effect of R&D and technology commercialization capabilities on the innovation performance of Korean IT SMEs: The case of direct and indirect receipts of public R&D funding*. IEEE Conference Publications, pp. 1531–1541.

Sheila Lam (2016). *Policy address '16 review: Reintroducing re-industrialization, computer world Hong Kong*. Retrieved from: http://cw.com.hk/feature/policy-address-16-review-reintroducing-re-industrialization. 27 Jan 2016.

Sofija Hotomski, Eya Ben Charrada, & Martin Glinz (2016). *An exploratory study on handling requirements and acceptance test documentation in industry*. IEEE Conference Publications, pp. 116–125.

Souhaib Besrour, Lukman Bin Ab Rahim, & P. D. D. Dominic (2015). *The study of the relation between requirement engineering techniques and challenges in software industry*. IEEE Conference Publications, pp. 49–53.

Suleiman K. Kassicieh, Steven T. Walsh, John C. Cummings, Paul J. McWhorter, Alton D. Romig, and W. David Williams (2002). Factors differentiating the commercialization of disruptive and sustaining technologies. *IEEE Transactions on Engineering Management, 49*(4), 375–387.

Uda Hashim (2002). *Development of as silicon sensor wafer fab for local industry requirement*. IEEE Conference Publications, pp. 145–149.

Venlakaisa Holtta, Taneli Eisto, & Katrine Mahlamaki (2009). *Benefits for cast product development through early supplier involvement*. IEEE Conference Publications, pp. 1–7.

Vittorio Chiesa (2001). Models of technological innovation process. *R&D Strategy and Organization Managing Technical Change in Dynamic Contexts, 4*, 6–7.

Wessner, C. W. (2005). Discussion Papers on entrepreneurship, Growth and Public Policy, "MYTHS and Realities about Government Support of Industry R&D". p. 6. Retrieved from https://papers.econ.mpg.de/egp/discussionpapers/2005-06.pdf.

Xiaohua Sun (2016). *Research on risk management of engineering project*. IEEE Conference Publications, pp. 1–5.

Yayoi Hirose (2012). *Knowledge transfer from researches to society: How to offset the cultural gap?* IEEE Conference Publications, pp. 1107–1116.

Yuji Miyashita, Yuriko Sawatani, & Kiyoshi Niwa (2006). *Improved market-creation R&D management through researcher-consultant collaboration*. IEEE Conference Publications, pp. 663–667.

Zhao, C. L., Ming, X. G., Wang, X. H., & Li, D. (2009). *A framework of supplier involved collaborative project management*. IEEE Conference Publications, pp. 4130–4135.

Chapter 3
A Meta-Analysis of Sustainable Supplier Selection Approaches

Felix Tuczek and Tina Wakolbinger

3.1 Introduction

A company is no more sustainable than the vendors from which it sources (Krause et al. 2009). Therefore, supplier selection plays a key role for achieving sustainable supply chains due to the decisive influence at the beginning of the material flow of the company (Preuss 2001). Moreover, the outsourcing trend is still ongoing (Van Weele 2014) emphasizing the importance of selecting sustainable suppliers. The supplier selection process is a multi-criteria decision-making process evoking trade-offs between competing aims such as quality and costs (e.g., Jain et al. 2007) considering qualitative and quantitative factors (e.g., Sarkis and Talluri 2002). Since the "traditional" supplier selection process is a complex task, the incorporation of environmental and social criteria in recent times increases the complexity of supplier selection decisions (Bai and Sarkis 2010a, Bai and Sarkis 2010b). Therefore, the sustainable supplier selection process can be defined as the traditional supplier selection process enhanced with environmental and social factors to compare the performance of potential suppliers (Azadnia et al. 2015, Genovese et al. 2013). In order to operationalize sustainability, the triple bottom-line (TBL) approach is widely applied in literature encompassing economic, environmental, and social dimensions (Elkington 1998). Several researchers published literature reviews on supplier selection methods focusing on the economic dimension (Aissaoui et al. 2007, De Boer et al. 2001, Wu and Barnes 2011). Furthermore, reviews on green supplier selection were conducted (Genovese et al. 2013, Govindan et al. 2015, Igarashi et al. 2013). Recently, Wetzstein et al. (2016) reviewed topics of sustainable supplier selection without an analysis of methods and criteria applied. Apart

F. Tuczek (✉) · T. Wakolbinger
Institute for Transport and Logistics Management, WU (Vienna University of Economics and Business), Vienna, Austria
e-mail: felix.tuczek@wu.ac.at; tina.wakolbinger@we.ac.at

H. Qudrat-Ullah (ed.), *Innovative Solutions for Sustainable Supply Chains*,
Understanding Complex Systems, https://doi.org/10.1007/978-3-319-94322-0_3

from thematically deviating objectives from this review paper, the paper of Wetzstein et al. (2016) is not included because the present review paper covers the time horizon until 2015.

Even though reviews on economic and environmental dimensions of the triple bottom line (TBL) exist, a comprehensive review on sustainable supplier selection is required for structuring this fast growing field of research. Drawing on previously published reviews on supplier selection permits to gain deeper insights into the interplay between methods and criteria applied to approaches for supplier selection. Moreover previous reviews on supplier selection focus on (1) the identification of criteria and (2) methods to select suppliers (Perçin 2006). In contrast, the objective of this review paper is to combine these two perspectives by linking the methods employed with economic, environmental, and social criteria and subsequently scrutinize whether or not methods applied to each dimension differ. Given the importance of industry-related factors (Chen et al. 2014), it is analyzed how they are reflected in the literature. The paper answers the following research questions:

1. Which criteria are applied to measure sustainability dimensions in papers on supplier selection?
2. Which methods are applied to supplier selection in these papers?
3. Are industry-related characteristics reflected in approaches on sustainable supplier selection?

The remainder of this paper is structured as follows. The paper starts with outlining the content analysis method as applied in the research process. Further, the findings of the content analysis are discussed with a particular focus on sustainability criteria and approaches applied to sustainable supplier selection. Then collected papers are scrutinized in terms of industry-specific requirements of sustainability efforts. Subsequently, certifiable environmental standards are discussed. This will be followed by an identification of research gaps and recommendations for future research as well as managerial implications.

3.2 Research Method

The present review aims to compile approaches for supplier selection and the corresponding sustainability criteria employed. Furthermore, our objective is to analyze industry-specific characteristics in the area of sustainable supplier selection. We employ content analysis based on two interconnected successive key searches. Since a considerable amount of literature reviews concerning methods and/or criteria for supplier selection are published, we start with a meta-analysis of previous literature reviews. We then complement it with a second search for recent literature.

Content analysis permits to condense existing literature reliably, objectively, and systematically (e.g., Brandenburg et al. 2014, Seuring and Gold 2012). Mayring (2010) developed a process model for content analysis consisting of the following steps:

1. Material collection: the material to be collected and the unit of analysis (e.g., single paper) are defined and delimited.
2. Descriptive analysis: formal aspects of the materials are assessed, e.g., number of publications per year and publications per journal.
3. Category selection: the structural dimensions and related analytic categories, which are applied to the collected material, are selected.
4. Material evaluation: the material is analyzed and sorted corresponding to the structural dimensions and related analytic categories. This should permit to identify relevant issues and to interpret the results.

3.2.1 Material Collection

The first literature screening aims at identifying reviews considering methods in combination with criteria applied to the supplier selection process. For this meta-analysis, we combine the keywords "supplier" OR "partner" OR "vendor" AND "selection" AND "review" OR "state of the art" OR "methods and criteria" yielding search strings such as "supplier selection review" or "partner selection methods and criteria" contained in the document title. These papers are searched in scientific databases including Emerald, ScienceDirect, ABI/Inform ProQuest, Springer-Link Journals, and Wiley.

After eliminating duplicate results, the search for literature reviews yielded 13 unique results. The papers were then screened to make sure that they are in line with the focus of this paper and that they list the methods used in each collected individual paper permitting the assignment of criteria applied to each paper. Seven papers did not fulfill these criteria. For instance, the review conducted by De Boer et al. (2001) does neither assess the methods applied numerically nor highlights the criteria considered in the collected papers. Hence, De Boer et al. (2001) are not included in this paper. Consequently, six review papers are taken into account for the meta-analysis in the present paper. Three of these papers address the environmental dimension of TBL (Genovese et al. 2013, Govindan et al. 2015, Igarashi et al. 2013); another three consider economic factors (Chai et al. 2013, Ho et al. 2010, Weber et al. 1991). Reviews focusing on social criteria for supplier selection have not been published so far. Furthermore, no evaluation of methods for supplier selection integrating all three sustainability dimension is available until 2015. The literature cited in these six reviews was checked in terms of duplication and alignment with respect to the aims pursued by this paper. Two hundred forty-four papers with a focus on economic criteria as well as 23 papers with a focus on environmental criteria were chosen for the review paper.

The second step aims at closing the gap to current state of the art. We conduct a key search covering the time interval from 2012 to 2015. We start in January 2012 in order to connect directly to the last reviews analyzing criteria and methods in a single review concerning environmental criteria. We searched the same databases as in the first search mentioned above. In order to identify appropriate keywords for

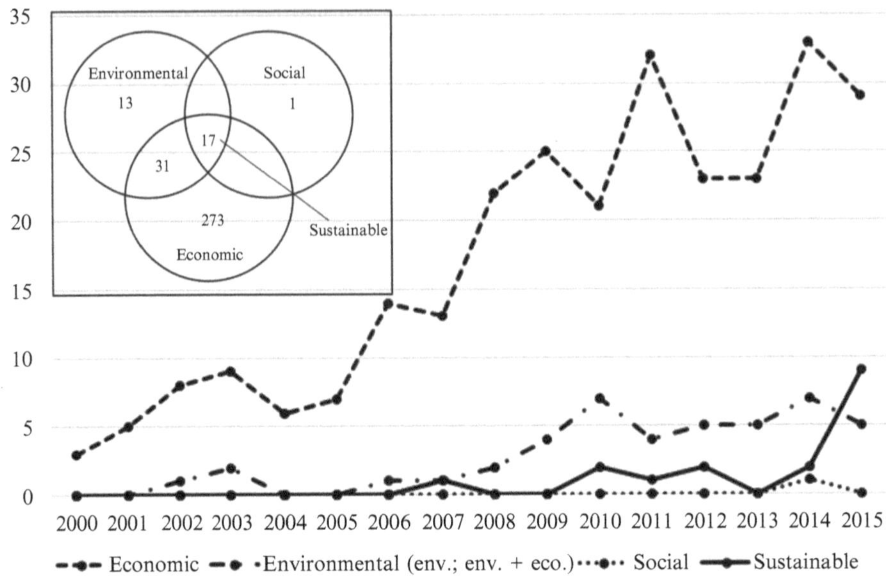

Fig. 3.1 Number of publications on supplier selection from 2000 to 2015 over time and summarized in Venn diagram. (Following Carter and Rogers 2008)

this second search, we orientate toward the search strings utilized by De Boer et al. (2001), Wu and Barnes (2011), as well as Bruno et al. (2012). This leads to a combination of the keywords "supplier" AND "select*" OR "eval*" within the document title. This search yielded 404 initial results, out of which 143 papers were chosen for the review.

3.2.2 *Descriptive Analysis*

Supplier selection has its roots in the study conducted by Dickson (1996) and the work of Weber et al. (1991). Green purchasing initiatives started during the 1980s and 1990s (Dowlatshahi 2000), but first papers on green purchasing appeared in the middle of the 1990s (Igarashi et al. 2013). As Weber et al. (1991) do not explicitly address environmental issues (Igarashi et al. 2013), we display the evolution of approaches for sustainable supplier selection starting in the year 2000 in accordance with the review conducted by Ho et al. (2010). In Fig. 3.1, the number of peer-reviewed papers on mathematical models for supplier selection that appeared in scientific journals is presented over time. According to the three dimensions of TBL, papers are divided into economic (dashed line), environmental (dot chain line), social (dotted line), and sustainable (solid line). Economic refers to models applying criteria such as price and quality. Environmental depicts papers encompassing at least one environmental criterion regardless of the integration of economic criteria or environmental criteria solely. Furthermore, social means the consideration of social criteria or a combination of social and economic criteria. Papers considering sustainable topics integrate the economic, environmental, and social dimension of TBL within a single paper. In total, the amount of papers on supplier selection rises. Economic criteria are addressed most often with an increasing number of publications. Environmental issues are tackled since 2009 with approximately five publications per year. Only one paper in this review addresses social issues exclusively. Papers on supplier selection approaches covering all three dimensions of TBL gain considerable attention in scientific literature. Nine out of 17 collected papers addressing sustainability criteria are published in 2015. Even more, in 2015 more papers considering all three dimensions of TBL are published than papers integrating only environmental criteria.

Further, the Venn diagram integrated in Fig. 3.1 summarizes the collected papers published between 2000 and 2015 with a focus on interfaces in terms of different dimensions of TBL. As already mentioned, 17 papers tackle economic, environmental, and social criteria simultaneously which is categorized as sustainable. Two hundred seventy-three papers take into account economic criteria for supplier selection, and one paper focuses on social criteria exclusively. Papers dealing with environmental criteria are covered by 17 publications integrating all 3 dimensions, 31 papers consider economic and environmental criteria in a single paper, and 13 approaches on supplier selection consider environmental issues exclusively.

Figure 3.2 displays the number of papers published by a journal. Moreover, the amount of papers in each dimension is presented. Only one paper addresses social topics exclusively. *Expert Systems with Applications* publishes the highest share of papers on sustainable supplier selection approaches regardless of the specific dimension, 68 on economic, 8 on environmental, and 6 on sustainable dimensions. Overall, most papers were published in journals with strong quantitative traditions. Regarding environmental issues covered by the collected papers, several journals with origin in environmental subjects publish papers on sustainable supplier selection. This is most obvious by means of *Journal of Cleaner Production*. This journal published four papers covering environmental criteria and three approaches focusing on the integration of all three dimensions of TBL simultaneously.

Interesting is the high share of papers published in *Expert Systems with Applications*. It is surprising since its scope lies not on purchasing and supplier selection and no special issue on this topic appears. This could be explained by the huge amount of articles published per year (12 issues and about 167 articles per issue). For comparison, the *European Journal of Operational Research* publishes 3 issues with approx. 26 papers per year, and the *International Journal of Production Economics* publishes around 10 volumes per year consisting of approx. 30 articles.

3.2.3 Category Selection

The categorization in this review refers to the TBL approach with economic, environmental, and social dimensions (Elkington 1998). This separation is widely applied in literature (Carter and Easton 2011, Pagell et al. 2010, Seuring 2013) and

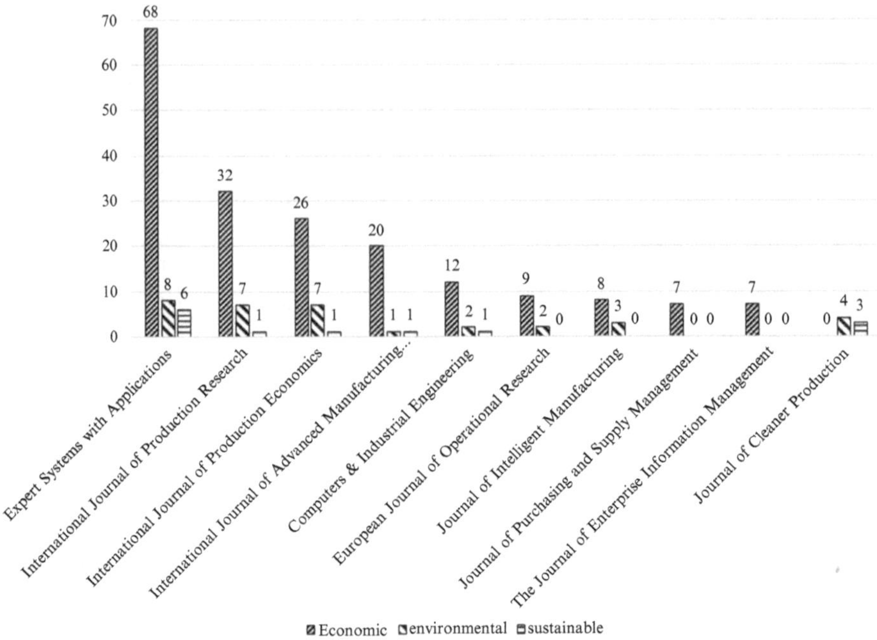

Fig. 3.2 Most frequently applied sources of papers on supplier selection in terms of triple bottom line. (Date from 2000 to 2015)

facilitates a consecutive research in the domain of supplier selection approaches. Thus, this categorization permits the comparison of traditional concepts with environmental as well as social perspectives receiving attention in recent years.

In Sect. 2.1, we show evidence for a multitude of papers summarizing economic criteria for supplier selection. In his seminal work, Dickson (1996) identified 23 selection criteria, Weber et al. (1991) reviewed these criteria, and even recent studies such as Ageron et al. (2013) refer to the Dickson study. Consequently, this classification is used in the paper.

In scientific literature several categorizations of environmental criteria exist for supplier selection (Awasthi et al. 2010, Bai and Sarkis 2010b, Huang and Keskar 2007). In this paper, environmental criteria applied to supplier selection refer to the seven subcategories according to the taxonomy developed by Humphreys et al. (2003a, b). This taxonomy considers qualitative as well as quantitative factors and is built upon well-known frameworks consolidating the studies of Noci (1995), Sarkis et al. (1996), Azzone and Noci (1996), Noci (1997), and Enarsson (1998). The resulting framework divides environmental criteria into qualitative and quantitative. In addition, Lloyd (1994) distinguishes between product-related and organization-related criteria, which are still prevailing in the literature on green supplier selection (Igarashi et al. 2013). Thus, we compile criteria according to this classification in the present paper.

In order to find a common denominator for classifying metrics, the application of well-established frameworks such as those from Global Reporting Initiative (GRI) is recommended (Miemczyk et al. 2012). Thus, we follow this suggestion and refer to the categories of GRI guidelines for assigning social criteria applied to supplier selection. The social criteria of the GRI guidelines are based on international recognized standards such as OECD Guidelines for Multinational Enterprises, International Labour Organization (ILO) Declaration, and United Nations Global Compact ten principles (GRI 2013). Four categories within total 30 corresponding subcategories represent the measures for the social pillar of the TBL.

Two main streams for categorizing methods for supplier selection can be observed in literature. Both streams distinguish between singular and hybrid approaches at the highest level. They differ with respect to the second level. The first group classifies methods corresponding to the phases of the supplier selection process (De Boer et al. 2001, Wu and Barnes 2011). The second stream divides approaches for supplier selection into mathematical and artificial intelligence (Chen 2011, Singh 2014). In this paper, the methods applied to supplier selection are categorized based on the second stream. Thus, in the first step, single and hybrid approaches are separated. Then individual approaches are further divided into mathematical and artificial intelligence. With regard to hybrid approaches, the existing scheme is extended. Previous taxonomies list several combinations of approaches without numerical assessment of occurrence (e.g., Ha and Krishnan 2008, Singh 2014). In this review, hybrid approaches are separated in terms of the type of combination of methods classified as mathematical or artificial intelligence.

3.3 Material Evaluation

In this section, papers collected are evaluated. The criteria applied to supplier selection are assessed numerically differentiating between economic, environmental, and social criteria. Subsequently, the methods applied to supplier selection are scrutinized in terms of each sustainability dimension. In order to compare stakeholder requirements in specific industries and criteria employed, the status quo of supplier selection criteria in examples of high-polluting industries and labor-intensive industries are analyzed. Furthermore, due to the importance of environmental standards in practice, we scrutinize the application of certifiable management systems as criteria for supplier selection.

3.3.1 Criteria Applied to Supplier Selection

A plethora of economic criteria are applied in the collected papers. For indicating the most important economic criteria, we compare the review papers of Weber et al. (1991) and Ho et al. (2010). "Price/cost," "delivery," and "quality" remain the top

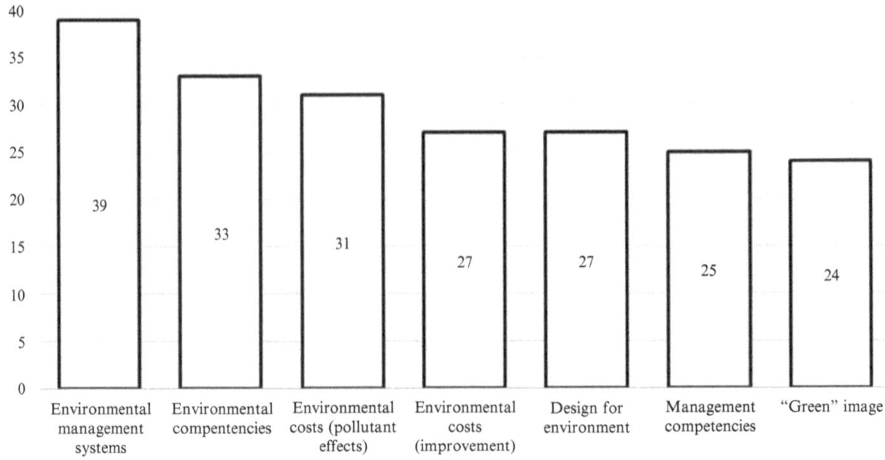

Fig. 3.3 Environmental criteria considered in the collected papers

three economic criteria. The importance of "quality" rises from 53% considered in the approaches between 1966 and 1990 (Weber et al. 1991) up to 89% (Ho et al. 2010) yielding "quality" the most important criterion. Ageron et al. (2013) confirm these findings, and Aissaoui et al. (2007, p. 3520) state that "Although the evolution of the industrial environment modified the degrees of the relative importance of supplier selection criteria since the 1960s, the 23 ones presented by Dickson still cover the majority of those presented in the literature until today."

In our review, we identify 62 papers published between 1997 and 2015 considering environmental criteria. Seventeen papers integrate economic, environmental, and social criteria in a single approach. Thirty-two papers combine economic and environmental criteria, and 13 papers employ environmental criteria exclusively. In this review paper, the aim is to assess these criteria quantitatively. Therefore, the criteria used are assigned to the scheme developed by Humphreys et al. (2003a, b). Figure 3.3 shows the number of each category considered by the collected papers. Thirty-one approaches refer to "environmental costs (pollutant effects)," and 27 take into account "environmental costs (improvements)." Qualitative criteria are applied more frequently than quantitative criteria. The most popular is the adoption of criteria belonging to "environmental management systems" with 39 occurrences. Due to this frequency, we analyze an in-depth certification as a criterion for green supplier selection in Sect. 3.3.2. More than half of the collected papers (33 approaches) utilize criteria of the category "environmental competencies," "design for the environment," "management competencies," and "green image." They are considered 27, 25, and 24 times, respectively.

The heterogeneity of single criteria employed is enormous. Qualitative criteria are more popular than quantitative ones for assessing candidate suppliers regarding environmental criteria. Thus, the selection of suppliers in terms of environmental issues is assessed predominately based on subjective assessments of the decision-maker of buying companies.

The categorization of criteria into product- and organization-related factors is widely recognized (Igarashi et al. 2013, Lloyd 1994). Thirty-five papers (56%) rely on both product-related and organization-related criteria. Twenty-one (34%) take into consideration organization-related criteria solely, and six approaches (10%) use product-related criteria exclusively.

The social dimension is addressed by 18 papers. Seventeen papers integrate social criteria into "real sustainable" approaches, and one paper focuses on the social dimension of sustainability exclusively. Similar to the evaluation of environmental criteria, the collected papers considering social criteria are assessed with the number of papers considering criteria of each category. Twelve papers integrate criteria addressing "society" into their approaches. "Labor practices and decent work" is the most popular category integrated into 15 of the 18 papers evaluated. "Product responsibility" and "human rights" are considered two times each.

As outlined in Sect. 2.3, the GRI guidelines are based on OECD Guidelines for MNE, ILO Declaration, and UNGC. Lozano and Boni (2002, p. 172) scrutinize these codes concluding that "Human Rights are considered transcultural and universal; they are the normative cohesion element that relates the international system as a whole." Thus, governments have transformed these requirements to national laws (e.g., GRI 2013). Another aspect for this low rate of integration of "human rights" is inherent to the category selected because "It is important to note that many aspects that provide insight into human rights performance and impacts can be found in other (sub-) Categories in the Guidelines, and are not limited to the Human Rights sub- Category" (GRI 2013, p. 71). "Product responsibility" is a more downstream-oriented measure and thus hardly considered yet. Taking into account the entire life cycle of a product, "product responsibility" is an important aspect for the OEM. Changing the perspective from end consumer to customer and considering a product as an intermediate item purchased, the importance of this category should rise. "Society" encompasses the responsibility of companies to local communities and philanthropic contributions which are assessed qualitatively. "Labor practices and decent work" contains predominantly quantifiable criteria such as incidents/accidents records. All social criteria are captured at the organizational level in the collected papers.

3.3.2 Methods Applied to Sustainable Supplier Selection

The comparison of approaches applied to supplier selection is conducted in line with the search design. First, results obtained from the meta-analysis are elucidated. Second, the approaches for supplier selection are analyzed for the time interval from 2012 to 2015.

3.3.2.1 Meta-Analysis

According to Sect. 2.3, the results are presented tackling economic, environmental, and "sustainable" dimensions of TBL subsequently. Further, we differentiate between single and hybrid approaches within each dimension.

Economic Dimension: Single Approaches

Weber et al. (1991) group quantitative approaches applied to supplier selection into linear weighting, mathematical programming, and statistical approaches. Ho et al. (2010) conclude that single approaches were slightly more popular than hybrid approaches between 2000 and 2008. The most popular single approaches are data envelopment analysis (DEA) (14 papers), followed by mathematical programming (9 papers) and AHP- and case-based reasoning (CBR) (7 papers, respectively). In contrast, only 28% of the papers considered in the review of Chai et al. (2013) are single approaches. Results show that the most popular single approach is DEA, followed by AHP, ANP, and MP techniques between 2008 and 2011.

The overview shows that DEA has received considerable attention for supplier selection. Also the application of AHP is accepted. Moreover, MP techniques are applied widely. With the progress in computer technology, more complex MP techniques such as nonlinear programming gain ground. The progress in IT facilitates the development of new approaches such as CBR. In general, an ongoing trend toward integrated approaches is observed.

Economic Dimension: Hybrid Approaches

Weber et al. (1991) do not separate between single and combined approaches. Ho et al. (2010) identify 20 different combinations of methods. Applying AHP with other methods receives attention in the period between 2000 and 2008. Seventeen papers use nine different combinations considering AHP. The most popular is to combine AHP with MP techniques. The combination of these techniques permits to integrate qualitative and quantitative criteria for supplier selection and integrating the supplier scores into the MP model and subsequently allocating orders to each supplier (e.g., Ghodsypour and O'Brien 1998, Ho et al. 2011). Further, the integration of methods subsumed to artificial intelligence is becoming more popular such as artificial neural networks (ANN). Evolving is the integration of fuzzy set theory (FST), which is combined with AHP and MP techniques mainly. Worth mentioning is the use of "borrowed" methods such as quality function deployment (QFD), which was initially developed in Japan in the late 1960s in the area of product development (Akao and Mazur 2003). Chai et al. (2013) identify 64 papers for supplier selection focusing on economic criteria covering 43 combinations of methods. The most popular is the integration of FST with other methods (53 approaches). Even the utilization of MP techniques (27 approaches) followed by the technique for order performance by similarity to ideal solution (TOPSIS) (14 approaches) and AHP combinations (13 approaches) is the most popular. TOPSIS is an upcoming method for supplier selection that can substitute AHP and ANP (Hwang and Masud 2012). Combining DEA with other methods is less applied to supplier selection compared to prior reviews of this meta-analysis. The application of AI methods is progressing with new methods developed.

Overall, the combination of methods has increased in importance. Methods based on pairwise comparison, mainly AHP and more recently TOPSIS, are combined with MP techniques. Thus, subjective evaluation of candidate suppliers is combined with quantitative factors. Recently, these approaches are extended using FST for including vague data or encompassing decision-making of groups. Thus, fuzzy AHP and fuzzy TOPSIS are gaining popularity.

Environmental Dimension: Single Approaches

Our meta-analysis of approaches considering environmental criteria covers the time between 1996 and 2011 including seven papers referring to single approaches. The most popular single approach is to apply AHP to green supplier selection (three papers). One paper, respectively, employs ANP, FST, genetic algorithm, and particle swarm optimization.

Environmental Dimension: Hybrid Approaches

Sixteen papers employ hybrid approaches using 12 different combinations of methods for green supplier selection. Integrating FST is the most popular (ten papers) followed by considering AHP (eight papers) in hybrid approaches. Considering FST and AHP together (two papers) or in combination with a third method is popular (three further papers). Just as for the economic dimension, the most popular combination is AHP with MP (three papers). In addition, AI methods are combined similar to the economic dimension. Even more, gray relation analysis is used for green supplier selection.

Compared to the economic dimension, DEA and TOPSIS are utilized less in hybrid approaches considering environmental criteria. DEA is considered only once and is applied in combination with FST, AHP, and TOPSIS, which can be traced back to the use of quantitative input data originally inherent to the DEA method. Only two papers address TOPSIS for selecting green suppliers.

The methods analyzed in this meta-analysis are facilitating the integration of qualitative data. Researchers use AI methods for quantifying qualitative criteria or employ OR methods such as AHP, ANP, or TOPSIS to consider qualitative and quantitative criteria.

"Sustainable" Approaches

The meta-analysis of the social dimension yields no results. However, four approaches consider all three dimensions simultaneously. The only single approach applies AHP to select sustainability-conscious suppliers. Three hybrid approaches use three different combinations containing mathematical and AI-based methods. Papers published on sustainable supplier selection until 2011 integrate approaches facilitating the use of qualitative as well as quantitative data.

Findings of the Meta-Analysis

The meta-analysis clearly confirms that several approaches are developed over time facilitating the integration of qualitative and quantitative criteria. A clear trend toward the use of combined methods is observed across sustainability dimensions. Interesting is the similarity of the first method applied to the respective dimension of TBL. AHP was applied to environmental (Handfield et al. 2002, Noci 1997) and "real" sustainable (Chan et al. 2007). Even the first approach considered in this review paper on social criteria uses AHP (Mani et al. 2014). Thus, researchers rely on a well-recognized methods if new criteria emerge. AHP can handle qualitative as well as quantitative factors. Across all dimensions of sustainability, at the beginning, single methods are applied to supplier selection, and then hybrid approaches are developed. We do not observe specific patterns for sequencing the adoption of combined methods.

3.3.2.2 Comparison of Approaches from 2012 to Present

In the subsequent section, the time interval ranging from 2012 to 2015 is analyzed in terms of methods applied to each pillar of TBL. In total 143 papers are assessed. Thirteen papers consider all 3 dimensions, 22 papers take into account environmental criteria, and 108 papers focus on economic criteria. Fifty-five papers use a single method for selecting suppliers, and 88 papers refer to hybrid approaches. This underscores the trend toward hybrid approaches indicated in the meta-analysis. Further one paper considers social criteria exclusively. Subsequently, single approaches are elucidated, and then hybrid approaches for supplier selection are scrutinized.

Single Approaches

Figure 3.4 summarizes the most popular approaches of this category. MP techniques are applied to the economic dimension predominantly. Twenty-six papers use MP techniques for selecting suppliers focusing on economic criteria. Four papers aiming at selection of green suppliers use MP techniques, and one paper uses it to select

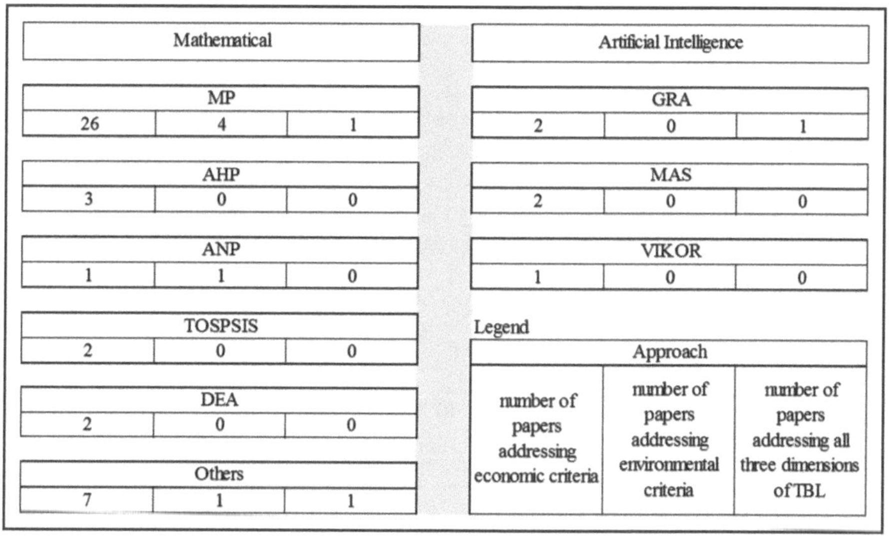

Fig. 3.4 Number of single approaches for supplier selection embracing distinction in terms of TBL (2012–2015)

sustainably sound suppliers. The high share of MP techniques for solving economic supplier selection problems is explained by the use of quantifiable criteria. Most popular economic criteria (see Sect. 3.1) are “quality,” “price/cost,” and “delivery.” Surprising is the dominance of the economic dimension regarding AHP and TOPSIS. The only approach considering social criteria exclusively employs AHP, but no approach for environmental and “real sustainable” issues applies to AHP solely. Congruent to the results of the meta-analysis, DEA is applied to the economic dimension exclusively. With regard to AI-based methods, six single approaches use these methods.

Hybrid Approaches

In total, 88 papers are analyzed employing 55 different combinations of methods in the time interval from 2012 to 2015. Nineteen hybrid approaches combine two mathematical methods. Thirty-five hybrid approaches combine mathematical and AI methods. One paper combines different AI methods.

To keep in line with the meta-analysis, we elucidate the most popular methods and their combinations. Thus, we elucidate MP techniques, AHP, DEA, and TOPSIS. Figure 3.5 displays the interconnected approaches among these four methods. Worth mentioning is the use of borrowed methods from engineering such as axiomatic design (AD) (two papers) and QFD (three papers). The latter is mentioned in the meta-analysis, and between 2012 and 2015 this method is applied to economic criteria. In contrast, AD (Suh 1990) is applied to the environmental pillar

Mathematical and mathematical			
AHP + others	3	0	0
MP + others	3	1	0
AHP + MP	2	0	0
DEA + others	2	1	0
AHP + DEA	2	0	0
MP + TOPSIS	1	0	0
AHP + TOPSIS	0	1	0
DEA + MP	0	1	0
TOPSIS + DEA + MP	1	0	0
TOPSIS + ANP + MP	1	0	0
Others	2	0	0

Mathematical and Artificial Intelligence			
AHP + others	3	0	0
FST + AHP + (others)	6 (1)	(1)	(2)
MP + others	3	0	1
FST + MP + (others)	6 (2)	1 (1)	1
FST + AHP + MP	1	1	1
FST + TOPSIS + (others)	1	2 (1)	2
FST + AHP + TOPSIS	4	0	1
FST + TOPSIS + MP	3	0	0
FST + AHP + TOPSIS + MP	0	1	0
FST + others	12	1	2
Other combinations	2	2	0

Legend

Approach		
number of papers addressing economic criteria	number of papers addressing environmental criteria	number of papers addressing all three dimensions of TBL

Fig. 3.5 Number of hybrid approaches for supplier selection embracing distinction in terms of TBL

of TBL (Büyüközkan 2012) and further in an approach for selecting all three dimensions simultaneously (Kannan et al. 2015).

Regarding AI-based methods, FST has received considerable attention. Therefore, we assess hybrid approaches integrating FST numerically. Integrating FST and AHP as well as combining FST and MP techniques is the most popular. This further strengthens the dominance of these combined methods as outlined in the meta-analysis. AHP and MP techniques are combined with AI-based methods such as ANN. Worth mentioning is the sophisticated combination of FST, TOPSIS, and system dynamics considering all three dimensions of sustainability simultaneously. New methods subsumed to AI such as rough set theory are emerging.

All hybrid approaches considering criteria of all three dimensions simultaneously include AI-based methods with mathematical methods. Nine out of ten papers use FST (90%). Even green supplier selection methods integrate FST (10 of 16 hybrid approaches, approx. 63%). Referring to economic criteria, 37 out of 62 (approx. 60%) use FST for solving supplier selection problems.

Generally, the trends observed in the meta-analysis also hold in primary search. The increasing popularity of FST that was shown in the meta-analysis was continuing between 2012 and 2015. The application of economic criteria is strongly dominating the field throughout the analyzed time period. MP techniques were starting to be applied to approaches combining all three dimensions since 2012, especially in combination with AI-based methods. Furthermore, some methods that were only applied with respect to the economic dimension pre-2012 were also integrating environmental and social criteria after 2012. For instance, TOPSIS was applied to consider all three dimensions of sustainability simultaneously since 2012.

3.3.3 *Further Analyses*

The collected papers on sustainable supplier selection refer to specific industries. Therefore, we will analyze the literature with respect to the industries they are considering. Moreover, we explore if the existence of environmental management systems is considered as a criterion in approaches for sustainable supplier selection.

3.3.3.1 Analysis with Regard to Industries

Each industry has a specific sustainability profile. Schneider and Wallenburg (2012) operationalize sustainable sourcing activities regarding the emphasis on each of the three dimensions yielding specific archetypes of sustainable sourcing. Based on stakeholder theory, the authors derive expected profiles for companies of a particular industry and profiles for specific industries. In the following section, we check the sustainability profiles of approaches for supplier selection. As an example of high-polluting industries, papers on supplier selection in the automotive industry are analyzed (Carbone et al. 2012). In order to scrutinize an example of labor-intensive industries, we analyze approaches conducted in textile/apparel industry.

According to Schneider and Wallenburg (2012, p. 251), "labor-intensive industries, such as the textile industry, might be primarily affected by social legislation and thus place special emphasis on social sustainability in sourcing." Nine of the collected papers of our review conduct studies for supplier selection in the textile or apparel industry. Eight of these papers consider economic criteria for supplier

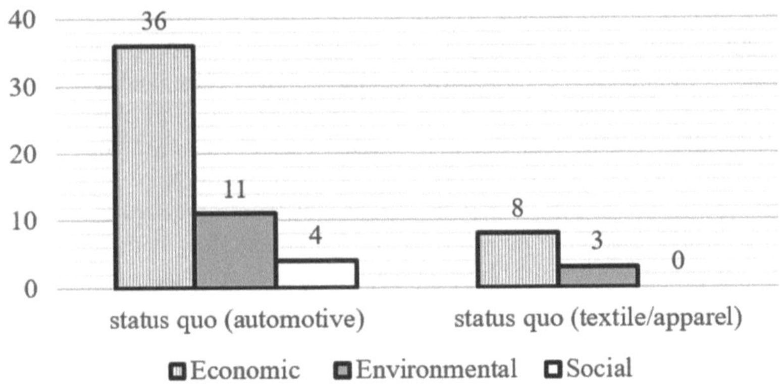

Fig. 3.6 Juxtaposition of sustainability criteria employed in collected papers for textile/apparel industry and automotive

selection, and four papers integrate environmental criteria. None of these papers include social criteria for supplier selection.

In contrast, environmental regulation impacts mainly manufacturing companies, yielding emphasis on environmental sustainability (Schneider and Wallenburg 2012). Subsequently, the profile of the automotive industry is analyzed for supplier selection. Expected is a focus on economic and environmental criteria considered in the corresponding papers. Forty collected papers deal with supplier selection for automotive items. Only 11 approaches integrate environmental criteria into the supplier selection process. Three papers consider all three dimensions of sustainability in the model.

Figure 3.6 displays the number of papers on automotive and textile/apparel industry considering sustainability criteria.

In conclusion, the expected results based on stakeholder theory deviate from the amount of sustainability criteria considered in scientific multi-criteria approaches. Cost pressures might lead to the dominance of economic criteria. Cost pressures are prevailing in both industries under consideration yet due to different reasons. Automotive industry has been struggling with market saturation leading to cost pressures being put on suppliers. Cost pressures in the textile/apparel industry becomes most obvious in terms of the emergence of fast fashion resulting in the neglect of ethical issues in developing countries (Bhardwaj and Fairhurst 2010, Turker and Altuntas 2014).

With respect to the use of environmental criteria in the automotive industry, we see that environmental regulations for automotive companies are manifold and becoming stricter in recent times (e.g., Thun and Müller 2010). Furthermore, new technological developments such as electronic propulsion are demanded by end consumers yielding increased awareness of environmental factors for supplier selection (Orsato and Wells 2007). Therefore, integrating environmental criteria into approaches on supplier selection conducted in the automotive industry receives increasing attention. However, this trend is not sufficiently reflected in the literature.

The missing representation of social criteria in the textile/apparel industry is surprising considering recent trends. Social misconduct in the textile/apparel industry is highlighted in media yet hardly discussed in literature on supplier selection in the past (Hoejmose and Adrien-Kirby 2012). Currently, companies are recognizing the need for improving working conditions on supplier sites located in developing countries. Recently, social and ethical factors are receiving increasing attention in scientific literature (Kozlowski et al. 2015).

3.3.3.2 Standards as Criteria for Supplier Selection

In 2002, Handfield et al. (2002, p. 81) conclude that until ISO 14001 management system "has diffused further, it may have limited applications in differentiating supplier environmental performance." By now, voluntary standards are receiving increasing attention across industrial sectors around the globe (Marimon Viadiu et al. 2006). Even for evaluating suppliers, certifiable environmental standards are important (Ageron et al. 2013) due to its easy assessment (Handfield et al. 2002). According to the ISO 14001 standard, an EMS is "the part of the overall management system that includes organizational structure, planning activities, responsibilities, practices, procedures, processes and resources for developing, implementing, achieving, reviewing, and maintaining the environmental policy" (ISO14001 2009, Article 3).

Subsequently, we scrutinize the development of the use of certifiable EMS as a criterion for supplier selection over time. Figure 3.7 displays the total amount of papers integrating environmental criteria and papers considering certifiable EMS as a criterion for supplier selection (beside further criteria). The dashed line shows the collected papers considering green criteria. The number of papers increases. From 2013 to 2015, the number grows rapidly with a clear peak in 2015. Even in 2010, an upward swing is observed. The solid line shows papers considering certifiable EMS as a criterion for supplier selection. The peak is in 2010 where six papers use certifiable EMS as a criterion. After a drop in 2011, the amount of papers integrating certifiable EMS keeps almost stable slightly altering in the range of three and four. The gap of papers considering certifiable EMS and total amount of papers integrating green criteria is increasing in recent times. In contrast, from 2002 to 2006, every paper dealing with environmental issues refers to certifiable EMS for supplier selection. In the time horizon from 2007 to 2009, the use of certifiable EMS grows steadily.

Remarkable is the growing drift between the total number of papers on green supplier selection and total number of papers integrating certifiable environmental standards. The most widely adopted environmental standard is ISO 14001 (Darnall 2006). ISO 14001 was first launched in 1996 (Darnall et al. 2008). Thus it can be assumed that using ISO 14001 as a criterion for supplier selection is signaling environmental-conscious behavior (for signaling effect, see Spence (1973); Terlaak and King (2006)). Over time, implementation of standards has become quasi-legal (Terlaak 2007) implying a binding effect (Beske et al. 2008). Thus, researchers see

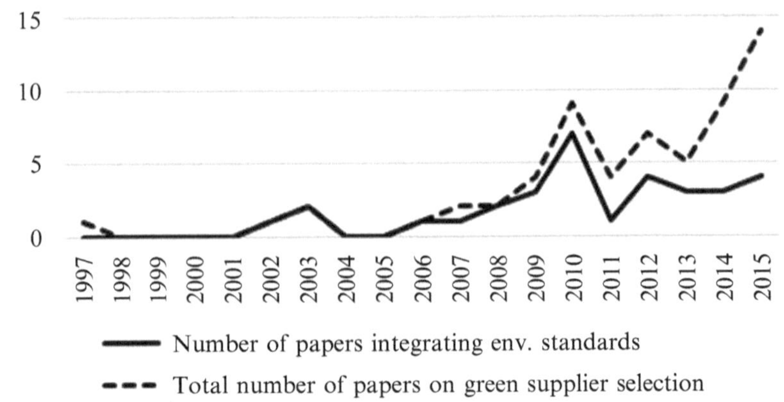

Fig. 3.7 Comparison of papers considering certifiable management systems as criterion for supplier selection and total amount of papers on green supplier selection over time

the implementation of standards for granted and do not consider, e.g., ISO 14001 as a criterion specifically. The adoption pattern of standards differs between industrial sectors and regions. Furthermore, certifiable management systems are often part of the supplier qualification phase. For example, OEMs require the adoption of ISO 14001 as a prerequisite for doing business (Ahsen 2014).

3.4 Insights and Research Gaps

Since the mid-1990s, environmental concerns are receiving increasing attention in literature on purchasing and supply chain management (Seuring and Müller 2008). Social issues are gaining ground in literature on purchasing since the 2000s (Carter and Jennings 2002, Muller-Camen et al. 2015). In this paper, we analyzed how the three dimensions of sustainability are reflected in the literature on supplier selection and which criteria and methods are applied.

The results confirm the call for more approaches integrating all three dimensions of TBL (Seuring and Müller 2008). Only few papers consider all three dimensions of sustainability. Additionally, these papers consider few criteria in terms of sustainability goals and scratch sustainability issues on the surface yielding demand for a taxonomy of criteria for supplier selection in order to ease integration of sustainability criteria into the respective approaches. Even though frameworks covering supplier selection criteria exist (e.g., Dou and Sarkis 2010), this review shows that a multitude of environmental criteria are applied without referring to taxonomies of criteria. Even more "the operationalization of these criteria into meaningful, practical and measurable variables often poses challenges, both for purchasers and suppliers" (Igarashi et al. 2013, p. 247 referring to Jabbour and Jabbour 2009; Lloyd 1994). Well-recognized taxonomies and frameworks should be adapted to sustainability concerns. Future research could develop a scheme for classifying methods linked to type of input data. Further insights could be gained by assigning methods to type of data input providing appropriate methods to a given set of criteria. For example, AHP can deal with qualitative as well as quantitative factors. In contrast, MP techniques require quantitative input.

With regard to criteria applied to each dimension of sustainability, quantitative and qualitative factors are considered. The number of qualitative criteria that can be integrated into qualitative methods is limited due to constraints of human's information processing capacity (Miller 1956, Sarkis and Talluri 2002). Thus, in order to consider a wide set of sustainability criteria, the use of quantitative criteria is mandatory. This can be achieved by combining methods considering qualitative criteria with methods processing quantitative data input. Our insights are in line with insights of Wu and Barnes (2011, p. 268) stating that "existing approaches to partner selection highlight the need to adopt and meet a combination of qualitative and quantitative objectives. Therefore, no single methodology is likely to be able to solve the partner selection problem, especially, when different organizations have different requirements. Further research is needed to work towards developing a new more mature combination of methods and models."

In terms of the use of hybrid approaches, the present review reveals changes since the statement of Sevkli et al. (2008) that hybrid approaches are scarce. The utilization of hybrid approaches gained ground yielding 88 papers using 55 different hybrid approaches in the time interval between 2012 and 2015. Even more, borrowing methods from other disciplines is becoming popular. Methods originating in engineering such as FMEA, QFD, and AD are applied to supplier selection. Applying additional methods from other disciplines could provide further insights to the domain of supplier selection. For instance, project management approaches have gained considerable attention. Transforming methods of six sigma (Zhang et al. 2016) could support the efficient selection process. Additionally, agile methods such as "scrum" can be implemented to improve responsiveness of processes such as selection, development, and monitoring of suppliers. Employing "Kano method" could help to structure criteria for supplier selection. Also, statistical methods such as conjoint analysis might be applicable for sorting importance of selection criteria.

Analyzing industry-specific characteristics, we determined that scientific literature lags behind requirements formulated by stakeholders in terms of sustainability. Approaches for supplier selection do not sufficiently integrate criteria demanded by stakeholders. In particular, past literature focusing on the automotive industry did not sufficiently include environmental criteria for supplier selection. Papers on the textile/apparel industry lack social criteria within the respective models for supplier selection.

The implementation of certifiable environmental standards is considered as a criterion for supplier selection. Until 2011, almost all papers on green supplier selection cover these standards. Since then, the number of papers on green supplier selection has grown rapidly, while the number of papers integrating certifiable environmental standards stays almost stable. Certifiable environmental standards are an appropriate criterion for preselecting a pool of candidate suppliers leading to avoidance of this criterion for final supplier selection approaches.

Most studies reviewed in our analysis are applied to situations buying a single component. Few studies develop methods for buying product bundles from a single supplier. This is not reflecting real-world demand occurring in many industries such as automotive (e.g., Yu and Wong 2015). Therefore, future research should consider buying multiple items from a single supplier yielding synergy effects. Even more, close buyer-supplier relationships are critical to long-term success. The collected papers lack the integration of supplier selection into further sourcing activities. Sourcing encompasses processes such as "supplier development and all other processes associated with a holistic supplier management" (Schneider and Wallenburg 2012, p. 245) additionally to the supplier selection process. Basing the selection process on an analysis of the R&D capabilities is a promising approach for enhancing sustainability performance. Using complementary resources and capabilities yields competitive advantages (Dyer and Singh 1998) through pooling of the firms' resources in R&D (Das and Teng 2000). Cooperative orientation decreases costs (Kefi and Bencherqui 2014). Thus, integrating the assessment of R&D capabilities and consequently potential synergy effects between buyers and suppliers into supplier selection approaches is recommended as a future research topic.

The paper provides implications for managers toward securing sustainability-conscious behavior of supply chain members. These implications refer to the use of an appropriate set of criteria as well as employing manageable methods for supplier selection. Referring to criteria, managers should consider economic, environmental, and social criteria simultaneously. For achieving this, managers have to consider qualitative criteria for supplier selection and incorporate these qualitative factors into the respective approaches. Advantages of integrating a broad set of criteria include that companies strive for robustness in terms of sustainability. This avoids performing in one subcategory superior while neglecting another subcategory of sustainability. Thus, the company is more robust with regard to unexpected changes in legislation and internal processes.

Regarding methods applied to supplier selection, many approaches are complex in nature yielding problems in terms of traceability and transparency of decision-making. However, methods for practical use exist. In order to achieve a traceable supplier selection process, managers should combine well-established methods such as SMART and AHP for evaluating suppliers. Further, companies should provide data of the process to suppliers facilitating areas for improvements at each supplier's site. The development of a comprehensive software system as basis for an AI-based decision-making is recommended.

3.5 Conclusions

The paper provides an analysis of approaches on sustainable supplier selection applying content analysis. First, a meta-analysis is conducted in order to examine papers on the economic, environmental, and social dimension of TBL as well as papers integrating all three dimensions of TBL simultaneously covering the time horizon until 2011. Second, we conducted a key search for identifying individual papers in order to bridge the gap from 2012 to 2015. This review highlights criteria applied to supplier selection as well as methods used in the respective papers. Economic criteria did not change dramatically since 1966. "Quality," "delivery," and "cost/price" remain the most popular economic criteria. With regard to environmental criteria, qualitative and quantitative criteria have to be integrated. Referring to the classification scheme developed by Humphreys et al. (2003a, b), integrating criteria belonging to the category "environmental management systems" in terms of qualitative criteria is the most popular. Regarding quantitative environmental criteria, the category "environmental costs (pollutant effects)" is applied most often. Consideration of social criteria is categorized according to GRI guidelines yielding "labor practices and decent work" most popular. Regarding methods applied to sustainable supplier selection, hybrid approaches are receiving increasing attention. In order to cope with uncertainty, the use of FST is popular across sustainability dimensions addressed in the collected papers. Moreover, studies conducted in high-polluting or labor-intensive industries do not match requirements of the respective sustainability dimension as predicted based on stakeholder theory. The integration of certifiable environmental standards is analyzed attesting a lower share of papers considering this criterion. Finally, research gaps and future research topics are discussed and managerial implications are provided.

References

Ageron, B., Gunasekaran, A., & Spalanzani, A. (2013). IS/IT as supplier selection criterion for upstream value chain. *Industrial Management & Data Systems, 113*, 443–460.

Ahsen, A. (2014). The integration of quality, environmental and health and safety management by car manufacturers – A long-term empirical study. *Business Strategy and the Environment, 23*, 395–416.

Aissaoui, N., Haouari, M., & Hassini, E. (2007). Supplier selection and order lot sizing modeling: A review. *Computers & Operations Research, 34*, 3516–3540.

Akao, Y., & Mazur, G. H. (2003). The leading edge in QFD: Past, present and future. *International Journal of Quality and Reliability Management, 20*, 20–35.

Awasthi, A., Chauhan, S. S., & Goyal, S. K. (2010). A fuzzy multicriteria approach for evaluating environmental performance of suppliers. *International Journal of Production Economics, 126*, 370–378.

Azadnia, A. H., Saman, M. Z. M., & Wong, K. Y. (2015). Sustainable supplier selection and order lot-sizing: An integrated multi-objective decision-making process. *International Journal of Production Research, 53*, 383–408.

Azzone, G., & Noci, G. (1996). Measuring the environmental performance of new products: An integrated approach. *International Journal of Production Research, 34*, 3055–3078.

Bai, C., & Sarkis, J. (2010a). Green supplier development: Analytical evaluation using rough set theory. *Journal of Cleaner Production, 18*, 1200–1210.

Bai, C., & Sarkis, J. (2010b). Integrating sustainability into supplier selection with grey system and rough set methodologies. *International Journal of Production Economics, 124*, 252–264.

Beske, P., Koplin, J., & Seuring, S. (2008). The use of environmental and social standards by German first-tier suppliers of the Volkswagen AG. *Corporate Social Responsibility and Environmental Management, 15*, 63–75.

Bhardwaj, V., & Fairhurst, A. (2010). Fast fashion: Response to changes in the fashion industry. *The International Review of Retail, Distribution and Consumer Research, 20*, 165–173.
Brandenburg, M., Govindan, K., Sarkis, J., & Seuring, S. (2014). Quantitative models for sustainable supply chain management: Developments and directions. *European Journal of Operational Research, 233*, 299–312.
Bruno, G., Esposito, E., Genovese, A., & Passaro, R. (2012). AHP-based approaches for supplier evaluation: Problems and perspectives. *Journal of Purchasing and Supply Management, 18*, 159–172.
Büyüközkan, G. (2012). An integrated fuzzy multi-criteria group decision-making approach for green supplier evaluation. *International Journal of Production Research, 50*, 2892–2909.
Carbone, V., Moatti, V., & Vinzi, V. E. (2012). Mapping corporate responsibility and sustainable supply chains: An exploratory perspective. *Business Strategy and the Environment, 21*, 475–494.
Carter, C. R., & Easton, L. P. (2011). Sustainable supply chain management: evolution and future directions. *International Journal of Physical Distribution and Logistics Management, 41*, 46–62.
Carter, C. R., & Jennings, M. M. (2002). Logistics social responsibility: An integrative framework. *Journal of Business Logistics, 23*, 145–180.
Carter, C. R., & Rogers, D. S. (2008). A framework of sustainable supply chain management: Moving toward new theory. *International Journal of Physical Distribution and Logistics Management, 38*, 360–387.
Chai, J., Liu, J. N., & Ngai, E. W. (2013). Application of decision-making techniques in supplier selection: A systematic review of literature. *Expert Systems with Applications, 40*, 3872–3885.
Chan, F., Chan, H., Ip, R., & Lau, H. (2007). A decision support system for supplier selection in the airline industry. *Proceedings of the Institution of Mechanical Engineers, Part B: Journal of Engineering Manufacture, 221*, 741–758.
Chen, L., Olhager, J., & Tang, O. (2014). Manufacturing facility location and sustainability: A literature review and research agenda. *International Journal of Production Economics, 149*, 154–163.
Chen, Y. J. (2011). Structured methodology for supplier selection and evaluation in a supply chain. *Information Sciences, 181*, 1651–1670.
Darnall, N. (2006). Why firms mandate ISO 14001 certification. *Business and Society, 45*, 354–381.
Darnall, N., Jolley, G. J., & Handfield, R. (2008). Environmental management systems and green supply chain management: Complements for sustainability? *Business Strategy and the Environment, 17*, 30–45.
Das, T. K., & Teng, B.-S. (2000). A resource-based theory of strategic alliances. *Journal of Management, 26*, 31–61.
De Boer, L., Labro, E., & Morlacchi, P. (2001). A review of methods supporting supplier selection. *European Journal of Purchasing and Supply Management, 7*, 75–89.
Dickson, G. W. (1996). *An analysis of vendor selection systems and decisions* (pp. 5–17). February: Journal of Purchasing.
Dou, Y., & Sarkis, J. (2010). A joint location and outsourcing sustainability analysis for a strategic offshoring decision. *International Journal of Production Research, 48*, 567–592.
Dowlatshahi, S. (2000). Designer–buyer–supplier interface: Theory versus practice. *International Journal of Production Economics, 63*, 111–130.
Dyer, J. H., & Singh, H. (1998). The relational view: Cooperative strategy and sources of interorganizational competitive advantage. *Academy of Management Review, 23*, 660–679.
Elkington, J. (1998). Partnerships from cannibals with forks: The triple bottom line of 21st-century business. *Environmental Quality Management, 8*, 37–51.
Enarsson, L. (1998). Evaluation of suppliers: How to consider the environment. *International Journal of Physical Distribution and Logistics Management, 28*, 5–17.
Genovese, A., Lenny Koh, S., Bruno, G., & Esposito, E. (2013). Greener supplier selection: State of the art and some empirical evidence. *International Journal of Production Research, 51*, 2868–2886.

Ghodsypour, S. H., & O'brien, C. (1998). A decision support system for supplier selection using an integrated analytic hierarchy process and linear programming. *International Journal of Production Economics, 56*, 199–212.

Govindan, K., Rajendran, S., Sarkis, J., & Murugesan, P. (2015). Multi criteria decision making approaches for green supplier evaluation and selection: A literature review. *Journal of Cleaner Production, 98*, 66–83.

Gri, G. R. I. (2013). *G4 sustainability reporting guidelines: Reporting principles and standards disclosures*. Amsterdam.

Ha, S. H., & Krishnan, R. (2008). A hybrid approach to supplier selection for the maintenance of a competitive supply chain. *Expert Systems with Applications, 34*, 1303–1311.

Handfield, R., Walton, S. V., Sroufe, R., & Melnyk, S. A. (2002). Applying environmental criteria to supplier assessment: A study in the application of the analytical hierarchy process. *European Journal of Operational Research, 141*, 70–87.

Ho, W., Dey, P. K., & Lockström, M. (2011). Strategic sourcing: A combined QFD and AHP approach in manufacturing. *Supply Chain Management: An International Journal, 16*, 446–461.

Ho, W., Xu, X., & Dey, P. K. (2010). Multi-criteria decision making approaches for supplier evaluation and selection: A literature review. *European Journal of Operational Research, 202*, 16–24.

Hoejmose, S. U., & Adrien-Kirby, A. J. (2012). Socially and environmentally responsible procurement: A literature review and future research agenda of a managerial issue in the 21st century. *Journal of Purchasing and Supply Management, 18*, 232–242.

Huang, S. H., & Keskar, H. (2007). Comprehensive and configurable metrics for supplier selection. *International Journal of Production Economics, 105*, 510–523.

Humphreys, P., Mcivor, R., & Chan, F. (2003a). Using case-based reasoning to evaluate supplier environmental management performance. *Expert Systems with Applications, 25*, 141–153.

Humphreys, P., Wong, Y., & Chan, F. (2003b). Integrating environmental criteria into the supplier selection process. *Journal of Materials Processing Technology, 138*, 349–356.

Hwang, C.-L. & Masud, A. S. M. 2012. Multiple objective decision making—Methods and applications: A state-of-the-art survey., Springer Science & Business Media.

Igarashi, M., De Boer, L., & Fet, A. M. (2013). What is required for greener supplier selection? A literature review and conceptual model development. *Journal of Purchasing and Supply Management, 19*, 247–263.

ISO14001. (2009). *Environmental management systems - requirements with guidance for use (ISO 14001:2004 + Cor 1:2009)*. Geneva: International Organization for Standardization.

Jabbour, A. B. L., & Jabbour, C. J. (2009). Are supplier selection criteria going green? Case studies of companies in Brazil. *Industrial Management & Data Systems, 109*, 477–495.

Jain, V., Wadhwa, S., & Deshmukh, S. (2007). Supplier selection using fuzzy association rules mining approach. *International Journal of Production Research, 45*, 1323–1353.

Kannan, D., Govindan, K., & Rajendran, S. (2015). Fuzzy axiomatic design approach based green supplier selection: A case study from Singapore. *Journal of Cleaner Production, 96*, 194–208.

Kefi, M. K., & Bencherqui, D. B. (2014). Inter-firm cooperative agreements and innovation-oriented behavior: An intersectional approach. *The Business and Management Review, 4*, 17.

Kozlowski, A., Searcy, C., & Bardecki, M. (2015). Corporate sustainability reporting in the apparel industry: An analysis of indicators disclosed. *International Journal of Productivity and Performance Management, 64*, 377–397.

Krause, D. R., Vachon, S., & Klassen, R. D. (2009). Special topic forum on sustainable supply chain management: Introduction and reflections on the role of purchasing management. *Journal of Supply Chain Management, 45*, 18–25.

Lloyd, M. (1994). *How green are my suppliers* (pp. 36–39). October: Purchasing and Supply Management.

Lozano, J. F., & Boni, A. (2002). The impact of the multinational in the development: An ethical challenge. *Journal of Business Ethics, 39*, 169–178.

Mani, V., Agrawal, R., & Sharma, V. (2014). Supplier selection using social sustainability: AHP based approach in India. *International Strategic Management Review, 2*, 98–112.

Marimon Viadiu, F., Casadesús Fa, M., & Heras Saizarbitoria, I. (2006). ISO 9000 and ISO 14000 standards: An international diffusion model. *International Journal of Operations & Production Management, 26*, 141–165.

Mayring, P. 2010. Qualitative Inhaltsanalyse *Handbuch qualitative Forschung in der Psychologie*, 601–613.

Miemczyk, J., Johnsen, T. E., & Macquet, M. (2012). Sustainable purchasing and supply management: A structured literature review of definitions and measures at the dyad, chain and network levels. *Supply Chain Management: An International Journal, 17*, 478–496.

Miller, G. (1956). The magical number seven, plus or minus two: Some limits on our capacity for processing information. *The Psychological Review, 63*, 81–97.

Muller-Camen, M., Rohr, C. & Maurer, I. 2015. Human rights: Employment relations and HRM in the supply chain. *SASE 2015 - mini-conference "Multinational Firms - Labour, Management and Society"*.

Noci, G. (1995). Supporting decision-making for recycling-based investments. *Business Strategy and the Environment, 4*, 62–72.

Noci, G. (1997). Designing 'green' vendor rating systems for the assessment of a supplier's environmental performance. *European Journal of Purchasing and Supply Management, 3*, 103–114.

Orsato, R. J., & Wells, P. (2007). U-turn: The rise and demise of the automobile industry. *Journal of Cleaner Production, 15*, 994–1006.

Pagell, M., Wu, Z., & Wasserman, M. E. (2010). Thinking differently about purchasing portfolios: An assessment of sustainable sourcing. *Journal of Supply Chain Management, 46*, 57–73.

Perçin, S. (2006). An application of the integrated AHP-PGP model in supplier selection. *Measuring Business Excellence, 10*, 34–49.

Preuss, L. (2001). In dirty chains? Purchasing and greener manufacturing. *Journal of Business Ethics, 34*, 345–359.

Sarkis, J., Nehman, G. & Priest, J. A systemic evaluation model for environmentally conscious business practices and strategy. Electronics and the Environment, 1996. ISEE-1996., Proceedings of the 1996 IEEE International Symposium on, 1996. IEEE, 281–286.

Sarkis, J., & Talluri, S. (2002). A model for strategic supplier selection. *Journal of Supply Chain Management, 38*, 18–28.

Schneider, L., & Wallenburg, C. M. (2012). Implementing sustainable sourcing—Does purchasing need to change? *Journal of Purchasing and Supply Management, 18*, 243–257.

Seuring, S. (2013). A review of modeling approaches for sustainable supply chain management. *Decision Support Systems, 54*, 1513–1520.

Seuring, S., & Gold, S. (2012). Conducting content-analysis based literature reviews in supply chain management. *Supply Chain Management: An International Journal, 17*, 544–555.

Seuring, S., & Müller, M. (2008). From a literature review to a conceptual framework for sustainable supply chain management. *Journal of Cleaner Production, 16*, 1699–1710.

Sevkli, M., Lenny Koh, S., Zaim, S., Demirbag, M., & Tatoglu, E. (2008). Hybrid analytical hierarchy process model for supplier selection. *Industrial Management & Data Systems, 108*, 122–142.

Singh, A. (2014). Supplier evaluation and demand allocation among suppliers in a supply chain. *Journal of Purchasing and Supply Management, 20*, 167–176.

Spence, M. (1973). Job market signaling. *The Quarterly Journal of Economics, 87*, 355–374.

Suh, N. P. (1990). *The principles of design*. Oxford University Press.

Terlaak, A. (2007). Order without law? The role of certified management standards in shaping socially desired firm behaviors. *Academy of Management Review, 32*, 968–985.

Terlaak, A., & King, A. A. (2006). The effect of certification with the ISO 9000 quality management standard: A signaling approach. *Journal of Economic Behavior & Organization, 60*, 579–602.

Thun, J. H., & Müller, A. (2010). An empirical analysis of green supply chain management in the German automotive industry. *Business Strategy and the Environment, 19*, 119–132.

Turker, D., & Altuntas, C. (2014). Sustainable supply chain management in the fast fashion industry: An analysis of corporate reports. *European Management Journal, 32*, 837–849.

Van Weele, A. J. 2014. Purchasing and supply chain management: Analysis, strategy, planning and practice., Cengage Learning EMEA.

Weber, C. A., Current, J. R., & Benton, W. (1991). Vendor selection criteria and methods. *European Journal of Operational Research, 50*, 2–18.

Wetzstein, A., Hartmann, E., Benton, W. C., Jr., & Hohenstein, N. O. (2016). A systematic assessment of supplier selection literature – State-of-the-art and future scope. *International Journal of Production Economics, 182*, 304–323.

Wu, C., & Barnes, D. (2011). A literature review of decision-making models and approaches for partner selection in agile supply chains. *Journal of Purchasing and Supply Management, 17*, 256–274.

Yu, C., & Wong, T. (2015). A multi-agent architecture for multi-product supplier selection in consideration of the synergy between products. *International Journal of Production Research, 53*, 6059–6082.

Zhang, A., Zhang, A., Luo, W., Luo, W., Shi, Y., Shi, Y., Chia, S. T., Chia, S. T., Sim, Z. H. X., & Sim, Z. H. X. (2016). Lean and six sigma in logistics: A pilot survey study in Singapore. *International Journal of Operations & Production Management, 36*, 1625–1643.

Chapter 4
Green Supply Chain Management, Green Innovations, and Green Practices

Umar Burki

4.1 Introduction

The globalization process provides immense growth in international trade, especially in the manufacturing sector. Simultaneously, this process resulted in environmental degradation and bigger carbon footprint. Business companies outsource an increasing part of their manufacturing processes globally that puts more pressure on cargo movement-related logistics services and thereby adversely affect the environment (WTO 2014). Consequently, the Paris Climate Agreement of 2016 (UNFCCC 2017) brought together all the world's stakeholders including national governments, international business firms and third-party logistics providers, environmental scientists, and global consumers to demonstrate their deep commitment to finding a clear solution to save our planet, Earth, from future environmental degradation.

Present-day consumers are becoming sensitive to environmental issues and hence demonstrate a stronger preference for green products. In response, business firms, marketing channels, and supply chains are adopting new practices and innovations to become greener and minimize their carbon footprint. When business firms become green and operate in an environmentally friendly manner, they reap economic and social benefits (Chiou et al. 2011). Furthermore, when business firms become green, they cultivate a sensible environmental image in the mind of consumers, which further enhances opportunities for commercial benefits. That is why some globally recognized business firms show a greater propensity to function as green/environmental-friendly business entities. For instance, Toyota, Google, and Starbucks are greening their respective supply chains and embracing green supply chain management (henceforward denote as GSCM) practices.

U. Burki (✉)
School of Business, University of South-Eastern Norway, Kongsberg, Norway
e-mail: umar.burki@usn.no

H. Qudrat-Ullah (ed.), *Innovative Solutions for Sustainable Supply Chains*,
Understanding Complex Systems, https://doi.org/10.1007/978-3-319-94322-0_4

Evident deteriorating in environmental conditions and rising environmental awareness among the world's consumers forced business firms to change actively and introduce, adopt, and implement sustainable environmental operations. Under this new approach, business firms are making green, sustainable practices an integral part of their everyday policies and strategies. For instance, Chinese manufacturing firms, which make the bulk of today's global suppliers, have aggressively started implementation of environmental practices and green supply chain management (Liu and Chang 2017). This new paradigm emphasizes that business firms should put emphasis on developing ecologically sensitive and strategic management policies and not solely prioritize on achieving economic profits (Luthra et al. 2013). In supply chain settings, this implies that business firms and especially manufacturing companies make their respective supply chains more sustainable and eco-friendly.

Developing GSCM implies a comprehensive integration of environmental thinking into supply chain management, which involves procurement and material sourcing, supplier selection, manufacturing processes, and delivery of final products to consumers (Srivastava 2007). In other words, GSCM practices refer to all those intra- and interorganizational practices that incorporate environmental rational into daily supply chain operations. Green supply chain practices are referred to as internal and external GSCM practices (Zhu et al. 2017). External GSCM practices focus on greening fundamental external supply chain activities such as procurement, supplier relationships, product design, innovation, and others, whereas internal GSCM practices focus on improving internal support systems and commitment to green management practices (Chiou et al. 2011). When adopting and implementing internal and external GSCM practices, it is prudent for supply chain partners to evaluate each step to ensure that the economic and operational performance of a supply chain is not compromised.

Literature (Chen et al. 2006; Chen 2008) suggests that business firms employ a number of hardware and software innovations to improve areas such as environmental management and environmental performance. Such innovations are known as green innovations, and business firms apply them to enhance environmental performance and competitive advantage (Chen et al. 2006; Van Den Berg et al. 2013). For instance, the Spanish fashion apparel giant Zara brought together green innovations and organic cotton to make ecological fabrics, which resulted in improving the company's green image and reducing energy consumption. Essentially, green innovations improve environmental management and deal with green products, green designs, green processes, green technological innovations (resulting in energy saving, pollution prevention, reduction in environmental waste), and the acceptance of environmental rational by top management levels. Similarly, business firms apply green practices for achieving environmental objectives. Therefore, the focus of this chapter is to provide an overview about the role and effect of green innovations on GSCM practices.

In the next section of this chapter, the author provides an overview of green supply chain management and its different dimensions. Next sections describe green innovation, types of green innovations, and key drivers and outcomes about green innovations. A literature review appendix provides an overview of green innovations in different research contexts. In addition, this section outlines the role of green practices in GSCM, followed by a conclusion.

4.2 Green Supply Chain Management (GSCM)

A simple supply chain is a set of dynamic networks of business organizations involved in sharing vital information and resources focused on moving a product or a service from one supplier/manufacturer to another and finally to the end consumer. A supply chain includes suppliers and/or manufacturers of raw material, first-tier and second-tier intermediate suppliers/manufacturers, final product manufacturers, and their connection with distributors, retailers, and consumers. Hence, a generic supply chain model composes of goal-oriented coordination and collaboration with all the supply chain partners such as suppliers, manufacturers, intermediaries, third-party service providers, distributors, and, finally, the end consumers (see Fig. 4.1).

Minor or major operational activity in a supply chain link involves consumption of natural resources and produces a carbon footprint. This consequence is principally unavoidable because no business activity in a supply chain could be labeled as a carbon neutral activity. Unsophisticatedly, many business companies keep on labeling their consumer products or manufacturing units as carbon neutral by emphasizing that neither their products nor their manufacturing units adversely affect the environment. Furthermore, all kind of operational activity produces zero carbon emissions and, hence, no carbon footprint. Business firms face an uphill task in justifying such carbon neutrality approach as consumers gain more awareness about environmental issues, greater international focus on environmental sustainability, and greater attention toward the compliance of dynamic environmental regulations at the national and international levels.

One such example is the Paris Climate Conference (held in 2016) where the international community came together to cut down carbon emissions in a concrete manner. Such kind of collective commitment to improve environment sustainable is beginning to have an impact on the way business firms will operate in their current and future activities. For instance, Starbucks, the coffee retail giant, adopted marketing strategies *green stores and green coffee cup* and initiated supply chain initiative known as *ethical sourcing* under which the company ensures that coffee, tea, cocoa, and manufactured goods follow an ethical production and purchasing process. By

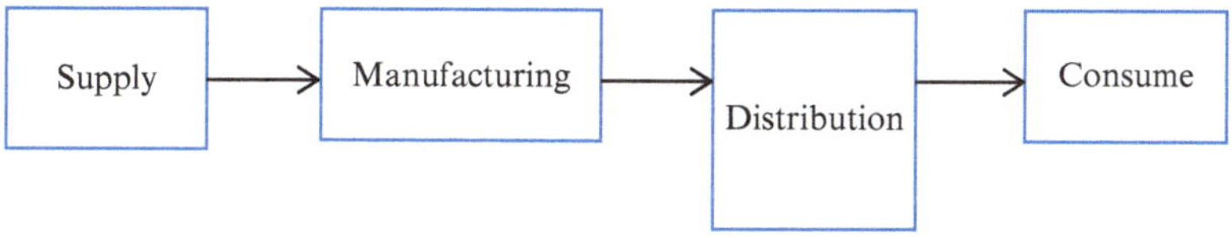

Fig. 4.1 Conventional supply chain management. (Adapted from Liu and Chang 2017)

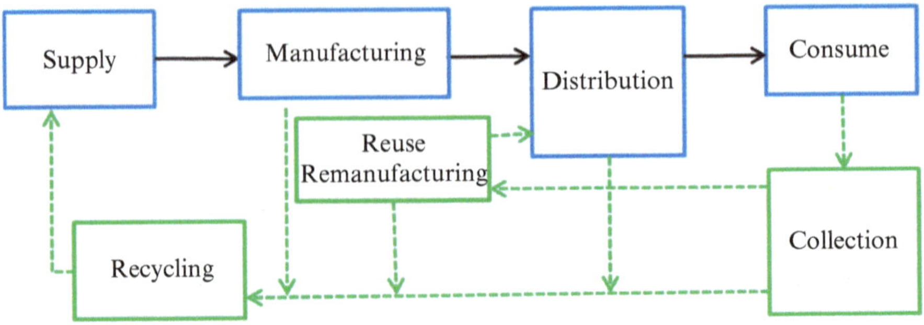

Fig. 4.2 A supply chain with an EMS. (Adapted from Liu and Chang 2017)

adopting and following such concrete steps, Starbucks is improving environmental management programs that mitigate the negative effect of climate change on farmers. Such kind of business initiatives results in improving environmental sustainability levels and corporate environmental image and hence creating a unique balance between the success of business objectives and sustainability.

Regarding contributions toward environmental sustainability, business firms usually implement an *environmental management system* (EMS), connected with the reverse logistics segment of their supply chains. Environmental management system (EMS) involves a number of successive processes and practices, which facilitate a business unit in reducing its environmental impact and contribute toward increasing its operating efficiency (EPA 2017). Under EMS framework, a business firm try achieving a number of targeted environmental goals through a set of reliable and ecologically regulated operations in a supply chain. In many ways, EMS initiates a recycling process directed toward reducing waste in a supply chain through reuse and remanufacturing maintainable processes (see Fig. 4.2).

Green supply chain management (GSCM) refers to as a sophisticated extension of EMS. Literature (Chin et al. 2015) reports GSCM as a management process in supply chains whose prime objective is to remove all kind of waste and reduce the environmental impact in various stages of a supply chain. When a supply chain holistically implements such kind of environmental thinking and management approach, that ordinary supply chain develops into a green supply chain (Chin et al. 2015). GSCM became particularly relevant when heightened commercial activity showed highly adverse impacts on the environment (Luthra et al. 2014). An increase in the environmental deterioration forced business firms to adopt ethical and social responsible measure in their everyday business operations and mindfulness about the role of green practices in their supply chains (Diabat and Govindan 2011).

Literature (Liu and Chang 2017) suggests that a business firm' supply chain becomes a green supply chain when it adopts and implements both environmental management system (EMS) and supply chain management (SCM) in their business practices. Further, the relationship between EMS and SCM explains the concept of sustainability. Achieving optimum sustainable performance levels, business firms need to focus on environmental and social performance as they work on economic

Table 4.1 Characteristics of an ordinary supply chain (SC) and a green supply chain (GSC)

Characteristics	SC	GSC
Objective/value/output	Economic	Economic and ecological
Ecological optimization	Integrated approach	Ecological impact
Industry/supplier relations	Short term	Long term
Adoption cost	Low	High
Speed/flexibility	High	Low
Focus	Supply chain	Whole environment

Adapted from Deshmukh and Vasudevan (2014)

performance (Carter and Rogers 2008). In other words, business firms need to focus on durable commercial profitable goals that can reduce the environmental and societal risks simultaneously.

Currently, business firms that integrate environmental sustainability and corporate social responsibility (CSR) elements to make their supply chains greener and sustainable adopt this attitude. This tactic provides then with a long-term competitive advantage as the worldwide consumer demand for products focuses more on environment-friendly elements. Achieving environment sustainability goals demands a sense of mutual dependence between organization, environment, and society because these results in achieving shared value and a win-win solution (Thoo et al. 2013).

Traditional supply chains focus on cost and delivery elements of the final product and service with minimal relevance to the ecological effects, whereas GSCM focuses on supply chains to become greener by integrating ecological and economic factors together. For instance, literature (Deshmukh and Vasudevan 2014) outlined the main difference between ordinary and green supply chains that the first type of supply chains is focused mainly on business and economic factors, whereas the latter supply chains combine business profitability with achieving long-term environmental goals. Literature (Sulistio and Rini 2015) shows that green supply chain management includes integrating greener management processes into key supply chain processes such as planning, procurement, production, consumption, and reverse logistics. Such requirements result in greener as well as sustainable supply chains (Singh and Trivedi 2016). Table 4.1 outlines the key differences between the two supply chains.

Green supply chain management to some extent follows the just-in-time (JIT) philosophy by minimizing waste in a supply chain. Similarly, GSCM focuses on eliminating waste that includes all kind of hazardous chemicals, emissions, energy, and solid waste with a given supply chain. For instance, eliminating all kind of waste associated with product design, material procurement, supplier selection, managing waste in a manufacturing process, delivery of final product, and reverse logistics connected with a product's end-of-life management (Srivastava 2007). Literature (Sarkis et al. 2011) considers GSCM as a holistic concept where supply chain operations of a company are aligned with the environmental elements. Further, a company focuses on integrating the environmental elements in interorganizational

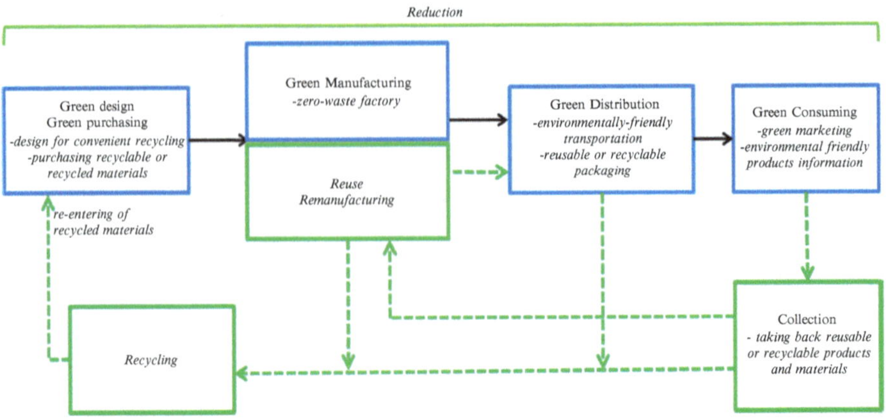

Fig. 4.3 Green supply chain management. (Adapted from Liu and Chang 2017)

practices for achieving better success. In this manner, the total supply chain functions as one green unit where the prime focus is on managing a greener environmental-friendly supply chain management system.

Implementing and achieving GSCM is not an easy task, as the concerned firms need to bring about all the actors within a supply chain. iPhone, the star product of Apple, has more than 500 first-tier suppliers located all over the world. Imagine how difficult it will be for Apple to monitor and ensure environmental compliance in this supply chain alone. This example underlines the enormity of turning an ordinary supply chain management into a tangible green supply chain management. Supply chain processes are dynamic processes and numerous obstacles. Hence, any slack in implementing a GSCM approach, whether in the management or other key competencies, is likely to augment the difficulty. Achieving GSCM requires time, resources, internal awareness, competence, and effective support mechanisms commitment from top management (DNV GL 2016).

A green supply chain has a number of direct and indirect actors that collaborate to eliminate any environmental harming waste in every single stage of a supply chain. However, realizing GSCM demands that all actors function harmoniously to achieve the defined environmental goals. Figure 4.3 depicts a supply chain where the focus of GSCM is on reducing overall waste.

4.3 Green Innovations and Green Supply Chain Management

Green innovation refers to any innovation that mitigates or reduces the environmental impact of business activities by reducing energy use, pollution, and waste via recycling, either by a better environmental management system, by environment-friendly product designs, or by adopting new green technologies (Chen et al. 2006). Recent literature (Seman et al. 2012) describes green innovation as an innovative approach that business firms adopt and implement to redesign their consumer products and services, and manufacturing processes continue to minimize the negative environmental impact. Such an approach creates an environment-friendly image and identifies a clear differentiation between competitors' products. Studies (Chen 2008; Chiou et al. 2011; Tseng et al. 2013) segment green innovations on features such as green technology, top managerial functions, green manufacturing processes, and green product design. The prime purpose of adopting and implementing green innovations is to enhance business firms' environmental image and competitive capabilities (Burki and Dahlstrom 2017).

Current literature (Chen et al. 2012) segments green innovations into two categories, namely, *proactive green innovation* and *reactive green innovation*. Proactive green innovation is those active environment-related innovations that relate to new practices or products developed ahead of competitors, to decrease cost, to seize opportunities, to lead in the market, or to obtain competitive advantages. On the other hand, reactive green innovation involves passive environment-related innovations that comply with environmental regulations, adapting to the environmental requests from stakeholders, responding to the changing environment, and/or reacting to competitors' challenges.

Within green supply chains, the role of green innovations is to how to minimize environmental damage caused by business activities. As the world's consumers show a greater inclination for environmentally friendly products, it is vital for business firms to integrate green innovations as a key game-changing component in their supply chains to become environmentally friendly. Business firms do realize that greening their supply chains via green innovations is likely to result in opening new business opportunities related to new green products and green-oriented consumer markets. This emerging perspective about green innovations is likely to have a positive effect on businesses as the increasing consumer demands for healthier products are on the rise. Hence, any investments associated with green innovations in making a supply chain greener are most likely to lead to an overall environmental efficacy and result in a win-win situation.

4.3.1 Types of Green Innovation

Green supply chain management literature identifies a number of green innovations. It is beyond the scope of this chapter to outline all types of green innovations. This section will highlight only the role of those green innovations applied recently by researchers in managing green supply chains.

1. *Green management innovation*

This innovation is the key innovation in initiating, adopting, and managing green supply chain management. Green management innovation demonstrates a firm's management capability as well as capacity to follow a green approach. It is a comprehensive innovation process that includes increasing a firm's capability in areas such as specific technical knowledge and skills, required operations, and profit-making suitability (Chen 2008; Lin et al. 2013). Literature (Van Den Berg et al. 2013; Qi et al. 2010) accentuates that when a firm adopts green management innovation, its top management must get aboard key supply chain partners regarding environmental management system (EMS) ISO 14000 certifications and reducing hazardous emissions and waste.

Literature (Tseng et al. 2013) identifies the role of this green innovation as holistic because it involves changing and restructuring internal production processes and operations, which facilitate in implementing and achieving internal green supply chain management. Furthermore, such internal changes also improve products or services by the newly established environmental criteria (Tseng et al. 2012). For instance, a firm needs to establish an internal environmental management system and follow specific ISO 14000 series certifications to minimize water, electricity, gas, and petrol usage. A firm should involve supply chain partners and regularly hold environmental awareness training, seminars, and training regarding disposal of hazardous waste, hazardous emissions, and ways to minimize consumption of natural and nonrenewable energy resources (Zhu et al. 2010).

2. *Green product innovation*

Green product innovation refers to the innovation of product that is suitable for the environment (Kushwaha and Sharma 2016). Green product innovation encompasses different modifications made in product design, quality, and safety features to reduce the impact of the entire product life cycle on the environment (Kammerer 2009). Furthermore, these modifications are directed toward reducing different environment-harming toxins generated during production stages, energy generation, and biodegradable packaging (Kammerer 2009).

Developing green products implies using minimum resources to cater to consumers' demand. Regarding attractiveness, green product innovations provide a number of business benefits such as lower prices, higher market penetration of new products, economical use of resources, and achieving the ultimate goal of sustainable business operations (Kushwaha and Sharma 2016). Simultaneously, tangible green product innovations help business firms to differentiate and create competitive advantage. Furthermore, it helps firms to improve their overall environmental performance and firm's green image before consumers and contribute to achieving general industrial-level environmental goals.

3. *Green process innovation*

Green process innovation includes all manufacturing processes that help in reducing the negative impact on the environment by concrete measures such as energy saving, pollution prevention, and waste recycling (Qi et al. 2012). By adopting this innovation, businesses modify their product manufacturing processes that result in producing environmentally friendly products that also meet set eco-targets (Abdullah et al. 2016). When firms adopt green innovations and green technologies in their manufacturing processes, it results in reducing usage of natural resources and energy consumption (Van Den Berg et al. 2013). In this manner, green process innovation improves a firm's environmental performance by reducing the production waste and enhances a firm's commercial performance by lowering costs associated with resource utilization and costs of production.

4. *Green design innovation*

Green design refers to as the self-improvement in an industrial design concept that demands continuous development and design improvements to overcome any shortcomings (Guo 2017). In other words, green design innovation also guarantees sustainability. A green design provides consumers a basic guarantee of a product's quality by following the principle of 3Rs (reduce, recycle, and reuse). Such kind of conformity ensures that a specific product fulfills all the required green and health standards, and therefore, the product design is not harmful to the environmental destruction and will not contribute to resource waste (Hu and Fu 2003). Furthermore, a green design product will not affect users adversely physically and psychologically.

Especially, green design is the future of enterprises and an effective method to eliminate the green trade barrier (Guo 2017). To develop green design innovations, a number of principles are applied, for instance, the principle of the best ecological and economic benefit, the principle of energy conservation, the zero pollution principle, the principle of advanced technology, and the principle of making full use of resources (Guo 2017).

5. *Green technological innovation*

Primarily, green technological innovation involves providing a competitive advantage for business firms. For instance, green technological innovations include investing in technological innovations related to energy saving, water saving, noise and waste reduction, and air pollution (Qi et al. 2010). Green technological innovation investment in green production technology has a strategic inclination and acts as a stimulus for supporting green innovations in a business environment. This stimulus has an interactive dynamic with internal and external business environment. Internally, firms innovate their organizations with well-defined rules and objectives. In this manner, green technological innovation simultaneously builds up a firm's environmental performance and competitive advantage by implementing innovative technological ideas in its daily processes and operations (Abdullah et al. 2016). Externally, especially in a supply chain context, green technological innovations are the technological measures that a business firm collaboratively embarks on with other business firms to achieve a set of well-defined environment goals. A green supply chain is one of such example.

4.4 Drivers and Outcome of Green Innovation

Literature (Díaz-García et al. 2015) identifies a number of *macro*- and microlevel factors. Macro-level factor includes policy instruments such as norm and regulations, subsidies, fiscal instruments, regulations, education policy, technological systems, country context, and regional factors. And *micro*level factors are structural factors, strategy, and business logic, resources, and capabilities. In addition, the study also identifies *meso*-level factors (market trends, pressure groups of stakeholders, networking and collaborations, and industry in which the firm operate) that drive green innovations. In addition, a number of market factors such as green brand image, political and institutional stakeholder pressure, external stakeholders, and a firm's organizational and technological capabilities are the key drivers in green innovations.

Regarding the positive results of green innovations, the key outcome areas are economic and financial performance, enhanced competitive advantage, overall improvement in marketing performance, employee performance, better company reputation, increased exports, and higher productivity. The most important outcomes specifically related to the environmental performance are the reduction in atmospheric emissions, solid waste, sewage waste, energy consumption, consumption of toxic materials, and use of inputs that also lowers the frequency of environmental accidents (Jabbour et al. 2015) (Fig. 4.4).

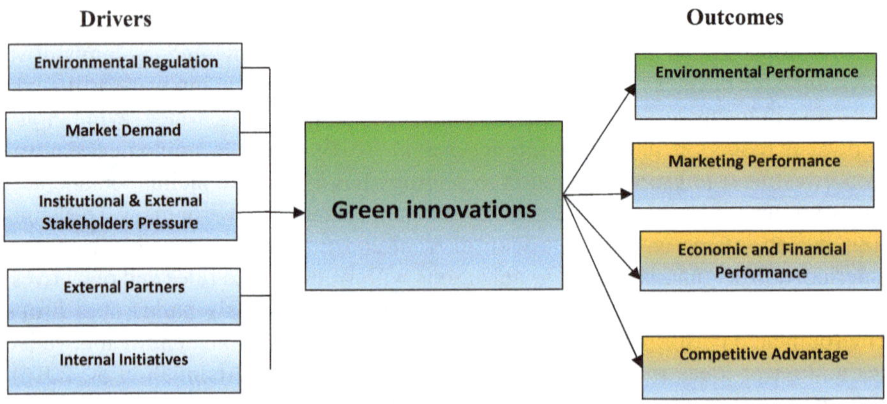

Fig. 4.4 Drivers and outcomes of green innovation

4.4.1 Drivers of Green Innovation

Literature (Qi et al. 2010) identifies government environmental regulations, environmental taxes, and subsidies as few of the key factors in adopting green practices. Similarly, regulations play a direct role regarding green innovation performance (Huang et al. 2016). Evidence shows that government regulations act as a support system for initiating green innovations (Abdullah et al. 2016). Market demand plays a vital role when it comes to introducing green products in a market and stimulating businesses to adopt innovation processes. Empirical evidence (Lin et al. 2013) show that market demand has a positive effect on green product innovation that leads to better firm performance. Consumer demand is a stimulus that facilitates to the adoption of green innovation (Abdullah et al. 2016). Literature (Zailani et al. 2014) reports that transportation companies adopted green innovations due to customer pressure. Hence, market demand is another facet of customer pressure and acts as the determinant factor that forces adoption and implementation of green innovations (Huang et al. 2016).

Stakeholder pressure is another key factor that enforces firms to adopt green innovations. Literature (Huang et al. 2009) shows that stakeholder pressure makes business firms to adopt green innovations. For business firms, stakeholder pressure assists business firms to adopt and integrate green innovation in their daily operations. Business firms are initially skeptical about the costs associated with adopting environmental management practices and investment in green innovations. However, the outcome benefits such as green product innovation and competitive advantage finally leave to firm profitability (Chan et al. 2016). Other external factors such media coverage on environmental issues also affect the adoption of green innovations. Literature (Li et al. 2017) shows that such kinds of external legitimate external pressure positively influence green product innovation and green process innovation.

Another driver of green innovation is the suppliers' role. To achieve green supply chain objective or environmental targets, supplier selection can play a critical role in the adoption of green innovations by a business firm (Gupta and Barua 2017). Concerning supplier selection, it is further suggested that factors such as green competencies, environmental management initiatives, green purchasing capabilities, environmental regulatory compliance, environmental investments, and end economic benefits play a critical role too. It is so because, in a supply chain context, supply-side factors play a more determining role in driving environmental processes and innovations necessary for any environmental innovations (Triguero et al. 2013). Evidence (Lee and Kim 2011) shows that suppliers play a key role in enhancing a firm's capability to accomplish and adopt environmental management and green innovations when a legal system supports it.

The preceding paragraph shows the significant role that external partners' collaboration has on the adoption of green product and process innovation (Dangelico 2016). For external collaboration, the involved partners should be knowledgeable about the relationship requirements regarding environmental requirements and green innovations (Wong 2013). Furthermore, the concerned partners should have a common goal toward achieving the well-desired environmental goals. It is also important for all the collaborating partners to work together with external research institutes such as universities to avoid any negative effect on green innovation adoption (Triguero et al. 2013; Abdullah et al. 2016).

A number of antecedents also play a key role in adopting green innovations Zailain et al. (2014). For instance, internal resources (Abdullah et al. 2016), environmental ethics (Chiou et al. 2011), internal environmental management and quality management (Leender and Chandra 2013), internal corporate profitability (Li et al. 2017), and firm size and the type of industry (Woo et al. 2014). All these internal factors somehow affect the attitudes and perceptions about the active adoption role of green innovations.

4.4.2 Outcomes of Green Innovation

Concerning outcomes of green innovations, firm performance and competitive capability are the top winners. When external driver such as green collaboration with external suppliers is adopted, it enhances firm's operational performance (Dai et al. 2017). Similarly, the green innovation related to green products also affects firm performance and competitive capabilities (Ar Ilker 2012). Literature also shows green innovations directed toward green products play a successful role in strengthening green product competitive advantage (Wong 2012).

Likewise, organizational factors such as environmental-friendly organizational innovations have a strong effect on a firm's business performance (Cheng et al. 2014). Emerging literature (Chan et al. 2016) supports the argument that the adoption of green innovation leads to cost-efficiency and better competitive advantage, resulting in better financial, environmental, and marketing performance. Studies (Xie et al. 2016) show that the adoption of green innovations improves a firm's absorptive capacity and financial performance. Other studies show that green innovations positively affect labor productivity (Woo et al. 2014), job intensity (Iranmanesh et al. 2017), and industry employment (Kunapatarawong and Martínez-Ros 2016).

4.4.3 Green Innovations, Green Practices, and Green Supply Chain Management

Green innovation plays a key role in achieving green supply chain management (GSCM). Different drivers of green innovation lead to a number of positive outcomes. Internal drivers take into account the main green initiatives and green processes involved in firm's operations. For instance, green initiatives start with top management commitment that later on translate into operations such as adopting

Table 4.2 Green practices and green innovations and green supply chain management (GSCM)

Studies	Green practices and green innovations
Zhu and Sarkis (2004)	Internal environmental management, green purchasing, customer cooperation with environmental concerns, investment recovery, eco-design
Zhu et al. (2008)	Internal environmental management, green purchasing, customer cooperation with environmental concerns, investment recovery, eco-design
Holt and Ghobadian (2009)	Internal environmental management practices, supplier education and mentoring, supplier assessment and evaluation, green procurement Green logistics policy, industrial networks
Ninlawan et al. (2010)	Green procurement, stakeholder collaboration Eco-design, green manufacturing/operations, green distribution, reverse logistics
Lin and Ho (2010)	Regulatory pressure, governmental support, organizational support, green human resource management
Large and Gimenez Thomsen (2011)	Green supply management capabilities, strategic purchasing, environmental commitment level, green supplier assessment, green collaboration
Arimura et al. (2011)	International standard compliance (ISO 14001), government programs
Azevedo et al. (2011)	Environmentally friendly purchasing practices, environmental collaboration with suppliers, eco-design, ISO 14001 certification, environmental collaboration with customers, minimizing waste, environmentally friendly packaging, reverse logistics
Chiou et al. (2011)	Greening the supply chain, green innovation
Wu et al. (2011)	Criteria for selecting suppliers for GSC
Kumar et al. (2012)	Product design, process design, supplier collaboration, green purchasing, green manufacturing, transportation and shipping, consumption and disposal
Green et al. (2012)	Internal environmental management, green information systems, green purchasing, cooperation with customers, eco-design, investment recovery
Lee et al. (2012)	Employee satisfaction, operational efficiency, eco-design, relational efficiency
Baines et al. (2012)	Green materials/products, green processes and operations, green use of product, green end-of-life management, regulation compliance, stakeholder collaboration, corporate acceptance, acceptance by customer base
Seman et al. (2012)	New green product
Dangelico and Pontrandolfo (2013)	Green manufacturing/operations (energy, materials, pollution), collaboration with business actors, collaboration with nonbusiness actors
Laosirihongthong et al. (2013)	International standard compliance (ISO 14001), threat of legislation and regulation
DNVGL (2013)	Customer collaboration, market collaboration

(continued)

Table 4.2 (continued)

Studies	Green practices and green innovations
Luthra et al. (2013)	Private governance, government regulations, supplier involvement and auditing, green procurement, green design, green operations, environment-friendly distribution, management commitment, customer awareness, end-of-life management
Thoo et al. (2013)	Green procurement, green manufacturing/operations, green distribution, green logistics
Van Den Berg et al. (2013)	Greening of suppliers
Lee et al. (2014)	Internal environmental management, eco-design, investment recovery, green purchasing, and cooperation with customers
Stackhouse (2014)	Green procurement, supplier relationship management
Blome et al. (2014)	Green procurement, green supplier development
Gunasekaran et al. (2015)	Green benefits, green relationship, green integration, green practices
Stackhouse (2015)	Innovation of materials, green manufacturing/operations, superior resource management
Jabbour et al. (2015)	Internal environmental management, cooperation with customers
DNVGL (2016)	Green operations, improving stakeholder relations
Bhardwaj (2016)	Environmental policy, green human resource management
Kadam et al. (2017)	Reuse, recycle, remanufacture, reverse logistics
Zhu et al. (2017)	Green innovation, green purchasing
Li and Huang (2017)	Internal environmental management, external activities (design specification to suppliers, cooperation with suppliers), investment recovery, eco-design

green innovations, green design, green procurement, and other green operations (Liu and Chang 2017; Burki and Dahlstrom 2017).

Green supply chain literature demonstrates (see Table 4.2) that green innovations belong to a broad set of green supply chain practices. Integration between appropriate green practices and green innovations results in developing an appropriate green supply chain and an appropriate management process. For instance, integrating green innovations with external processes of a green supply chain such as logistics, regulations, supplier collaboration, and customer cooperation leads to green supply chain management (GSCM) (Liu and Chang 2017). In this manner, manufacturing firms adopt and implement green innovations alongside green practices for achieving GSCM (Zhu et al. 2012).

In a macro-perspective, Table 4.2 shows that both green practices and green innovations are interdependent and play a pivotal role in achieving green supply chains. Adoption of green practices initiates green innovations, and in the end, both factors facilitate in achieving green supply chain management. Green innovations are one of the vital mechanisms of green practices. Appendix A presents a comprehensive review of the international literature regarding the research question, research context regarding country or industry, and findings. The literature review also demonstrates a high degree of interdependence between green practices and green innovations.

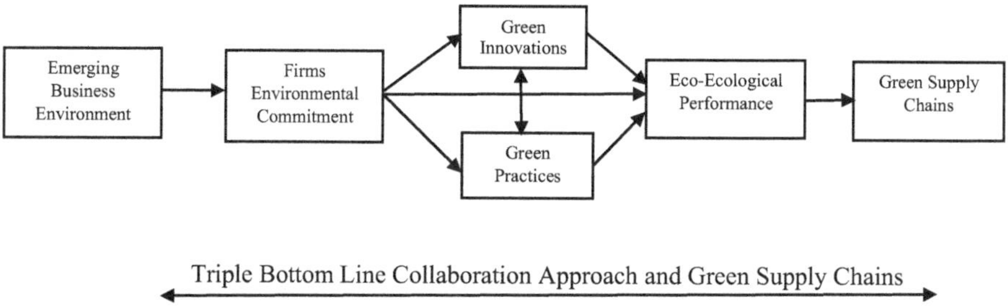

Fig. 4.5 Key antecedents of green supply chains

Based on the literature reviewed, the conceptual model (Fig. 4.5) identifies the triple bottom-line approach as key element in formulating sustainable and ecological-friendly supply chains. Current business environment is highly environmental conscious, and ecological issues such as reducing carbon foot are an integral part of firms' business strategies. Similarly, governments are making serious efforts to formulating and then implementing environmental regulations, for example, enforcement of strict regulations to improve the declining quality of air in the world major cities.

The model suggests that when businesses transform their ordinary supply chains to greener supply chains, a long-term sustainable triple bottom-line (TBL) approach will enforce businesses to conform to the environmental, economic, and social requirements. When it comes to the designing of green supply chains, they should conform to the concept of *green stewardship*. The green stewardship concept is concerned with integrating the economic, environmental, and social components in all the supply chain operations, ranging from production to the end-of-life process management. It is important to remember that supply chains, either ordinary or ecological, do not function in a vacuum but involve a number of economic, governmental, and community stakeholders.

Emerging ecological tendencies among consumers and supply chain stakeholders are forcing businesses and governments to show a clear commitment toward sustainability. When all the stakeholders show a clear commitment toward green stewardship, supply chains willingly adopt and implement green innovations and green practices in their supply chain operations. Once better economic and ecological results become evident, it strengthens firms' attitudes and commitments toward green supply chains. Green innovations and practice are one of the vital components of a functional green supply chain. However, adopting them into an ordinary supply chain demands a green stewardship resolve that should also fulfill the triple bottom-line benchmarks.

4.5 Conclusion

The importance of internal environmental management is the starting point of implementing and achieving GSCM. Under the IEM approach, a firm's top management needs to demonstrate a solid environmental commitment because staring and implementing environmental initiatives require monetary and organizational support. A well-established IEM embarking on adopting green practices and green innovations guarantees success. Implementing GSCM requires a combination of internal and external green practices and innovations. It is important to realize for supply chain partners that implementing green practices or green innovations does not always require new technological; rather improving existing business operations could facilitate in minimizing negative environmental impact and achieving TBL goals.

A firm and its management adopt green practices or green innovations in response to fulfill external stakeholder pressures. Business firms' choice of green strategies depends on the kinds of pressures and influences they experience from their stakeholders (Mander 2017; Baines et al. 2012). At times, the operational environment they work in also dictates the green initiatives they plan to implement. For instance, manufacturing industries focus more on initiating on internal green dimensions (Green et al. 2012). In another example, firm coordinates with its upstream and downstream supply chain partners to fulfill social compliance issues when it introduces eco-efficiency methods in its product design (Li et al. 2017). Achieving environmental practices is possible when firms develop a clear and strong association between enforcing international environmental regulations and achieving sustainable goals (Poulsen et al. 2016). For instance, when a third-party logistics-providing company adopts a green logistics model and starts following green regulation compliance, the initial support has to come from the company's top management and its employees under the internal environmental management system. Once this catalyst is in place, implementing green operations and processes, developing appropriate skills and training to employees, and understanding green business dimensions of partners become easier.

Collaboration with upstream and downstream supply chain stakeholders is another vital aspect that a firm has to take into consideration to achieve the transition from regular supply chain to a green supply chain. A supply chain cannot become a green supply chain if its key stakeholders fail to collaborate with each other regarding green supply chain objectives related to different levels of planning, operations, and international environmental regulations' compliance. Such coordination entails the transfer of required knowledge, skills, and functional support to achieve environmental efficiency. Better collaboration provides better opportunities to integrate green approaches in a supply chain to yield positive environmental and economic outcomes. Implementing a combination of green practices and green innovation to achieve GSCM is also critical for a firm to attain environmental, economic, and social gains and better supply chain performance.

When lawmaking organizations and regulatory authorities cannot keep up with the changing environment and stakeholder demands, industries and groups start collaborating at a local level to fill the gap and maintain the environment in which they operate. Although international regulatory bodies are actively working and introducing green initiatives and programs that help companies reduce their environmental impacts, their membership is low, the adoption is slow, and the control mechanisms are weak.

Further research should focus on extending the role of GSCM drivers about green practices and green innovations and their role in achieving triple bottom-line goals. It would be useful to know what factors companies and their top management consider as actual incentives to adopt GSCM. Additional information about business firms' environmental strategy for GSCM would also be useful. Environmental regulation compliance and governance issues with green supply chain partners are other areas that demand examination.

Appendix A: International Focus on Green Innovations

Study (author)	Research context	Country/ industry	Findings
The influence of green innovation performance on corporate advantage in Taiwan (Chen et al. 2006)	The correct evaluation and position for green innovation. The influence of green innovation performance on the corporate advantage	Taiwan/ information and electronic industries	Significant differences in the performance of green product innovation and the performance of green process innovation in information and electronics industries in Taiwan
The driver of green innovation and green image-green core competence (Chen 2008)	The issues of environmental management and core competence, which respond to the new concept of "green management" carrying both aspects of environmental protection and economy development	Taiwan/ information and electronic industries	Green core competence, green product innovation performance, green process innovation performance, and green images of SMEs depend on firm size
Salient stakeholder voices: family business and green innovation adoption (Huang et al. 2009)	The relationship between various stakeholder pressure perceived by the managers and their decision to adopt green technical and administrative innovations	Taiwan	The positive relationship between the degree of natural environmental pressure from stakeholders and a firm's decision to adopt green innovations

Study (author)	Research context	Country/ industry	Findings
The drivers for contractors' green innovation: an industry perspective (Qi et al. 2010)	Key factors influencing contractors to adopt green construction practices: government environmental regulations, managerial concern, project stakeholder pressure, and size of firms	China/ contractor industry	Managerial concerns and government regulatory pressures are the two most important driving forces for the adoption of green construction practices Significant relationships between government regulations and business size with the adoption of green construction practices
Mainstreaming green product innovation: why and how companies integrate environmental sustainability (Dangelico and Devashish 2013)	Insights on multifaceted nature of green product innovation process involving both micro- and macro-environmental issues	Italy, Canada/ small- to medium-sized manufacturing companies	Three key environmental dimensions of green product innovation such as energy minimization, material reduction, and pollution prevention. The green product development process, innovativeness of green products (radical versus incremental), and life cycle analysis as a tool for measuring environmental impact at each of stage of green products
Integrating suppliers into green product innovation development: an empirical case study in the semiconductor industry (Lee and Kim 2011)	The role of suppliers in enhancing the manufacturer's ability to carry out green innovation in product development successfully	Korea/ semiconductor industry	Regulations and legislation are major factors behind the promotion of environmental management and green innovation by companies. Two important factors for GPI: coordination and alignment with project teams and effective communication with suppliers
The influence of greening the suppliers and green innovation on environmental performance and competitive advantage in Taiwan (Chiou et al. 2011)	Relations between greening the supply chain, green innovation, environmental performance, and competitive advantage that helps to encourage companies to improve their environmental performance and to enhance their competitive advantage in the global market	Taiwan/eight industry sectors	Greening the supplier through green innovation contributes significant benefits to the environmental performance and competitive advantage of the firm

Study (author)	Research context	Country/ industry	Findings
The influence of corporate environmental ethics on competitive advantage: the mediation role of green innovation (Chiou 2011)	The relations between corporate environmental ethics, competitive advantage, and green product and process innovation that plays a mediation role	Taiwan/ manufacturing industry	The corporate environmental ethics can not only affect competitive advantage directly but also influence it indirectly via green product innovation in the Taiwanese manufacturing industry
The impact of green product innovation on firm performance and competitive capability: the moderating role of managerial, environmental concern (Ar Ilker 2012)	Relations between green product innovation, firm performance, and competitive capability and on the moderating role of managerial, environmental concern in this relation	Turkey/ manufacturer firms from various sectors	Green product innovation significantly positively affects both firm performance and competitive capability. Managerial, environmental concern only moderates the relationship between green product innovation and firm performance
Origins of green innovations: the differences between proactive and reactive green innovations (Chen et al. 2012)	The origins of the two types of green innovations: proactive and reactive green innovations	Taiwan/ Optronics Corp.	Two types of green innovations: proactive and reactive green innovations, considering their origins. Both of the internal origins (environmental leadership, environmental culture, and environmental capability and the external origins) and the environmental regulations and the environmentalism of investors and clients can generate reactive green innovation. Only the internal origins can facilitate proactive green innovation
The influence of green product competitiveness on the success of green product innovation: empirical evidence from the Chinese electrical and electronics industry (Wong 2012)	The article addresses the gap in green innovation theory concerning the associations among the key constructs of green innovation causal chain	China/electrical and electronics industry	Green product innovation has a positively stronger influence on both green product competitive advantage and green new product success than green process innovation

Study (author)	Research context	Country/ industry	Findings
Market demand, green product innovation, and firm performance: evidence from Vietnam motorcycle industry (Lin et al. 2013)	Correlations between market demand, green product innovation, and firm performance in the context of Vietnamese motorcycle industry	Vietnam/ motorcycle industry	Market demand is positively correlated to both green product innovation and firm performance, while green product innovation performance is also positively correlated with firm performance. Effect of three types of green product innovation on market demand and firm performance
Antecedents and consequences of green innovation in the wine industry: the role of channel structure (Leender and Chandra 2013)	The internal vs. external drivers of green innovation, the effect of green innovation on business performance, and the role of channel structures in making green innovation more productive regarding business performance	USA and Canada, South Africa, and Australia and New Zealand/ wine industry	Internal drivers, i.e., environmental management and quality management, in particular, play a greater role than external drivers (e.g., government and regulatory pressures) on the adoption of green innovation strategies. Producing and using organic products and processes and recycling activities have a significant direct positive impact on business performance
Improving performance of green innovation practices under uncertainty (Tseng et al. 2013)	The green managerial, process, product, and technology innovation aspects in Taiwanese computer products industry	Taiwan/ computer products industry	The appropriate green innovation aspects and criteria for the case firms evaluate the weights of the aspects and criteria as described by linguistic preferences and use an analytical network process with entropy weights to evaluate the proposed framework

Study (author)	Research context	Country/ industry	Findings
The effects of greening the supplier and innovation on environmental performance and competitive advantage: original research (Van Den Berg et al. 2013)	The relationship between greening the supplier, green innovation, environmental performance, and competitive advantages in companies in the City of Johannesburg metropolitan municipality region	South Africa/ companies in the City of Johannesburg	The green innovative process has a significant effect on environmental performance. Greening the supplier will initiate greener product innovation, and green managerial innovation has a significant correlation with a competitive advantage. The greening of suppliers using green innovation leads to enhanced environmental performance and competitive advantages
Environmental requirements, knowledge sharing and green innovation: empirical evidence from the electronics industry in China (Wong 2013)	Developing an integrative model that explores the determinants of green innovation with a focus being placed on knowledge sharing	China/ electronics industry	Knowledge sharing mediates the relationship between green requirements and new green product success as well as that of green requirements and green product and process innovations. The direct positive impact of knowledge sharing is the strongest on green process innovation
Drivers of different types of eco-innovation in European SMEs (Triguero et al. 2013)	The influence of several factors as collaboration with research institutes, supply-side, demand-side, and regulatory factors. The multivariate estimations considered allow showing potential complementarities across the three types of eco-innovation as well as differences across sectors and countries	27 European countries	Collaboration with research institutes, agencies, and universities is a significant factor. Supply-side factors are a more important driver for environmental processes and organizational innovations than for environmental product innovations. Market share has a significant positive influence on eco-product and eco-organizational innovations, while cost-savings are significant for eco-process innovations

Study (author)	Research context	Country/ industry	Findings
The link between eco-innovation and business performance: a Taiwanese industry context (Cheng et al. 2014)	Interrelationships among three types of eco-innovation and their relative impact on business performance	Taiwan	Eco-organizational innovation has the strongest effect on business performance. Eco-process and eco-product innovations partially mediate the effects of eco-organizational innovation, and eco-product innovation mediates eco-process innovations' effects on business performance. Business performance is directly and indirectly affected by eco-organizational, eco-process, and eco-product innovations
Impact of green innovation on labor productivity and its determinants: an analysis of the Korean manufacturing industry (Woo et al. 2014)	The impact of green innovation on labor productivity and the determinants of green innovation	Korea/ manufacturing industry	GI intended for both firm and customer benefits has a positive effect on labor productivity. There are significant differences in aggregate green innovations depending on different firm sizes and industries
Can political capital drive corporate green innovation? Lessons from China (Lin et al. 2013)	The factors influencing firms' GI under a new theoretical framework, combining the stakeholder theory and the RBV, and providing new insights that help clarify relationships among them	China/private manufacturing firms	Political capital plays a significant but negative role in firms' green product and process innovation performance; both regulations and suppliers positively promote GI in product and process; consumers are positively related to green product innovation but negatively related to green process innovation; and competitions do not have any significant effects on both green product and process innovation

Study (author)	Research context	Country/ industry	Findings
Drivers of green innovation and non-green innovation: empirical evidence in low-tech SMEs (Cuerva et al. 2014)	The main drivers of eco-innovation in SMEs. The differences between the factors influencing "green" and/or "non-green" innovations for SME in a low-tech sector	Spain/low-tech sector	Technological capabilities such as R&D and human capital foster the conventional innovation but not the GI. The implementation of QMS and differentiation explain only the adoption of innovative green activities. Higher implementation of voluntary scheme certifications such QMS would be more effective to enhance eco-innovation than public subsidies
Creating technological innovation via green supply chain management: an empirical analysis (Lee et al. 2014)	The relationship between GSCM practices and TI from the perspective of a developing nation	Malaysia/ manufacturing firms	The significant positive relationship between three GSCM practices (i.e., internal environmental management, eco-design, and investment recovery) and TI, green purchasing, and cooperation with customers does not have a significant positive correlation with TI. The importance of GSCM practices has been proven to enhance firms' TI, in addition to improving the environment, bringing about a positive impact on the manufacturing establishment
Green innovation adoption in automotive supply chain: the Malaysian case (Zaelani et al. 2015)	The determinants of green innovation adoption and its effect on firm performance	Malaysia/ automotive industry	Environmental regulations, market demand, and firm internal initiatives have a positive effect on green innovation initiative (GII), while GIIs have a positive effect on the three categories of sustainable performance (i.e., environmental, social, and economic)

Study (author)	Research context	Country/ industry	Findings
Taking risks in the face of uncertainty: an exploratory analysis of green innovation (Roper and Tapinos 2016)	The article focuses on how firms' perceptions of environmental uncertainty and their perceptions of the risks involved impact on their willingness to undertake GI	UK/food industry	Firms' probability of undertaking green innovating is positively related to both environmental uncertainty and the market-related risks of innovation
Barriers to green innovation initiatives among manufacturers: the Malaysian case (Abdullah et al. 2016)	The internal and external barriers to green innovation initiatives among Malaysian manufacturers	Malaysia/ manufacturing industry	The barriers to GI are a useful concept to prioritize innovation policies
The impact of legitimacy pressure and corporate profitability on green innovation: Evidence from China top 100 (Li et al. 2017)	The influence of external legitimacy pressure and internal corporate profitability and their interaction on green innovation	China/the top 100 listed companies	Legitimacy pressure from stakeholders has a significantly positive influence on both corporate green product innovation and process innovation. Corporate profitability positively affects green product innovation, while there was found to be no significant influence on green process innovation. Moreover, corporate profitability positively moderates the relationship between legitimacy pressure and green product innovation

References

Abdullah, M., Zailani, S., Iranmanesh, M., & Jayaraman, K. (2016). Barriers to green innovation initiatives among manufacturers: the Malaysian case. *Review of Managerial Science, 10*(4), 683–709 https://ezproxy1.usn.no:3528/10.1007/s11846-015-0173-9.

Ar Ilker. (2012). The impact of green product innovation on firm performance and competitive capability: The moderating role of managerial environmental concern. *Procedia – Social and Behavioral Sciences*, 854–864 https://doi.org/10.1016/j.sbspro.2012.09.144.

Arimura, T., Darnall, N., & Katayama, H. (2011). Is ISO 14001 a gateway to more advanced voluntary action? A case for green supply chain management. *Journal of Environmental Economics and Management, 61*(2). https://doi.org/10.2139/ssrn.1368393.

Azevedo, S. G., Carvalho, H., & Machado, V. C. (2011). The influence of green practices on supply chain performance: A case study approach. *Transportation Research Part E: Logistics and Transportation Review, 47*(6), 850–871 https://doi.org/10.1016/j.tre.2011.05.017.

Baines, T., Brown, S., Benedettini, O., & Ball, P. (2012). Examining green production and its role within the competitive strategy of manufacturers. *Journal of Industrial Engineering and Management, 5*(1). https://doi.org/10.3926/jiem.405.

Bhardwaj, B. R. (2016). Role of green policy on sustainable supply chain management: A model for implementing corporate social responsibility (CSR). *Benchmarking, 23*(2), 456–468. https://doi.org/10.1108/BIJ-08-2013-0077.

Blome, C., Hollos, D., & Paulraj, A. (2014). Green procurement and green supplier development: Antecedents and effects on supplier performance. *International Journal of Production Research, 52*(1), 32–49. http://ezproxy1.usn.no:2092/10.1080/00207543.2013.825748.

Burki, U., & Dahlstrom, R. (2017). Mediating effects of green innovations on interfirm cooperation. *Australasian Marketing Journal: AMJ, 25*(2), 149–156 https://doi.org/10.1016/j.ausmj.2017.05.001.

Carter, C. R., & Rogers, D. S. (2008). A framework of sustainable supply chain management: Moving toward new theory. *International Journal of Physical Distribution and Logistics Management, 39*(5), 360–387.

Chan, H. K., Yee, R. W. Y., Dai, J., & Lim, M. K. (2016). The moderating effect of environmental dynamism on green product innovation and performance. *International Journal of Production Economics, 181*, 384–391. https://doi.org/10.1016/j.ijpe.2015.12.006.

Chen, Y.-S., Chang, C.-H., & Wu, F.-S. (2012). Origins of green innovations: the differences between proactive and reactive green innovations. *Management Decision, 50*(3), 368–398 http://ezproxy1.usn.no:2092/10.1108/00251741211216197.

Chen, Y.-S., Lai, S.-B., & Wen, C.-T. (2006). The influence of green innovation performance on corporate advantage in Taiwan. *Journal of Business Ethics, 67*(4), 331–339 https://ezproxy2.usn.no:3481/10.1007/s10551-006-9025-5.

Chen, Y. S. (2008). The driver of green innovation and green image-green core competence. *Journal of Business Ethics, 81*(3), 531–543 https://ezproxy2.usn.no:3481/10.1007/s10551-007-9522-1.

Cheng, C. C. J., Yang, C.-L., & Sheu, C. (2014). The link between eco-innovation and business performance: a Taiwanese industry context. *Journal of Cleaner Production, 64*, 81–90 https://doi.org/10.1016/j.jclepro.2013.09.050.

Chin, T., Tat, H., & Sulaiman, Z. (2015). Green supply chain management, environmental collaboration and sustainability performance. *Procedia CIRP, 26*, 695–699 https://doi.org/10.1016/j.procir.2014.07.035.

Chiou, T.-Y., Chan, H. K., Lettice, F., & Sai Ho Chung, S. H. (2011). The influence of greening the suppliers and green innovation on environmental performance and competitive advantage in Taiwan. *Transportation Research Part E: Logistics and Transportation Review, 47*(6), 822–836 https://doi.org/10.1016/j.tre.2011.05.016.

Cuerva, M. C., Triguero-Cano, A., & Córcoles, D. (2014). Drivers of green and non-green innovation: empirical evidence in Low-Tech SMEs. *Journal of Cleaner Production*, 68, 104–113.

Dai, J., Cantor, D. E., & Montabon, F. L. (2017). Examining corporate environmental proactivity and operational performance: A strategy-structure-capabilities-performance perspective within a green context. *International Journal of Production Economics, 193*, 272–280. https://doi.org/10.1016/j.ijpe.2017.07.023.

Dangelico, R., & Pontrandolfo, P. (2013). Being 'green and competitive': The impact of environmental actions and collaborations on firm performance. *Business Strategy and the Environment, 24*(6), 413–430 https://doi.org/10.1002/bse.1828.

Dangelico, R. M. (2016). Green product innovation: Where we are and where we are going. *Business Strategy and the Environment, 25*(8), 560–576 https://doi.org/10.1002/bse.1886.

Deshmukh, A., & Vasudevan, H. (2014). Emerging supplier selection criteria in the context of traditional VS green supply chain management. *International Journal of Managing Value and Supply Chains, 5*(1), 19–33 https://doi.org/10.5121/ijmvsc.2014.5103.
Diabat, A., & Govindan, K. (2011). An analysis of the drivers affecting the implementation of green supply chain management. *Resources, Conservation and Recycling, 55*(6), 659–667 https://doi.org/10.1016/j.resconrec.2010.12.002.
Díaz-García, C., González-Moreno, Á., & Sáez-Martínez, F. J. (2015). Eco-innovation: Insights from a literature review. *Innovation: Management, Policy and Practice, 17*(1), 6–23 http://ezproxy2.usn.no:2087/10.1080/14479338.2015.1011060.
DNV GL. (2013). Sustainable product: What's your approach? Retrieved from https://www.dnvgl.com/assurance/viewpoint/viewpoint-surveys/sustainable-products.html
DNV GL. (2016). Integrating sustainability into the core business. Seizing the change. Retrieved from https://www.dnvgl.com/assurance/viewpoint/viewpoint-surveys/integrated-sustainability.html
Epa.gov. (2017). Environmental Management Systems (EMS) I US EPA. [online] Available at: https://www.epa.gov/ems. Accessed 21 Jan 2017.
Green, K., Zelbst, P., Meacham, J., & Bhadauria, V. (2012). Green supply chain management practices: Impact on performance. *Supply Chain Management: An International Journal, 17*(3), 290–305 https://doi.org/10.1108/13598541211227126.
Gunasekaran, A., Subramanian, N., & Rahman, S. (2015). Green supply chain collaboration and incentives: Current trends and future directions. *Transportation Research Part E: Logistics and Transportation Review, 74*, 1–10 https://doi.org/10.1016/j.tre.2015.01.002.
Guo, Y.-F. (2017). Green innovation design of products under the perspective of sustainable development. *IOP Conference Series: Earth and Environmental Science, 51*(1), 012011 (5pp).
Gupta, H., & Barua, M. K. (2017). Supplier selection among SMEs on the basis of their green innovation ability using BWM and fuzzy TOPSIS. *Journal of Cleaner Production, 152*, 242–258 https://doi.org/10.1016/j.jclepro.2017.03.125.
Holt, D., & Ghobadian, A. (2009). An empirical study of green supply chain management practices amongst UK manufacturers. *Journal of Manufacturing Technology Management, 20*(7), 933–956 https://doi.org/10.1108/17410380910984212.
Hu, A.-W., & Fu, Z.-H. (2003). Green Design for Products. *Journal of Zhuzhou Institute of Technology, 17*(5), 21–23.
Huang, Y. C., Ding, H. B., & Kao, M. R. (2009). Salient stakeholder voices: Family business and green innovation adoption. *Journal of Management & Organization, 15*, 309–326 https://search.proquest.com/docview/233256299?accountid=43239.
Huang, X.-x., Hu, Z.-p., Liu, C.-s., Yu, D.-j., & Yu, L.-f. (2016). The relationships between regulatory and customer pressure, green organizational responses, and green innovation performance. *Journal of Cleaner Production, 112*, 3423–3433 https://doi.org/10.1016/j.jclepro.2015.10.106.
Iranmanesh, M., Zailani, S., Moeinzadeh, S., & Nikbin, D. (2017). Effect of green innovation on job satisfaction of electronic and electrical manufacturers' employees through job intensity: personal innovativeness as moderator. *Review of Managerial Science, 11*(2), 299–313 https://doi.org/10.1007/s11846-015-0184-6.
Jabbour, A. B. L. D. S., Frascareli, F. C. D. O., & Jabbour, C. J. C. (2015). Green supply chain management and firms' performance: Understanding potential relationships and the role of green sourcing and some other green practices. *Resources, Conservation and Recycling, 104*, 366–374 https://doi.org/10.1016/j.resconrec.2015.07.017.
Kadam, S., Karvekar, A., & Kumbhar, V. (2017). Traditional & green supply chain management – A review. *IARJSET, 4*(1), 38–41 https://doi.org/10.17148/iarjset/ncdmete.2017.11.
Kammerer, D. (2009). The effects of customer benefit and regulation on environmental product innovation: Empirical evidence from appliance manufacturers in Germany. *Ecological Economics, 68*(8), 2285–2295 https://doi.org/10.1016/j.ecolecon.2009.02.016.
Kumar, S., Teichman, S., & Timpernagel, T. (2012). A green supply chain is a requirement for profitability. *International Journal of Production Research, 50*(5), 1278–1296 https://doi.org/10.1080/00207543.2011.571924.

Kunapatarawong, R., & Martínez-Ros, E. (2016). Towards green growth: How does green innovation affect employment? *Research Policy, 45*(6), 1218–1232.

Kushwaha, G. S., & Sharma, N. K. (2016). Green initiatives: A step towards sustainable development and firm's performance in the automobile industry. *Journal of Cleaner Production, 121*, 116–129 https://doi.org/10.1016/j.jclepro.2015.07.072.

Laosirihongthong, T., Adebanjo, D., & Choon Tan, K. (2013). Green supply chain management practices and performance. *Industrial Management & Data Systems, 113*(8), 1088–1109 https://doi.org/10.1108/imds-04-2013-0164.

Large, R., & Gimenez Thomsen, C. (2011). Drivers of green supply management performance: Evidence from Germany. *Journal of Purchasing and Supply Management, 17*(3), 176–184 https://doi.org/10.1016/j.pursup.2011.04.006.

Lee, S., Tae Kim, S., & Choi, D. (2012). Green supply chain management and organizational performance. *Industrial Management & Data Systems, 112*(8), 1148–1180 https://doi.org/10.1108/02635571211264609.

Lee, K.-H., & Kim, J.-W. (2011). Integrating Suppliers into Green Product Innovation Development: an Empirical Case Study in the Semiconductor Industry. *Business Strategy and the Environment, 20*(8), 527–538 https://doi.org/10.1002/bse.714.

Lee, V.-H., Ooi, K.-B., Chong, A. Y.-L., & Seow, C. (2014). Creating technological innovation via green supply chain management: An empirical analysis. *Expert Systems with Applications, 41*(16), 6983–6994 https://doi.org/10.1016/j.eswa.2014.05.022.

Leender, M. A. A. M., & Chandra, Y. (2013). Antecedents and consequences of green innovation in the wine industry: The role of channel structure. *Technology Analysis & Strategic Management, 25*, 203–218. https://doi.org/10.1080/09537325.2012.759203.

Li, D., Zheng, M., Cao, C., Chen, X., Ren, S., & Huang, M. (2017). The impact of legitimacy pressure and corporate profitability on green innovation: Evidence from China top 100. *Journal of Cleaner Production, 141*, 41–49 https://doi.org/10.1016/j.jclepro.2016.08.123.

Li, Y-H., & Huang, J.-W. (2017). The moderating role of relational bonding in green supply chain practices and performance. Journal of Purchasing and Supply Management, In Press, Corrected Proof. Available online 8 June 2017. https://doi.org/10.1016/j.pursup.2017.06.001.

Lin, R.-J., Tan, K.-H., & Geng, Y. (2013). Market demand, green product innovation, and firm performance: evidence from Vietnam motorcycle industry. *Journal of Cleaner Production, 40*, 101–107 https://doi.org/10.1016/j.jclepro.2012.01.001.

Lin, C., & Ho, Y. (2010). Determinants of green practice adoption for logistics companies in China. *Journal of Business Ethics, 98*(1), 67–83 https://doi.org/10.1007/s10551-010-0535-9.

Liu, S., & Chang, Y. (2017). Manufacturers' closed-loop orientation for green supply chain management. *Sustainability, 9*(2), 222 https://doi.org/10.3390/su9020222.

Luthra, S., Garg, D., & Haleem, A. (2013). Identifying and ranking of strategies to implement green supply chain management in Indian manufacturing industry using analytical hierarchy process. *Journal of Industrial Engineering and Management, 6*(4). https://doi.org/10.3926/jiem.693.

Luthra, S., Garg, D., & Haleem, A. (2014). Green supply chain management. *Journal of Advances in Management Research, 11*(1), 20–46 https://doi.org/10.1108/jamr-07-2012-0027.

Mander, S. (2017). Slow steaming and a new dawn for wind propulsion: A multi-level analysis of two low carbon shipping transitions. *Marine Policy, 75*, 210–216 https://doi.org/10.1016/j.marpol.2016.03.018.

Ninlawan, C., Seksan, P., Tossapol, K., & Pilada, W. (2010). The implementation of green supply chain management practices in electronics industry. In Proceedings of the International MultiConference of Engineers and Computer Scientists; March 17–19: Hong Kong. Retrieved from http://www.iaeng.org/publication/IMECS2010/IMECS2010_pp1563-1568.pdf

Poulsen, R. T., Ponte, S., & Lister, J. (2016). Buyer-driven greening? Cargo-owners and environmental upgrading in maritime shipping. *Geoforum*, 68, 57–68.

Qi, G., Zeng, S., Li, X., & Tam, C. (2012). Role of internalization process in defining the relationship between ISO s1 certification and corporate environmental performance. *Corporate Social Responsibility and Environmental Management, 19*(3), 129–140 https://doi.org/10.1002/csr.258.

Qi, G. Y., Shen, L. Y., Zeng, S. X., & Jorge, O. J. (2010). The drivers for contractors' green innovation: an industry perspective. *Journal of Cleaner Production, 18*(14), 1358–1365 https://doi.org/10.1016/j.jclepro.2010.04.017.

Roper, S., & Tapinos, E. (2016). Taking risks in the face of uncertainty: An exploratory analysis of green innovation. *Technological Forecasting and Social Change, 112*, 357–363.

Sarkis, J., Zhu, Q., & Lai, K.-H. (2011). An organizational theoretic review of green supply chain management literature. *International Journal of Production Economics, 130*, 1–15.

Seman, N. A. A., Zakuan, N., Jusoh, A., Arif, M. S. M., & Saman, M. Z. M. (2012). The relationship of green supply chain management and green innovation concept. *Procedia - Social and Behavioral Sciences, 57*, 453–457 https://doi.org/10.1016/j.sbspro.2012.09.1211.

Singh, A., & Trivedi, A. (2016). Sustainable green supply chain management: trends and current practices. *Competitiveness Review, 26*(3), 265–288 https://search.proquest.com/docview/1826443162?accountid=43239.

Srivastava, S. K. (2007). Green supply-chain management: A state-of-the-art literature review. *International Journal of Management Reviews, 9*(1), 53–80 https://doi.org/10.1111/j.1468-2370.2007.00202.x.

Stackhouse, L. (2014). Developed supplier relationship management brings in the big bucks. Marine Trader. Retrieved from http://readmt.com/news/article/2014/11/07/developed-supplier-relationship-management-brings-in-the-big-bucks/

Stackhouse, L. (2015). 3 key sustainable maritime supply chain trends for 2015. Marine Trader. Retrieved from http://readmt.com/sustainability/article/2015/02/03/3-key-sustainable-maritime-supply-chain-trends-for-2015/

Sulistio, J., & Rini, T. A. (2015). A Structural Literature Review on Models and Methods Analysis of Green Supply Chain Management. *Procedia Manufacturing, 4*, 291–299 https://doi.org/10.1016/j.promfg.2015.11.043.

Tarig, A., Badir, Y. F., Tarig, W., & Bhutta, U. S. (2017). Drivers and consequences of green product innovation a systematic review, conceptual framework, and future outlook. *Technology in Society, 51*, 8–23 https://doi.org/10.1016/j.techsoc.2017.06.002.

Thoo, A., Abdul Hamid, A., Rasli, A., & Zhang, D. (2013). The moderating effect of Enviropreneurship on green supply chain management practices and sustainability performance. *Advanced Materials Research, 869-870*, 773–776 https://doi.org/10.4028/www.scientific.net/amr.869-870.773.

Triguero, A., Moreno-Mondéjar, L., & Davia, M. A. (2013). Drivers of different types of eco-innovation in European SMEs. *Ecological Economics, 92*, 25–33 https://doi.org/10.1016/j.ecolecon.2013.04.009.

Tseng, M.-L., Huang, F.-H., & Chiu, A. S. F. (2012). Performance drivers of green innovation under incomplete information. *Procedia - Social and Behavioral Sciences, 40*, 234–250 https://doi.org/10.1016/j.sbspro.2012.03.186.

Tseng, M.-L., Wang, R., Chiu, A. S. F., Geng, Y., & Lin, Y. H. (2013). Improving performance of green innovation practices under uncertainty. *Journal of Cleaner Production, 40*, 71–82 https://doi.org/10.1016/j.jclepro.2011.10.009.

UNFCCC (2017). The Paris agreement. Retrieved 18 August from http://unfccc.int/paris_agreement/items/9485.php

Van Den Berg, U., Labuschagne, J.-P., & Van Den Berg, H. (2013). The effects of greening the supplier and innovation on environmental performance and competitive advantage: original research. *Journal of Transport and Supply Chain Management, 7*(1), 1–7 https://doi.org/10.4102/jtscm.v7i1.121.

Wong, S. K. S. (2012). The influence of green product competitiveness on the success of green product innovation: Empirical evidence from the Chinese electrical and electronics industry. *European Journal of Innovation Management, 1*, 468–490.

Wong, S. K. S. (2013). Environmental Requirements, Knowledge Sharing and Green Innovation: Empirical Evidence from the Electronics Industry in China. *Business Strategy and the Environment, 22*(5), 321–338 https://doi.org/10.1002/bse.1746.

Woo, C., Chung, Y., Chun, D., Han, S., & Lee, D. (2014). Impact of Green Innovation on Labor Productivity and its Determinants: an Analysis of the Korean Manufacturing Industry. *Business Strategy and the Environment, 23*(8), 567–576 https://doi.org/10.1002/bse.1807.
WTO. (2014). *Trade and development: Recent trends and the role of the WTO*. WTO Publications Retrieved from https://www.wto.org/english/res_e/booksp_e/world_trade_report14_e.pdf.
Wu, K.-J., Tseng, M.-L., & Vy, T. (2011). Evaluation the drivers of green supply chain management practices in uncertainty. *Procedia - Social and Behavioral Sciences, 25*, 384–397 https://doi.org/10.1016/j.sbspro.2012.02.049.
Xie, X., Huo, J., Qi, G., & Zhu, K. X. (2016). Green process innovation and financial performance in emerging economies: Moderating effects of absorptive capacity and green subsidies. *IEEE Transactions on Engineering Management, 63*(1), 101–112 https://doi.org/10.1109/TEM.2015.2507585.
Zailani, S., Govindan, K., Iranmanesh, M., Shaharudin, M. R., & Sia Chong, Y. (2015). Green innovation adoption in automotive supply chain: the Malaysian case. *Journal of Cleaner Production, 108*, 1115–1122 https://doi.org/10.1016/j.jclepro.2015.06.039.
Zailani, S., Iranmanesh, M., Nikbin, D., & Jumadi, H. B. (2014). Determinants and environmental outcome of green technology innovation adoption in the transportation industry in Malaysia. *Asian Journal of Technology Innovation, 22*(2), 286–301 https://doi.org/10.1080/19761597.2014.973167.
Zhu, Q., & Sarkis, J. (2004). Relationships between operational practices and performance among early adopters of green supply chain management practices in Chinese manufacturing enterprises. *Journal of Operations Management, 22*(3), 265–289. https://doi.org/10.1016/j.jom.2004.01.005.
Zhu, Q., Feng, Y., & Choi, S.-B. (2017). The role of customer relational governance in environmental and economic performance improvement through green supply chain management. *Journal of Cleaner Production, 155*, 46–53 https://doi.org/10.1016/j.jclepro.2016.02.124.
Zhu, Q., Geng, Y., & Lai, K.-H. (2010). Circular economy practices among Chinese manufacturers varying in environmental-oriented supply chain cooperation and the performance implications. *Journal of Environmental Management, 91*(6), 1324–1331.
Zhu, Q., Joseph Sarkis, J., & Lai, K.-H. (2008). Green supply chain management implications for "closing the loop". *Transportation Research Part E: Logistics and Transportation Review, 44*(1), 1–18 https://doi.org/10.1016/j.tre.2006.06.003.
Zhu, Q., Sarkis, J., Lai, K. H. (2012). Green supply chain management innovation diffusion and its relationship to organizational improvement: An ecological modernization perspective. *Journal of Engineering and Technology Management, 29*(1), 168–185. https://doi.org/10.1016/j.jengtecman.2011.09.012.

Chapter 5
The Road Toward a Circular Economy: The Role of Modular Product Designs in Supply Chains

Thomas Nowak, Fuminori Toyasaki, and Tina Wakolbinger

5.1 Introduction

Globalization, rapid technological advances, demanding customers, and rising environmental concerns lead to increasing challenges for companies when creating new products and designing their interlinked supply chains. Coping with these issues, modular product design has proven to be a successful design paradigm that may pave the way toward a circular economy by fostering product recovery operations. This is of specific relevance, since returning and recovering end-of-use or end-of-life returns will lead to a significant reduction of waste, energy, and raw material consumption. According to the Ellen MacArthur Foundation, a circular economy is "restorative and regenerative by design, and aims to keep products, components, and materials at their highest utility and value at all times" (Ellen Mac Arthur Foundation 2017). The EU acknowledges that product design plays an important role as facilitator of a circular economy. "Better design can make products more durable or easier to repair, upgrade or remanufacture. It can help recyclers to disassemble products in order to recover valuable materials and components" (European Commission 2015). The EU also points out that facilitating better design

T. Nowak (✉)
WU (Vienna University of Economics and Business), Institute for Production Management, Vienna, Austria
e-mail: thomas.nowak@wu.ac.at

F. Toyasaki
York University, School of Administrative Studies, Toronto, ON, Canada
e-mail: toyasaki@yorku.ca

T. Wakolbinger
WU (Vienna University of Economics and Business), Institute for Transport and Logistics Management, Vienna, Austria
e-mail: tina.wakolbinger@wu.ac.at

H. Qudrat-Ullah (ed.), *Innovative Solutions for Sustainable Supply Chains*, Understanding Complex Systems, https://doi.org/10.1007/978-3-319-94322-0_5

is not an easy task and that the right incentives need to be set in place. "However, current market signals appear insufficient to make this happen, in particular because the interests of producers, users and recyclers are not aligned. It is therefore essential to provide incentives for improved product design, while preserving the single market and competition, and enabling innovation" (European Commission 2015). Hence, a good understanding of how product design decisions affect and are affected by decisions of producers and consumers is necessary. This article wants to pursue this goal.

The concept of modularity is applied in many industries where complex products and systems are designed, including the automotive, the computer, and the airline industries. IBM's System/360 and the Boeing 777 are two famous examples how modular designs can be beneficial (Baldwin and Clark 2006; Kusiak 2002). In general, modular product designs are characterized by a decomposition of a product into subassemblies, i.e., modules, such that parts within a module are highly dependent and there is little dependency across modules. The concept of a modular product architecture is not new and has been heavily investigated, as highlighted in Sect. 5.2. However, research only hesitantly began to analyze the impacts of modular product designs on sustainability aspects associated with the production and usage phase of products. The importance of these product-related life cycle implications are pointed out by the EU Commission (European Commission 2012) by stating that "more than 80% of the environmental impact of a product is determined at the design stage." Product design and hence a product's disassembly abilities are a crucial factor in determining a product's life cycle costs as well as its effects on efficient reverse logistics operations. Huang et al. (2012) note that modular designs improve a product's 3R abilities (reduce, reuse, recycle) and can be seen as a strategy in mitigating a product's negative environmental impact.

The alignment of product and supply chain design is known to be an essential factor for the effectiveness of a supply chain and, hence, a company's success. As Fine (1998) argues, the production of highly modular products favors a modular and only loosely coupled supply chain. Pero et al. (2015) and Hilletofth and Eriksson (2011) are examples of studies investigating this relationship. As Ülkü et al. (2012) and Abbey et al. (2015) point out, consumers' awareness toward integral and modular as well as new and refurbished products will be major drivers for a company's optimal product modularity decision. While a large part of the literature focuses on product recovery operations by only considering implications for manufacturers, research only hesitantly began to analyze consumers' perception toward new or remanufactured and integral or modular products.

Therefore, the aim of this article is to shed more light on the interrelationships between supply chain strategies, consumer perceptions, and product modularity (for an index of product modularity, see, e.g., Hölttä-Otto and de Weck (2007)) in a circular economy context. Specifically, we compare anticipatory and reactive strategies using two-period models that reflect the expected life cycle profits of a product.

In such models, sales from the first period may generate returns in the second period. Using this modeling framework, the present work answers the following three research questions:

- What is the influence of increasing return rates on a manufacturer's preferred supply chain strategy?
- What is the impact of supply chain strategies and associated costs on modularity levels? How is this relationship influenced by consumer behavior?
- What is the effect of changes in product lifetime duration on optimal modularity levels? How is this relationship influenced by supply chain strategies and consumer behavior?

The numerical analysis indicates that the reactive strategy tends to achieve higher profits for manufacturers. The influence of the return rate on the relative advantage of the reactive strategy depends on consumers' consumption share of remanufactured products. The preferability of a strategy in terms of its ability to achieve a higher modularity level critically depends on the impact of modularity on consumers' willingness to pay. Furthermore, our research analysis indicates that longer lifetime leads to reduced modularity levels from manufacturers' perspective. Previous research by Ülkü et al. (2012) indicates that for products with longer lifetime, consumers prefer higher modularity levels. Due to this conflict of interest between stakeholders, policy-makers play an important role as intermediaries.

This article is organized as follows: Sect. 5.2 summarizes aspects of modular product designs for manufacturers as well as consumers. These findings are then reflected by the model formulation and its implied assumptions that are presented in Sect. 5.3. Section 5.4 presents the model outcomes using numerical analysis and interprets the findings. Section 5.5 discusses managerial insights and future research avenues.

5.2 Implications of Modular Product Designs

In this section, we briefly summarize implications of product modularity for the design of integrated supply chains, by distinguishing between consumers' and manufacturers' effects of modular product designs. The findings from this section are then used for the formulation of the optimization problems in Sect. 5.3. Following Baldwin and Clark (2006), we differentiate in our analysis between *modularity in use* and *modularity in production*.

5.2.1 Effects of Modularity on Production

Concerning the forward supply chain, i.e., the way from a point of origin to consumers, modular designs allow for the efficient organization of a product's components. Modular products enable an increased feasibility for mass customization that helps in reacting to changes in product requirements or demand (Swaminathan and Lee 2003). Simple assembly operations facilitate the usage of low-skilled labor (see, for instance, Sako 2005). Modular-designed products will also make service operations easier compared to their integral counterparts, as modules may simply be replaced. Summarizing these effects, Knittig et al. (2004) point out that production costs will tend to fall when a product's degree of modularity is increased. However, when the design of a product becomes highly modular, production costs may begin to rise when the degree of modularity would be raised further due to an increased coordination effort that is necessary for the modules to work together properly (Hohnen et al. 2013). According to Lau et al. (2011), modular products may also have a negative impact on product innovation, since modular designs may be easier to copy for competing firms and, hence, need to be protected against reverse engineering.

With respect to reverse logistics, Campagnolo and Camuffo (2010) as well as Mazahir et al. (2011) note that product modularity will, generally, increase the ease of disassembly operations. Modular designs may, therefore, increase the opportunity to recover parts of the product (Fernández and Käkele 2005), have positive implications for a product's life cycle costs (Newcomb et al. 1998), and may diminish a product's carbon footprint (Kremer et al. 2013). Chang and Yeh (2013) explore the effect of order quantities, customer's return probability, and refund policy on the optimal modularity level in a supply chain for a returnable modular product.

5.2.2 *Effects of Modularity in Use*

While the benefits and drawbacks of modular designs concerning manufacturers are well discussed in the literature, the possible effects for consumers often tend to be neglected. Concerning the effects of *modularity in use*, standardized modules enable an increased interconnectivity with other devices and, therefore, may increase a product's usability. A well-known example here is the widespread usage of USB modules, which allow personal computers, for instance, to communicate and supply power to peripheral devices. The usability for consumers may also be increased due to an easier replacement of old or outdated modules by new ones. By replacing these modules, incremental functional improvements can then be included with the new modules. As King et al. (2007) point out, XEROX, for instance, is known for using modular designs in order to decrease repair costs and give its consumers the chance to keep their products up-to-date. The aspect of modular upgradability is investigated in Ülkü et al. (2012), who find in their survey on upgradability of desktop computers that, counterintuitively, modular upgradability is valued by consumers for slowly improving goods more than for faster improving goods. The authors argue that this result is mainly based on consumers' hyperbolic discounting of their future savings when upgrading their products versus buying new ones.

Modular products may also have negative implications for customers. Modular products may contain more parts than integral products and may hence be bulkier. Additionally, the interfaces between modules represent a possible weakness. This may cause the danger of a potential interface loss and a suboptimal use of space, mass, and energy during operations (Hölttä-Otto and de Weck 2007). Since, by definition, function sharing across modules is limited; Newcomb et al. (2003) state that modular products tend to suffer from redundant physical architecture that can lead to a lower performance of the product and, therefore, to a decreased usability.

Based on benefits and drawbacks of product modularity for manufacturers as well as consumers, the following section describes our model setup, its assumptions, and the formulation of our optimization problems.

5.3 Model

In order to analyze the linkage between the chosen supply chain strategy, consumers' awareness toward new and remanufactured or integral and modular products, as well as reverse logistics operations on optimal product modularity levels, we consider a two-period profit maximization problem of a single manufacturer who decides on optimal production quantities of new and remanufactured products as well as the degree of modularity. We formulate two optimization problems, one for a company following an anticipatory (Sect. 5.3.1) and one for a company following a reactive supply chain strategy (Sect. 5.3.2). While in the reactive model, production decisions are delayed which cancels demand uncertainty, the anticipatory model considers demand uncertainty assuming a news vendor setup.

Nowak et al. (2014) study a production/remanufacturing and product design decision problem that is related to the focus of the study at hand. However, the model presented in this article differs from the latter one in several aspects: While Nowak et al. (2014) study an inter-temporal production/product design decision problem, this study optimizes a products expected life cycle profit. Furthermore, the latter study assumes demand to be a uniformly distributed random variables, while the present study considers a more realistic demand structure with a multiplicative, Weibull-distributed error term. Table 5.1 outlines the structure, the similarities, and the main differences of the two optimization problems.

Table 5.1 Structure of the two models" setup

	Reactive strategy model	Anticipatory strategy model
Demand	Delayed finalization of products causes production quantities to equal demand quantities	Demand uncertainty incorporated by a news vendor setting
Decision variables	Degree of modularity m_r (production quantities are set equal to demand, i.e., $q_r = d(m_r)$ and $\hat{q}_r = \hat{d}(m_r)$)	Production quantity of new products q_a Remanufacturing quantity $\hat{q}_a$ Degree of modularity m_a
Cost structure	Flexibility increases production costs, $\alpha c(m_r)$ with $\alpha \geq 1$	Prescient capacity planning causes lower production costs, $c(m_a)$
Prices consumers are willing to pay	Prices for new products p and for remanufactured products $\hat{p}$ are considered as given	

In our models, we use an index on the unit interval to describe a product's degree of modularity. Such an index is calculated for various products, for instance, by Hölttä-Otto and de Weck (2007), who calculate their singular value modularity index based on design structure matrices that reflect the interconnectedness of products. Using this methodology, a highly integral product (e.g., a jet engine) would be assigned a value close to 0, while a highly modular product would be assigned a value in the proximity of one (e.g., a chain). Based on the effects of modularity in production that are described in Sect. 5.2.1, we model production costs for new products using a quadratic convex function (see Hohnen et al. 2013). With such a functional form, production costs first become smaller when a product becomes more modular due to economies of scale in the production of modules. From a certain point on, the coordination effort that is necessary for the proper functioning of the product becomes increasingly large, such that production costs will begin to rise when the product becomes even more modular. In concordance with Hohnen et al. (2013), we model unit costs for new production as $c(m_i) = a_1 + a_2(m_i - a_3)^2, i \in \{a, r\}$, where a_1 determines the minimum production costs, a_2 is a slope parameter, and a_3 determines the degree of modularity with minimum production costs. The effects of modular designs on product recovery operations, as described in Sect. 5.2.1, indicate that remanufacturing costs will tend to decrease in the degree of modularity in many situations. Therefore, we model unit remanufacturing costs as a linearly decreasing function, $\hat{c}(m_i) = a_4 - a_5 m_i, i \in \{a,r\}$, where a_4 and a_5 are positive parameters. Since $\hat{c}(m_i) \geq 0$ has to hold for all $0 \leq m_i \leq 1$, $i \in \{a, r\}$, we require $a_4 \geq a_5$. However, to capture situations where the degree of modularity might lead to increased remanufacturing costs, we distinguish in our analysis between situations where a_5 might be positive or negative. Throughout the article, we assume that $p > c(m_i)$ and $\hat{p} > \hat{c}(m_i), i \in \{a,r\}$ in order to exclude trivial and unrealistic cases. In our models, we consider that the degree of modularity has also a direct impact on consumers' willingness to pay θ. Using a similar approach as in Atasu and Souza (2013), consumers' utility from buying new products is $U = \phi(w_1 + w_2 m_i) - p$, $i \in \{a, r\}$ and from buying remanufactured products is $\hat{U} = \delta\phi(w_1 + w_2 m_i) - \hat{p}, i \in \{a,r\}$, where δ reflects consumers' depreciation in their willingness to pay for the remanufactured product. In these utility functions, the term $w_1 + w_2 m_i$ reflects the possible impact product modularity might have on consumers' willingness to pay. Under the assumption of a uniform distribution of consumers' willingness to pay, $\phi \sim U(0, 1)$, the demand functions for new products and remanufactured products are given by

$$d(m_i) = 1 - \frac{p - \hat{p}}{(w_1 + w_2 m_i)(1 - \delta)}$$

$$\hat{d}(m_i) = \frac{p - \hat{p}}{(w_1 + w_2 m_i)(1 - \delta)} - \frac{\hat{p}}{(w_1 + w_2 m_i)\delta}, \text{ where } i \in \{a,r\}.$$

Clearly, in order for $d(m_i) \geq 0$ and $\hat{d}(m_i) \geq 0$ to hold, $\hat{p}/\delta \geq p \geq \hat{p} + (w_1 + w_2 m_i)(1-\delta)$, where $0 < \delta < 1$ and $w_1 + w_2 m_i \neq 0$. While Zhang et al. (2014) find that product modularity will increase a company's capability for mass customization and Franke et al. (2009) confirm an overall positive impact of customized products on consumer's willingness to pay, w_2 might be positive in many cases and, hence, increases consumers' willingness to pay for a product. However, as discussed in Sect. 5.2.2, modular products could also entail certain drawbacks for consumers. In order to also reflect cases where these drawback play a dominant role, we distinguish in our analysis between situations where w_2 takes positive and negative values.

5.3.1 Anticipatory Strategy Model

A company following an anticipatory strategy will make prescient capacity and production planning decisions. Therefore, we assume that these companies face uncertainties concerning future realizations of demand quantities. In order to reflect these uncertainties, we assume a news vendor-like setup. In such a modeling framework, consumers are assumed to take the price of a product as given and then decide whether they are willing to accept the price or not. News vendor models in the context of reverse logistics can be found, for instance, in Mostard et al. (2005), Drake et al. (2012), and Reimann and Lechner (2012). Given a quantity q_a ($\hat{q}_a$) of new (remanufactured) products, expected sales of new (remanufactured) products $S(q_a, m_a)$ and $\hat{S}(\hat{q}_a, m_a)$ are described by the following conditional expressions, where $d(m_i)$ and $\hat{d}(m_a)$ denote demand for new and remanufactured products, respectively:

$$
\begin{aligned}
S(q_a, m_a) &= \mathbb{E}\left[\begin{cases} q_a, & q_a \leq d(m_a)\cdot\epsilon \\ d(m_a)\cdot\epsilon, & q_a > d(m_a)\cdot\epsilon \end{cases}\right] \\
\hat{S}(\hat{q}_a, m_a) &= \mathbb{E}\left[\begin{cases} \hat{q}_a, & \hat{q}_a \leq \hat{d}(m_a)\cdot\epsilon \\ \hat{d}(m_a)\cdot\epsilon, & \hat{q}_a > \hat{d}(m_a)\cdot\epsilon \end{cases}\right]
\end{aligned}
\tag{5.1}
$$

In the expected sales quantities given by (5.1), ε is a random variable that reflects potential deviations from the demand function. In order to find an expression for these quantities, we assume ε to have a Weibull distribution with $\mathbb{E}(\epsilon) \approx 1$ and Skewness(ε) ≈ 0. The Weibull distribution with a cdf of $1-\exp\left\{-(\epsilon/\xi_2)^{\xi_1}\right\}$ has a mean of 1 and a diminishing skewness whenever $\xi_1 \approx 3.6023$ and $\xi_2 \approx 1.1098$. For new product sales, for example, we can then derive a closed-form expression for $S(q_a, m_a)$:

$$S(q_a,m_a) = d(m_a)\int_0^{q_a/d(m_a)} f_\epsilon(\epsilon)\cdot\epsilon \; d\epsilon + q_a \int_{q_a/d(m_a)}^{\infty} f_\epsilon(\epsilon)d\epsilon$$

$$= q_a - d(m_a)\int_0^{q/d(m_a)} F_\epsilon(\epsilon)\; d\epsilon$$

$$= \frac{\xi_2}{\xi_1} d(m_a)\left[\Gamma\left(\frac{1}{\xi_1}\right) - \Gamma\left(\frac{1}{\xi_1},\left(\frac{q_a}{\xi_2\cdot d(m_a)}\right)^{\xi_1}\right)\right]$$

The objective function of the optimization problem reflects expected life cycle profits that arise from selling a product as new in the first period and selling it as remanufactured in the second period. Earnings from future product recovery operations are discounted by a factor β that reflects the timing of product returns and hence the durability of the product. Parameter γ denotes the return rate, i.e., the fraction of products from period 1 that are returned for remanufacturing purposes. Then, the company's profit maximization problem with respect to the production of new and remanufactured products as well as the degree of modularity is given as follows:

$$\begin{aligned} &\max_{q_a,\hat{q}_a,m_a} \Pi_a = S(q_a,m_a)p - q_a c(m_a) + \beta\cdot\left[\hat{S}(\hat{q}_a,m_a)\hat{p} - \hat{q}_a\hat{c}(m_a)\right] \\ &s.t. \hat{q}_a \le \gamma S(q_a,m_a) \end{aligned} \tag{5.2}$$

The objective function is comprised of earnings from new product sales in the first period and discounted earnings that arise from sales of remanufactured products in the second period. Parameter β is then a discount factor that will be determined by a product's return period, i.e., the time between initial sale and remanufacturing. While sales are limited by demand using the expression $S(q_a, m_a)$ and $\hat{S}(\hat{q},m_a)$, recovery quantities are additionally limited by the availability of product returns, where γ denotes the return rate. The solution of optimization problem (5.2) is given below.

- Case 1: limited remanufacturing, $\lambda = 0$

$$\begin{aligned} &q_a = \xi_2 d(m_a)\log\left(\frac{p}{c(m_a)}\right)^{\frac{1}{\xi_1}} \\ &\hat{q}_a = \xi_2 \hat{d}(m_a)\log\left(\frac{\hat{p}}{\hat{c}(m_a)}\right)^{\frac{1}{\xi_1}} \\ &\frac{1}{\gamma}\left[-p + \left(\frac{\partial S(q_a,m_a)}{\partial m_a}\right)^{-1}\left(qc'(m_a) - \beta\hat{p}\frac{\partial\hat{S}(\hat{q}_a,m_a)}{\partial m_a} + \beta\hat{q}\hat{c}'(m_a)\right)\right] = 0 \end{aligned} \tag{5.3}$$

Table 5.2 Parameterization of the two optimization problems

Parameter	Base case value/ range for the numerical analysis	Description
α	1.2 $1 \le \alpha \le 2$	Factor that increases unit production cost under reactive strategy compared to the anticipatory strategy
β	0.672 $0 < \beta \le 1$	Discount factor for earnings from the second period where returned products from the first period are remanufactured; this parameter reflects the time between sale and return of a product, i.e., the return period. With an average life of a mobile PC of about 40 months (Daniel Research Group 2013), and a monthly discount rate of 1% (Ovchinnikov et al. 2014), $\beta = 1/(1 + 0.01)^{40} \approx 0.672$
γ	0.6 $0 \le \gamma \le 1$	Fraction of period 1 sales that are returned in period 2; the return rate for the base case scenario of 60% is taken from Ferguson and Toktay (2006).
w_1, w_2	$w_1 = 1.2$ $w_2 \in \{-0.1, 0.1\}$	Parameter that determine consumers' preference toward modular new $\phi(w_1 + w_2 m_i)$ and remanufactured $\phi\delta(w_1 + w_2 m_i)$ products, where $i \in \{a, r\}$. In the simulations, we test a positive impact $(1.1 + 0.1m_i)$ as well as a negative impact $(1.1 - 0.1m_i)$ of modularity on consumers' willingness to pay
$p, \hat{p}$	$p = 0.5$ $\hat{p} = 0.225$	Prices for new and remanufactured products. Hauser and Lund (2003) report that remanufactured products are around 35–55% cheaper than new products. When assuming the price for a new product to be 0.5, the price for a remanufactured product is 45% lower, i.e., $\hat{p} = 0.225$
a_1, a_2, a_3	$a_1 = 0.16$ $a_2 = 0.4$ $a_3 = 0.7$	Parameters for the unit production cost function. The values for the slope parameter a_2 and the cost minimizing level of modularity a_3 are assumptions. Based on the examples from Ahiska and Kurtul (2014), we set a_1 such that $c(a_3)/p \approx 0.32$
a_4, a_5	$a_4 = 0.16$ $a_5 = 0.1$ $a_5 \in \{-0.1, 0.1\}$	Intercept and slope in the unit remanufacturing cost function. When assuming $a_5 = 0.1$, we set a_4 $\hat{c}(a_3)/\hat{p} \approx 0.26$ (Ahiska and Kurtul 2014)
δ	0.63	Discount in consumers' willingness to pay for the remanufactured product. When assuming the discount in consumers' willingness to pay to be around 80%, $\delta = 0.8/(w_1 + w_2 a_3) \approx 0.63$
ξ_1, ξ_2	$\xi_1 = 3.6$ $\xi_2 = 1.11$	Parameters from the Weibull distribution of the error term in the anticipatory model. The parameters are chosen such that $\mathbb{E}(\epsilon) \approx 1$ and Skewness(ε) ≈ 0.

these increasingly large swings would hinder accurate forecasting of product returns, prescient capacity planning would be impeded. Additionally, seasonal influences or changes concerning the legal framework for returning products represent an additional source for unforeseeable developments. Corum et al. (2014) confirm this finding in their simulation study and state that a pull strategy, where remanufacturing operations are scheduled as late as possible, outperforms the anticipatory strategy in terms of total inventory costs and the potential for dampening the bullwhip effect. The results shown in Fig. 5.1 demonstrate that the relative

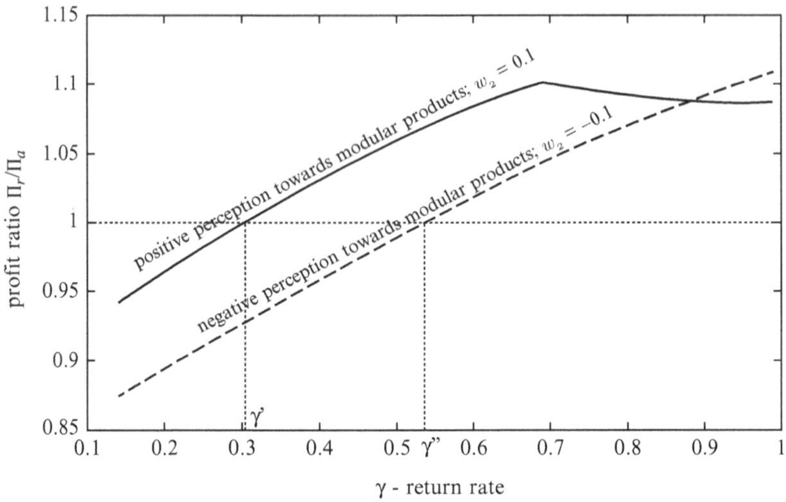

Fig. 5.1 Ratio between optimal profits from the reactive and from the anticipatory case

advantage of the reactive strategy compared to an anticipatory strategy mainly depends on the level of the return rate γ and consumers' preferences. Based on the models, we observe a higher relative profitability of the reactive strategy in cases when consumers have a positive perception toward modular products and the return rate is approximately less than 89%. On the contrary, when the drawbacks of modular designs play a dominant role and consumers' willingness to pay ($w_2 = -0.1$), our models indicate a lower relative profitability of the reactive strategy.

The main driver behind the effects that can be observed in Fig. 5.1 is that optimal degrees of modularity rise when the availability of product returns increases. While this effect causes a decrease in unit remanufacturing costs ($a_5 = 0.1$), unit production costs for new products $c(m_i)$ will increase in m_i whenever $m_i > a_3$, $i \in \{a, r\}$. In the case of the anticipatory model, an increase in γ starting from a low level would cause an increase in new production. By this behavior, a manufacturer generates future product returns and becomes able to offer remanufactured products. While new production gets more and more costly, starting from a certain level of the return rate γ, new production q_a starts to decrease in order to minimize costs. These effects cause optimized profits in the anticipatory case to be u-shaped: for low levels of γ, an increase in the return rate has a negative impact on profits since the negative effects from a decline in new production play a dominant role. However, when the return rate is increased even further, earnings from remanufacturing increase and the increase in profits from remanufacturing is even amplified by smaller remanufacturing costs that comes along an increase in the degree of modularity m_a. At the reactive model, we observe an increase in new production $q_r = d(m_r)$, remanufacturing $\hat{q}_r = \hat{d}(m_r)$, and the degree of modularity m_r up to a point where, in our experimental setting, the return rate is around 70%. From this point on, the market for remanufactured products is saturated, which causes m_r and, thus, q_r and $\hat{q}_r$ to stagnate.

The behavior of optimized profits in our anticipatory and reactive model then results in profit ratios that, first, increase and then, depending on the market's saturation for remanufactured products, begin to decrease. Based on calculations shown by Fig. 5.1, the reactive strategy has the highest relative advantage over the anticipatory strategy when the return rate is around 70%.

5.4.2 Impacts of Supply Chain Strategies on the Optimal Level of Modularity

Product design plays a major role for the circular economy, since it directly determines a product's ability for recovery operations. Additionally, as noted by Khan et al. (2012), aligning product design with the supply chain structure and strategy is a crucial success factor for a company. Hence, from a policy-maker's perspective, the question remains which supply chain strategy results in higher modularity levels. In order to shed more light on this relationship, Fig. 5.2 illustrates the results of the solutions to optimization problems (5.2) and (5.5) by showing the impact of a variation of α on optimal modularity levels.

Figure 5.2 highlights the importance of the supply chain strategy, the cost differential between the reactive and the anticipatory strategy, and consumers' perception of modular products for determining the optimal degree of modularity. Figure 5.2 indicates that when the reactive strategy becomes costly compared to the anticipatory strategy, i.e., α increases, the optimal degree of modularity may increase or decrease, depending on consumers' awareness toward modular products. When consumers are willing to pay higher prices for more modular products (see left graph of Fig. 5.2 where $w_2 = 0.1$), the optimal degree of modularity is higher than the degree of modularity that achieves minimal production costs determined by parameter a_3. Due to the convexity of the cost function, production costs could be decreased by decreasing the optimal degree of modularity. Whenever overall production costs are increased by a factor α, the company will then try to lower its production costs by decreasing the degree of modularity. This effect is illustrated by the left graph of Fig. 5.2. The contrary is true for a situation where consumers have a negative perception of modular products (the right graph of Fig. 5.2, where $w_2 = -0.1$). In such a situation, the optimal degree of modularity is lower than the level of a_3 that achieves the minimum production cost of a new product, and an increase in α would, therefore, increase the optimal degree of modularity. While the optimal product design is more modular under a reactive strategy than under an anticipatory strategy when consumers have a positive perception toward modular products (and when α is lower than approximately 1.28), the contrary is true when consumers have a negative perception toward modular products (and α is lower than approximately 1.34).

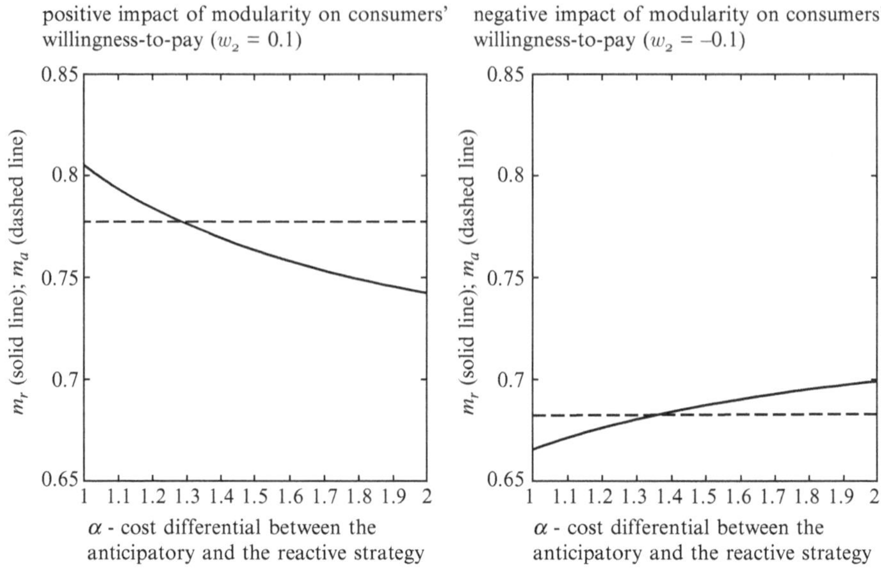

Fig. 5.2 Optimal degree of modularity for a reactive and an anticipatory strategy, i.e., m_r and m_a, calculated for various values for the parameter α that determines how much the unit production costs under a reactive strategy are higher compared to an anticipatory strategy

5.4.3 Effects of Changes in the Return Period

The return period of a product is determined by a product's lifetime duration and the time until consumers return the product for recovery operations. While companies have very limited control over consumers' behavior regarding if and when they return their products (Wakolbinger et al. 2014), a product's lifetime duration can often be controlled quite well. As product modularity can be seen as a way to extend a products' expected lifetime duration because of an increased potential for upgradability of the products, modular product designs might have a positive environmental impact (Ülkü et al. 2012). On the one hand, this positive effect may be dampened by an early replacement of modules which would increase waste (Ülkü et al. 2012), but on the other hand, this could offer the possibility to replace old modules by new ones with a lower environmental impact during their usage phase (Agrawal and Ülkü 2013). This section explores how a product's lifetime duration impacts optimal modularity levels and how this relationship is influenced by consumer behavior and supply chain policies.

The analysis of variations of parameter β, the discount factor for future profits from product recovery operations that reflects a product's return period, sheds light on the effect of the length of the return period on optimal product modularity levels. When considering a non-durable good with little time between sale and return, β will be near to 1. Parameter β will decrease when the duration of the return period is extended.

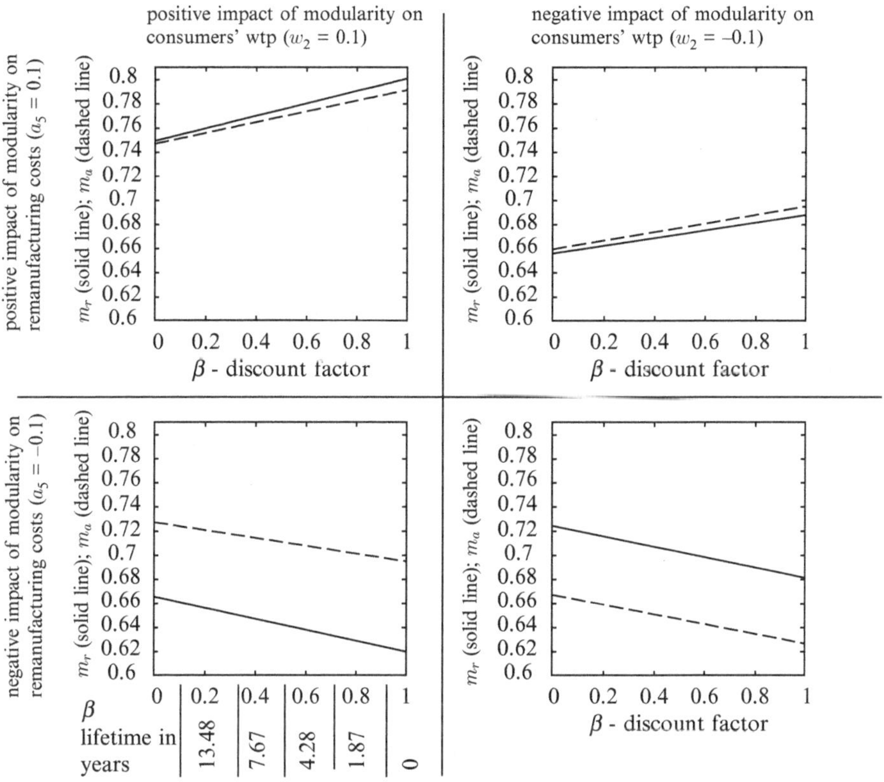

Fig. 5.3 Optimal degrees of modularity for a company following a reactive (m_r^* , solid line) and an anticipatory (m_a^* , dashed line) strategy for various values of the parameter β. With a monthly discount rate of 1%, the corresponding lifetime duration is calculated by $n = -\log(\beta)/\log(1 + 1\%)$

Based on our literature review, we expect modularity to decrease remanufacturing costs for many situations. However, as highlighted in Zhang and Gershenson (2003), cases exist where modularity might also have an increasing effect on unit remanufacturing costs. According to the two optimization problems, increase in a product's return period (β approaches 0) will reduce the optimal level of modularity and thus the optimal product design will become more integral. While this result is true when modularity decreases unit remanufacturing costs (a_5 is positive), we observe an opposite effect in situations where modularity would increase remanufacturing costs (a_5 is negative). From Fig. 5.3, it can be seen that a decrease in a product's lifetime duration (an increase in β) increases the optimal degree of modularity for both situations when consumers have a positive or a negative perception of product modularity (when $w_2 = 0.1$ or $w_2 = -0.1$). From a modeling perspective, the reason for this effect is that with a decrease in the time between the first sale of a product and the time when the product gets remanufactured, parameter β will approach 1. Then, earnings from remanufacturing receive a higher weight in the objective function and the optimal degree of modularity will increase as $\hat{c}(m_i)$ is a decreasing function of m_i, $i \in \{a, r\}$.

From simulating the impact of changes in parameter α, it can be seen that optimal modularity levels are higher under the reactive strategy than under the anticipatory strategy whenever $\alpha = 1.2$ and consumers have a positive perception toward modular products (and vice versa). From Fig. 5.3, it can be seen that this result is robust against changes in a product's return period.

Based on the resulting profits from these simulations, it is obvious that both anticipatory and reactive profits increase when the product becomes less durable (β approaches 1). Interestingly, when $w_2 = 0.1$, the increase in reactive profits is higher than the increase in anticipatory profits (vice versa when $w_2 = -0.1$). Thus, when $w_2 = 0.1$, companies under a reactive strategy would have a higher incentive for reducing the expected lifetime duration of their products than companies under an anticipatory strategy. Again, this result is reversed when an increase in a product's degree of modularity would have a negative impact on consumers' willingness to pay.

5.5 Managerial Insights and Conclusions

Table 5.3 summarizes the main results of the numerical analysis.

The numerical analysis derives several insights on sustainable operations and product design in a circular economy. Our results indicate that either the reactive or the anticipatory strategy can be more preferable for manufacturers, depending on consumers' preferences for modular product and the level of the return rate. Thus, when product return flows are considered in the decision-making process, the reactive strategy turns out to be economically beneficial for high return rates, while the anticipatory strategy appears beneficial for relatively low return rates. However, the relative advantage of the reactive strategy might be weakened when a high product return rate is achieved and consumers' preference for modular products is low. The recast of the Waste Electrical and Electronic Equipment (WEEE) Directive intends to increase the minimum collection rate from 45% to 65% by 2019. Our numerical results suggest that these changes could, depending on consumers' preference, directly impact manufacturers' optimal strategies.

From a manufacturer's viewpoint, consumers' willingness to pay for new or remanufactured integral or modular products and the chosen strategy are key determinants for the optimal modularity level of a product. Lastly, our analysis also highlights that coordination between consumers and manufacturers would be indispensable for pursuing the circular economy through modular product designs. As mentioned in Sect. 5.4, Ülkü et al. (2012) discuss the relationship between a product's lifetime duration and its degree of modularity from consumers' viewpoints. The study concludes that consumers appreciate modular upgradability for products with a longer lifetime more than for those with a shorter lifetime. Our results shows that manufacturers should decrease the modularity level for products with a longer product lifetime. To avoid this conflict of interests between consumers and manufacturers, policy-makers may need to coordinate.

Table 5.3 Summary of results

	Consumers have a positive perception toward modular products $w_2 = 0.1$	Consumers have a negative perception toward modular products $w_2 = -0.1$
Effects of a changing return rate of products	Profits under the reactive strategy are higher when the return rate is approximately greater than 31%	Profits under the reactive strategy are higher when the return rate is approximately greater than 54%
	The relative benefit of the reactive strategy is higher when $w_2 = 0.1$ compared to situations where $w_2 = -0.1$	
Effects of a changing cost differential between the anticipatory and the reactive strategy	The product should become more integral when the reactive strategy becomes costly compared to the anticipatory strategy (α is increased)	The product should become more modular when the reactive strategy becomes costly compared to the anticipatory strategy (α is increased)
Effects of a changing duration of the return period	The optimal product design should become more modular when the desired lifetime duration is decreased	
	$m_r > m_a$	$m_a > m_r$

Our analysis is subject to certain limitations with respect to the model setup: In the anticipatory strategy model, where companies try to anticipate future demand, we do not incorporate possible gains from effective demand forecasting. Forecasting future demand quantities, however, is a crucial success factor for companies following an anticipatory strategy. Furthermore, in the anticipatory model, we assume that a manufacturer optimizes life cycle profits of a product once, and decisions in the first period are not changed in the second period. While this assumption appears plausible, since product design cannot be changed easily, there might exist cases, where adaptations could be possible.

Another limitation of this model concerns the interactions between supply chain members. In this article, we assume that the company can decide freely on the optimal level of modularity and that the company is not restricted by other supply chain members' product design decisions. These interdependencies could effectively be analyzed by a variational inequality formulation of the optimization problems of all supply chain members, similar to the work of Wakolbinger et al. (2014). The analysis of these effects is subject to future research. Our research shows that consumer behavior plays a major role for manufacturers' production and product design decisions as well as the associated supply chain design decisions. Hence, we recommend future empirical research that explores consumers' willingness to pay for modular products and how this differs for different sectors and products.

Appendix A: Anticipatory Model

The system of equations given by (5.3) and (5.4) are directly obtained from the Karush-Kuhn-Tucker conditions of the optimization problem. In this optimization problem, we can rule out cases where $q_a = 0$, since in those cases

$$\frac{\partial S(q_a, m_a)}{\partial q_a}[p + \lambda\gamma] - c(m_a) \le 0$$

would have to hold (the first partial derivative of the Lagrangian with respect to q_a). Since $\partial S(q_a, m_a)/\partial q_a = 1$ in the case of $q_a = 0$, the inequality reduces to $p + \gamma\lambda \le c(m_a)$ and, hence, contradicts the assumption of $p > c(m_a)$. The same argument can be applied to the situation where $\hat{q}_a = 0$ and $\lambda = 0$: From the first partial derivative of the Lagrangian with respect to $\hat{q}_a$, we would obtain the condition $\hat{p} \le \hat{c}(m_a)$ that contradicts the assumption of $\hat{p} > \hat{c}(m_a)$. Additionally, if there would be no remanufacturing and the constraint was binding ($\hat{q}_a = 0$ and $\lambda > 0$), $\gamma S(q_a, m_a) = 0$ would have to hold. This would require $\Gamma(1/\xi_1) = \Gamma\left(1/\xi_1, \left(q_a / \left[\xi_2 \cdot d(m_a)\right]\right)^{\xi_1}\right)$, which is only true in cases where $q_a = 0$. Therefore, the only case where the Karush-Kuhn-Tucker conditions have to be considered is where $q_a, \hat{q}_a > 0$ holds and the shadow price of the constraint that limits the availability of returned cores is strictly positive.

Sufficient Conditions

Since the analysis of the eigenvalues of the Hessian is very cumbersome, checking if a solution to either (5.3) or (5.4) satisfies the sufficient conditions for a global maximum cannot be done in closed form. Therefore, we use the following procedure to identify a global maximum to problem (5.2):

1. Generate a number of starting solutions for $q_a, \hat{q}_a$ and m_a using shifted sine (or cosine) functions in order to capture as many combinations of $q_a, \hat{q}_a$ and m_a as possible.

   In order to avoid irrelevant solutions, the starting values for q_a and $\hat{q}_a$ are bounded on [0, 1] and the starting solutions for m_a are bounded below by $u = \max\left\{0, \min\left\{1, \left(a_4 - \hat{p}\right)/a_5\right\}, \min\left\{1, \left(\left(p - \hat{p}\right)/(1-\delta) - w_1\right)/w_2\right\}\right\}$ and above by $v = \min\left\{1, \max\left\{0, \sqrt{(p - a_1)/a_2} - a_3\right\}\right\}$ in order to make sure that $d(m_a) \ge 0$, $c(m_a) \le p$ and $\hat{c}(m_a) \le \hat{p}$.
2. Try to solve each system of equations using the Levenberg-Marquardt algorithm in MATLAB with each starting solution from step 1.
3. Discard all solutions from step 2 where the algorithm did not converge and where regularity conditions (check if $\lambda > 0$ in case two, if the constraint is fulfilled, if $0 \le q_a, \hat{q}_a, m_a \le 1$) are not met. Proceed only when all remaining solutions are the same.

   At this stage, the remaining solution must either satisfy the requirements for case 1 or the requirements for case 2.
4. Check if the solution found in step 3 satisfies the sufficient conditions:

Check if the eigenvalues of the objective function's Hessian are all smaller or equal to 0 (for the objective function to be concave) and the constraint is convex.

In order to check the convexity of the constraint $\hat{q}_a - S(q_a, m_a)$, it is sufficient to analyze the leading principal minors of its Hessian matrix. The first principal minor is

$$\frac{\exp\left\{-\left(\frac{q}{\xi_2 d(m_a)}\right)^{\xi_1}\right\}\gamma\xi_1\xi_2\left(\frac{q}{\xi_2 d(m_a)}\right)^{1+\xi_1}\left((w_1+w_2m_a)(1-\delta)-p+\hat{p}\right)}{(1-\delta)(w_1+w_2m_a)q_a^2}$$

and the second and third principal minor are zero. Thus, for the constraint to be convex, the first principal minor has to be positive. Considering that ξ_1, ξ_2, q_a, δ, γ, $w_1 + w_2m_a > 0$ and $\delta, \gamma < 1$, the expression is positive when $(w_1 + w_2m_a)(1-\delta) > p - \hat{p}$ holds, which is equivalent to the condition for positive demand $d(m_a)$. Checking the concavity of the objective function using eigenvalues of its Hessian can only be done numerically. For the base case parameter setting with $w_2 = 0.1$ and $a_5 = 0.1$, the objective function is concave, when the decision variables satisfy conditions illustrated by Fig. 5.4:

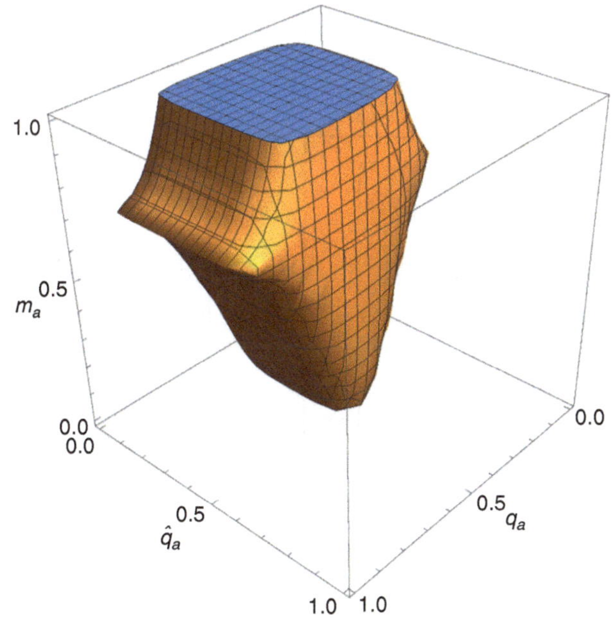

Fig. 5.4 Conditions for the objective function of the anticipatory model to be concave

Appendix B: Reactive Model

Using the base case scenario described in Table 5.2, the second-order derivative of the objective function is negative when m_r and $\hat{p}_r$ satisfy the conditions illustrated in Fig. 5.5

As can be seen from Fig. 5.5, the sufficient conditions for a global maximum are satisfied for all feasible regions of m_r, since $\hat{p} = 0.225$ in the base case scenario. Therefore, solutions to the optimization problem can be simply calculated by setting the first-order conditions to zero, i.e., solving the system of Eqs. (5.6) and (5.7).

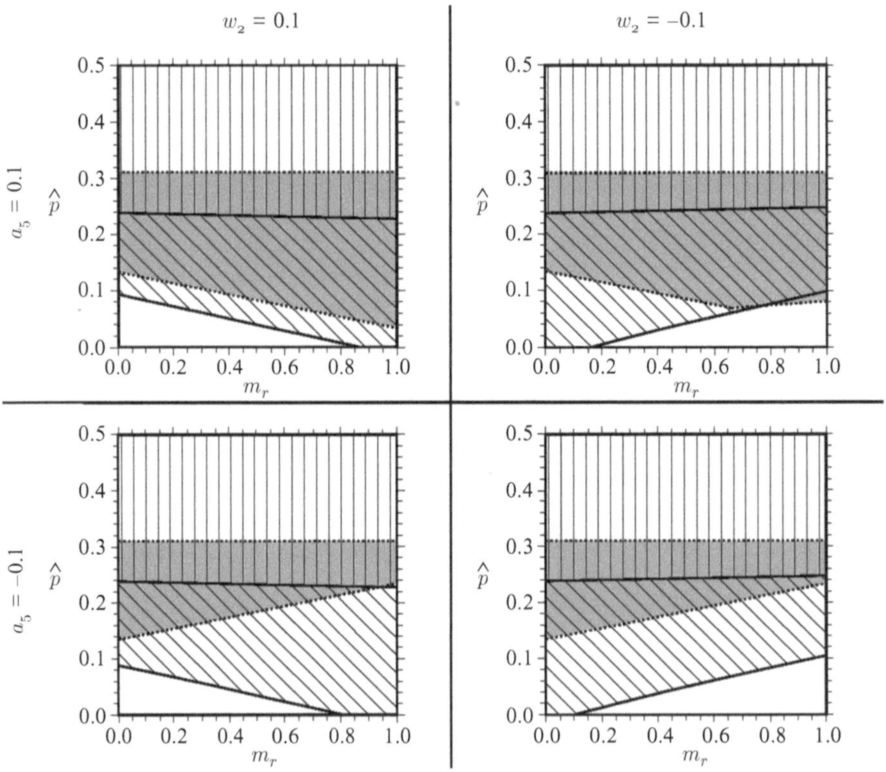

Fig. 5.5 Depending on situations where consumers have a positive or a negative perception toward modular products *($w_2 = 0.1$ or $w_2 = -0.1$)* and situations where higher levels of modularity would increase of decrease unit remanufacturing costs *($a_5 = 0.1$ or $a_5 = -0.1$)*, the graphs show regions where the second-order derivative of the objective function (5.5) is negative. While the vertical shaded area depicts parameter constellations with negative second-order derivative when $\hat{d}(m_r) \leq \gamma d(m_r)$, the diagonal shaded areas are drawn for parameter constellations where $\hat{d}(m_r) > \gamma d(m_r)$ holds true. The gray shaded areas indicate feasible regions, where $d(m_r), \hat{d}(m_r) \geq 0$, $p > c(m_r)$ and $\hat{p} > \hat{c}(m_r)$

References

Abbey, J. D., Meloy, M. G., Guide, V. D. R., Jr., & Atalay, S. (2015). Remanufactured products in closed-loop supply chains for consumer goods. *Production and Operations Management, 24*(3), 488–503.

Adenso-Dìaz, B., Moreno, P., Gutierrez, E., & Lozano, S. (2012). An analysis of the main factors affecting bullwhip in reverse supply chains. *International Journal of Production Economics, 135*, 917–928.

Agrawal, V., & Ülkü, S. (2013). The role of modular upgradability as a green design strategy. *Manufacturing & Service Operations Management, 15*(4), 640–648.

Ahiska, S., & Kurtul, E. (2014). Model and analysis of a product substitution strategy for a stochastic manufacturing/remanufacturing system. *Computers & Industrial Engineering, 72*, 1–11.

Atasu, A., & Souza, G. C. (2013). How does product recovery affect quality choice? *Production and Operations Management, 22*, 991–1010.

Baldwin, C. Y., & Clark, K. B. (2006). Modularity in the design of complex engineering systems. In D. Braha, A. Minai, & Y. Bar-Yam (Eds.), *Complex engineered systems, understanding complex systems* (pp. 175–205). Berlin/Heidelberg: Springer.

Campagnolo, D., & Camuffo, A. (2010). The concept of modularity in management studies: A literature review. *International Journal of Management Reviews, 12*(3), 259–283.

Chang, S. Y., & Yeh, T. Y. (2013). Optimal order quantity and product modularity for a two-echelon returnable supply chain. *International Journal of Production Research, 51*(17), 5210–5220.

Corum, A., Vayvay, O., & Bayraktar, E. (2014). The impact of remanufacturing on total inventory cost and order variance. *Journal of Cleaner Production, 85*, 442–452.

Daniel Research Group. (2013). United States PC, tablet, & mobile phone market size and forecast. URL http://ww1.prweb.com/prfiles/2013/06/20/10857238/DRG%20US%20PC%20Tablet%20Mobile%20Phone%20Market%20Forecast%20June%202013.pdf. Last accessed February 13, 2018.

Drake, M. J., Pentico, D. W., & Sroufe, R. P. (2012). The remanufacturing newsvendor problem. In T. M. Choi (Ed.), *Handbook of newsvendor problems, no. 176 in international series in operations research & management science* (pp. 249–262). New York: Springer Science + Business Media.

Ellen Mac Arthur Foundation. (2017). Circular economy overview. URL https://www.ellenmacarthurfoundation.org/circular-economy/overview/concept. Last accessed 13 Feb 2018.

European Commission. (2012). Ecodesign your future. How ecodesign can help the environment by making products smarter. URL http://ec.europa.eu/DocsRoom/documents/5187/attachments/1/translations/en/renditions/native. Llast accessed 13 Feb 2018.

European Commission. (2015). Communication from the commission to the European parliament, the council, the European economic and social committee and the committee of the regions. Closing the loop - an EU action plan for the circular economy. URL http://eur-lex.europa.eu/legal-content/EN/TXT/?uri=CELEX%3A52015DC0614. Last accessed 13 Feb 2018.

Ferguson, M. E., & Toktay, B. (2006). The effect of competition on recovery strategies. *Production and Operations Management, 15*(3), 351–368.

Fernández, I., & Kekäle, T. (2005). The influence of modularity and industry clockspeed on reverse logistics strategy: Implications for the purchasing function. *Journal of Purchasing and Supply Management, 11*(4), 193–205.

Fine, C. H. (1998). *Clockspeed: winning industry control in the age of temporary advantage*. Perseus Books. Amsterdam

Franke, N., Keinz, P., & Steger, C. J. (2009). Testing the value of customization: when do customers really prefer products tailored to their preferences? *Journal of Marketing, 73*(5), 103–121.

Hauser, W., & Lund, R. (2003). *The remanufacturing industry: Anatomy of a giant*. Boston: Boston University.

Hilletofth, P., & Eriksson, D. (2011). Coordinating new product development with supply chain management. *Industrial Management & Data Systems, 111*(2), 264–281.

Hohnen, T., Pollmanns, J., & Feldhusen, J. (2013). Cost-effects of product modularity – an approach to describe manufacturing costs as a function of modularity. In M. Abramovici & R. Stark (Eds.), *Smart product engineering, lecture notes in production engineering* (pp. 745–754). Berlin/Heidelberg: Springer-Verlag.

Hölttä-Otto, K., & de Weck, O. (2007). Degree of modularity in engineering systems and products with technical and business constraints. *Concurrent Engineering: Research and Applications, 15*(2), 113–126.

Huang, C. C., Liang, W. Y., Chuang, H. F., & Chang, Z. Y. (2012). A novel approach to product modularity and product disassembly with consideration of 3R-abilities. *Computers & Industrial Engineering, 62*(1), 97–107.

Khan, O., Christopher, M., & Creazza, A. (2012). Aligning product design with the supply chain: a case study. *Supply Chain Management: An International Journal, 17*(3), 323–336.

King, A., Barker, S., & Cosgrove, A. (2007). Remanufacturing and XEROX: Evaluating the process to establish principles for better design. In *International Conference on Engineering Design, ICED '07.*

Knittig, P. R., Shimizu, S., & Ballon, R. J. (2004). Modularisation and its limitations in the automotive industry. In *Second world conference on POM and 15th POM conference*. Cancun.

Kremer, G. E. O., Ma, J., Chiu, M. C., & Lin, T. K. (2013). Product modularity and implications for the reverse supply chain. *Supply Chain Forum: An International Journal, 14*(2), 54–68.

Kusiak, A. (2002). Integrated product and process design: a modularity perspective. *Journal of Engineering Design, 13*(3), 223–231.

Lau, A. K., Yam, R. C., & Tang, E. (2011). The impact of product modularity on new product performance: Mediation by product innovativeness. *Journal of Product Innovation Management, 28*(2), 270–284.

Mazahir, S., Lassagne, M., & Kerbache, L. (2011). Reverse logistics and push-pull manufacturing systems: The case of electronic products. *Supply Chain Forum: An International Journal, 12*(2), 92–102.

Mostard, J., de Koster, R., & Teunter, R. (2005). The distribution-free newsboy problem with resalable returns. *International Journal of Production Economics, 97*(3), 329–342.

Newcomb, P., Bras, B., & Rosen, D. (1998). Implications of modularity on product design for the life cycle. *Journal of Mechanical Design, 120*(3), 483–490.

Newcomb, P. J., Rosen, D. W., & Bras, B. (2003). Life cycle modularity metrics for product design. In *Proceedings of EcoDesign2003: Third international symposium on environmentally conscious design and inverse manufacturing.*

Nowak, T., Toyasaki, F., Wakolbinger, T., & Ng, D. (2014). Reverse logistics decision making for modular products: The impact of supply chain strategies. In *Lecture notes in management science, Tadbir Operational Research Group Ltd., Proceedings of the 6th international conference on applied operational research.*

Ovchinnikov, A., Blass, V., & Raz, G. (2014). Economic and environmental assessment of remanufacturing strategies for product + service firms. *Production and Operations Management, 23*(5), 744–761.

Pero, M., Stößlein, M., & Cigolini, R. (2015). Linking product modularity to supply chain integration in the construction and shipbuilding industries. *International Journal of Production, 170*(Part B), 602–615.

Reimann, M., & Lechner, G. (2012). Product and remanufacturing strategies in a closed- loop supply chain: A two-period newsvendor problem. In T. M. Choi (Ed.), *Hand- book of newsvendor problems, international series in operations research & management science* (Vol. 176, pp. 219–247). New York: Springer Science + Business Media.

Sako, M. (2005). Modularity and outsourcing: the nature of co-evolution of product architecture and organization architecture in the global automotive industry. In A. Prencipe, A. Davies, & M. Hobday (Eds.), (pp. 229–253). The Business of Systems Integration: Oxford University Press.

Swaminathan, J. M., & Lee, H. L. (2003). Design for postponement. In S. Graves & A. Kok (Eds.), *Supply chain management: Design, coordination and operation, hand- books in operations research & management science* (Vol. 11, pp. 199–226). North-Holland Publishing/Elsevier.

Ülkü, S., Dimofte, C. V., & Schmidt, G. M. (2012). Consumer valuation of modularly upgradeable products. *Management Science, 58*(9), 1761–1776.

Wakolbinger, T., Toyasaki, F., Nowak, T., & Nagurney, A. (2014). When and for whom would e-waste be a treasure trove? insights from a network equilibrium model of e-waste flows. *International Journal of Production Economics, 154*, 263–273.

Zhang, Y., & Gershenson, J. (2003). An initial study of direct relationships between life- cycle modularity and life-cycle cost. *Concurrent Engineering: Research and Applications, 11*(2), 121.

Zhang, M., Zhao, X., & Qi, Y. (2014). The effects of organizational flatness, coordination, and product modularity on mass customization capability. *International Journal of Production Economics, 158*, 145–155.

Chapter 6
Policy Design for Sustainable Supply Chain Through Training

The Case of XYZ Packaging Company

Ijaz Yusuf and Tashfeen M. Azhar

6.1 Introduction

System dynamics is a computer-aided simulation technique that addresses the complex, nonlinear, time-delayed, feedback-dominant, and dynamic models of various disciplines (Yusuf and Azhr 2017). System dynamics has the capability to handle the soft and hard systems simultaneously comprehending the complexity of the models (Forrester 1961). Broadly speaking the systems thinking, feedback notion, and control engineering are applied to economics, business, and organizational systems (Towill 1996) to gain the better insight of the problems under study. The aim of the chapter is to investigate the dynamics of the trainings related to soft skills and hard skills in order to reduce the waste level and cost of failures that ultimately lead to sustainable supply chain performance.

System dynamics masterly attempts to combine the key concepts like feedback controls, mutual causality, nonlinearity in the functions, cybernetics, complexity, counterintuitive behavior, and deviation-correcting and deviation-amplifying processes like goal-seeking, external resource production process, and many more to the organizational systems (Sterman 2007). Journey of system dynamics model begins with the problem statement and the problem under study; causal loop diagrams are developed to investigate the archetypes of the dynamic hypothesis using cognitive schemes and mental maps. Conceptual phase of the modeling demands the technical phase on the spectrum of the system thinking forming rates and levels and programming protocol for computer modeling (Richmond 1993). Father of System Dynamics, Jay Forrester (1961) defines system dynamics as a methodology

I. Yusuf (✉) · T. M. Azhar
School of Business and Economics, University of Management and Technology, Lahore, Pakistan
e-mail: ijaz.yusuf@umt.edu.pk

H. Qudrat-Ullah (ed.), *Innovative Solutions for Sustainable Supply Chains*, Understanding Complex Systems, https://doi.org/10.1007/978-3-319-94322-0_6

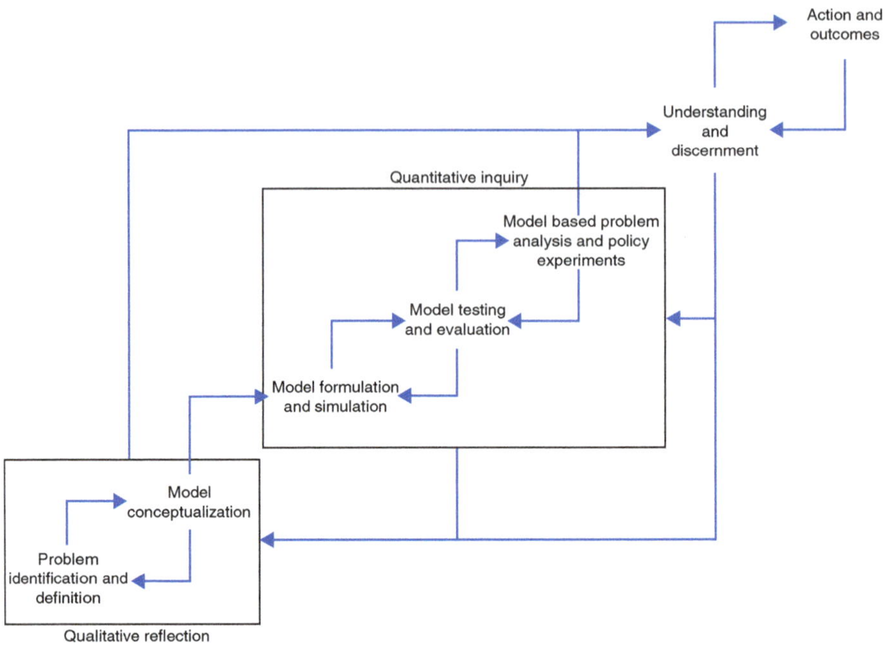

Fig. 6.1 Steps in system dynamics modeling. (Source: Adapted from Zagonel 2002)

that starts with important problems, comes to understand the structures that produce undesirable symptoms, and moves on to finding changes in structure and policy that would make a system better behaved. System dynamics is a method of dealing with questions about dynamic tendencies of complex systems, that is, the behavioral patterns they generate over time (Meadows et al. 1974). Zagonel (2002) has mentioned steps in system dynamics sharing the blend of qualitative reflection as well as quantitative inquire leads to policy design. The details of all steps in system dynamics are shown in Fig. 6.1.

Simulated system dynamics model includes the feedback-dominant structures, levels, flows, and their interactions to represent the world view. Experimentation with the model requires sleepless nights and constant interaction with the computer screen to learn, unlearn, and relearn the model phenomena. Many surprises in the model come up unveiling the hidden dimensions and enhancing the problem insight. Expectations may be wrong, and unexpected results explore the new avenues, that is, the real beauty of system dynamics modeling. Testing the model assumptions, validating the model against standard tests of verification and validating protocols, evaluating the real-life business scenarios, and altering the management policies are designed either based on parametric changes (decision stream seems to be easy and short run) or based on structural changes (incorporating the structures in the model to have sustained solutions on long-term basis). The purpose is to gain understanding so that problem to which the model is addressed could be solved. How the model will look like in the end depends a lot on the selection of dynamic variables, transparent goals, clear and unambiguous description of the situation, and closed boundary of the model. Without a purpose, there can be no answer to the question of what system components are important (Forrester 1968). Without the system boundary, it is impossible to define the modes of behavior which are created by the interaction of system components.

6.1.1 System Dynamics Paradigm

Paradigm is a set of basic philosophical assumptions to understand the nature of the world (ontological assumption), how we can understand the world (epistemological assumption), and how we can improve the understanding of the world (methodological assumption). A paradigm therefore represents shared philosophical understandings, terminology, rules, and research approaches adopted by a particular community that continuously evolve as they are negotiated and debated within the communities Morgan (2007). Meadows et al. (1974) in his debate about system dynamics paradigm states the world is composed of nonlinear, time-delayed, feedback-dominant structures and dynamic variables, and their interactions cause complexity to form a world (ontological frame), whereas causal loop diagrams on positive and negative loops enhance the understanding of the world (epistemological frame). Decision stream and the ability to analyze the behavior of models heavily rely on simple models linking feedback loops to system behavior (Oliva et al. 2006). Influence diagram and simple causal loops lead to conceptualization of real-life scenario during the model formulation. Behavior of the system is the result of interaction of positive and negative feedback control loops.

A positive loop deals with the amplification and generates runaway growth or collapse behavior that creates virtuous or vicious cycles: a change in one variable brings change in the next variable (Richardson 1986). Compounding and reinforcement are the outcomes of the positive loop. Negative feedback loop is a goal-seeking loop that maintains the status quo and generates honing in behavior. Negative loops resist the change and push them one way, and they come back while maintaining the equilibrium state. System thinkers define it as: "When a feedback loop response to a variable opposes the original perturbation, the loop is negative or goal-seeking. The definition of a negative loop is usually interpreted to mean that … a change in one element is propagated around the circle until it comes back to change that element in a direction opposite to the initial change" (Meadows et al. 1974).

Technical phase of the modeling protocol starts with block and flow diagram that has another name the level and rate diagram which consists of variables that depict the epistemology of the SD paradigm. Associated variables like auxiliary or convertor, graphical function, exogenous variable, source or sink of material, material and information flows, and constants all are building blocks to complete the block diagram. Feedback structure can be portrayed by equations or stock and flow diagrams (Richardson 1986).

Levels reflect on conditions within the system at a point in time. Stocks are what we would see if we suddenly freeze the activity within the system. Levels are just like the bathtubs in the sense that they accumulate or collect flows. Rates represent the stream of activity associated with particular stocks. Flows are depicted by a pipe which represents the conduit through which goal-seeking activities flow. They are just as valves that feed and drain stock (Yusuf and Ramish 2011). Policy statements in the system define how the existing conditions of the system produce a decision stream controlling actions that emerge from the rate equations.

6.2 Model Structure and Behavior

6.2.1 Case Study

XYZ Packaging Company was established in 1956 in collaboration with Akerlund and Rausing Sweden, the company with the largest printing complex in South Asia. At that time the company comprised of various production lines like carton line, flexible line, corrugated line, tissue line, and paper and board mill. The company has more than 20 departments to look at the various functions; one of the departments was human resource department which was fully responsible for capacity building and finding the right person for the right job. Management focus is to establish the company culture and ensure the proper placement of company values. That is why company has a recruitment policy to hire fresh graduates and develop them as per the job design for required organizational hierarchy. This approach was quite successful, and they developed many professional managers over the period of time.

6.2.2 Employee Recruitment Process

XYZ Packaging is the largest manufacturing unit in printing and packaging and believes on sound value systems. Employees are treated as family members, and selection of the company employees and hiring of new staff have a transparent and rigorous recruitment process. People may approach the head HR either through company web portal or advertisement in the print media or may have some referrals; all new recruits have to pass through the written test and initial interview. On the basis of test result, candidates are short listed, and a call for the initial interview is initiated. The candidates who qualified during the first interview with good feedback are chosen for the second interview with the top management. On the basis of the final feedback, the candidate has been recurited and passed through the company orientation training program. This process may take on average 2–3 months.

6.2.3 Training Types

Every newly hired person has been sent to company orientation for a 3-month period. Company orientation is the first training that is given to every new employee to get acquainted with the new department and associated staff. After that the individual is deployed to the related department on the basis of department need and employee aptitude. The first training which is provided to the employee is company orientation in which the employee visits every department, has a meeting with the department executives and departmental head, and gets sufficient operational knowledge of the department doings. Figure 6.2 shows various categories of trainings.

Table 6.1 shows the list of training which is provided to the company employees during their whole stay within the company depending upon the training need analysis, technology competency requirement, and any other operational need. For the

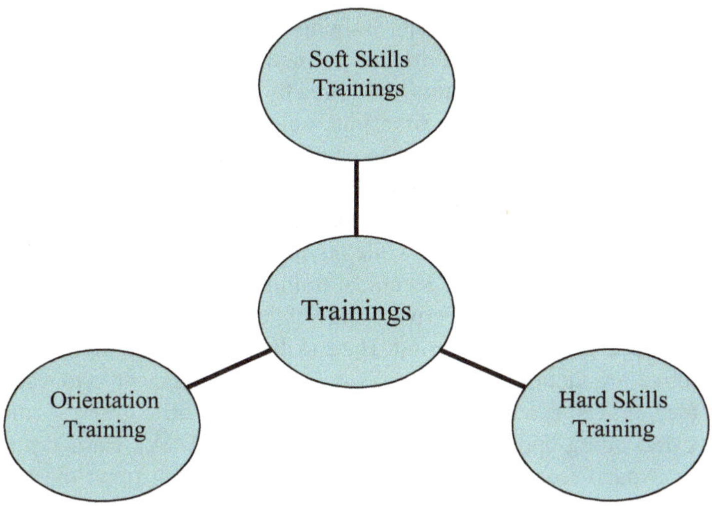

Fig. 6.2 Various categories of trainings

Table 6.1 List of trainings

Sr. No	Types of trainings	Category of training
1	Soft skills training	Soft
2	Hard skills training	Hard
3	Orientation training	Both
4	On-job training	Hard
5	Aboard training (outside country)	Hard
6	In-house training	Soft
7	Open enrollment training	Hard
8	Customized training	Soft
9	Outbound training	Soft

Table 6.2 List of model validation tests

Tests of model structure	Tests of model behavior	Tests of policy implications
Structure verification	Behavior reproduction	System improvement
Parametric verification	Behavior prediction	Changed behavior predictions
Extreme condition verification	Behavior anomaly	Boundary adequacy (policy)
Boundary adequacy (structure)	Family member	Policy sensitivity
Dimensional consistency	Surprise behavior	
	Extreme policy	
	Boundary adequacy	
	Behavior sensitivity	

sake of the simplicity, all sort of trainings have been categorized into soft skills training and hard skills trainings. Soft skills training time is calculated as product of number of trainers per month multiplied by number of training days per trainer multiplied by number of participants multiplied by sitting time of the participant per training day, whereas the hard skills training time is determined as product of number of trainers per month multiplied by number of training days per trainer multiplied by number of participants multiplied by sitting time of the participant per training day plus on-job training time plus outside country training. Hard skills training can be within country when the employee is doing work on machine (on-job training) or it can be from abroad at the time of procurement of new machinery categorized as outside country training. It is part of the agreement with equipment supplier and machine manufacturer while buying the new equipment or machinery that the supplier will provide the technical training to the company employees so that they can run the machine effectively and efficiently and train the other staff who are not exposed to the training period. Hard skills training like inventory management, effective production techniques, warehouse management, and quality management (see in Table 6.2 shown in Appendix A) is also given to the related employees depending upon the training need analysis (TNA). Hard skills training period may expand from few days to 3 months depending upon the operational complexity of the machinery and technical especially. Hard skills training time is usually from 8 h per month per trainer.

6.2.4 Training Process

Training need analysis (TNA) has been conducted in each department of the company and requests for nature of training either soft or hard; the proposed topic of training and number of persons to be trained from each department have been prepared by the department and sent to the HR department for execution purpose. On the basis of the TNA and using then input of other sources like competency gap, requirement of the new initiatives, and adoption of the new technology or emerging areas, the HR department prepares the budget and sends it to the management for approval. During a thorough discussion with the HR manager and concerned business unit head, training budget is approved. The next step is to select the appropriate trainers, and the HR department has a liaison with the training institutes and has the details of trainings to be conducted for the whole year. If there are around 20 or more participants for the particular training topic, an in-house training is designed worked jointly with the trainer and the company personals. If there are one or two participants for the said topic, then open enrollment courses are preferred. As per the training calendar, those one or two employees have been sent for those open enrollment workshops or training courses. Training outside the country whether it is hard skills training or soft skills training has been decided by the top management and business unit head; HR people are usually informed about the decision to maintain the training record. The training conduction hours per day per trainer are at least 8 h. Soft skills training is usually planned for 1 day per month. Few in-house workshops are planned for 2 days. So far as the hard skills trainings are concerned, employees are usually sent for open enrollment courses which may prolong for 1 day to 5 days depending upon the course topic. Trainings outside the country are mostly technical trainings which have been taken as a constant value for training hours on the basis of history. The trainer whether the expert is soft skills trainer or hard skills trainer, it took 8 h to conduct the training but the employee has taken 8 h training once in a month for just attending a 1-day training workshop. On-job trainings are part of the total stock of training hours and definitely contribute in the improvement of product quality and process improvement. Currently company has 2000 employees involved in the manufacturing activities, and they spend 8 eight hours per day and 25 days in a month as an average in the workplace. Total stock hours per month are around 400,000 hours, and as per the perception of the production manager, the learning rate is approximately 1% of the stock of monthly work hours. So total hours learned will be 4000 h and dividing by 2000 employees, we get 2 h per person per month are the hours which contribute as on job learning hours for the up gradation of the technical skill.

So far as the outside country trainings are concerned, as per the company history, two to three employees go outside the country at the time of purchase of new equipment and machinery and spend around 10–15 days to learn its operations and maintenance work spending 8 h per day including the lunch hours with a total of 160–360 h max per year. That indicates around 30 h per month divided by 3 persons. (10 h per person per month can be taken as a value for outside country training).

6.2.5 *Feedback Structure*

Stock and flow diagram of the model is captured in STELLA simulation software where the equations of the model are written automatically on the basis of the level and rate representation is shown in Fig. 6.6, whereas the listing of the equations written in STELLA has been shown in Appendix D.

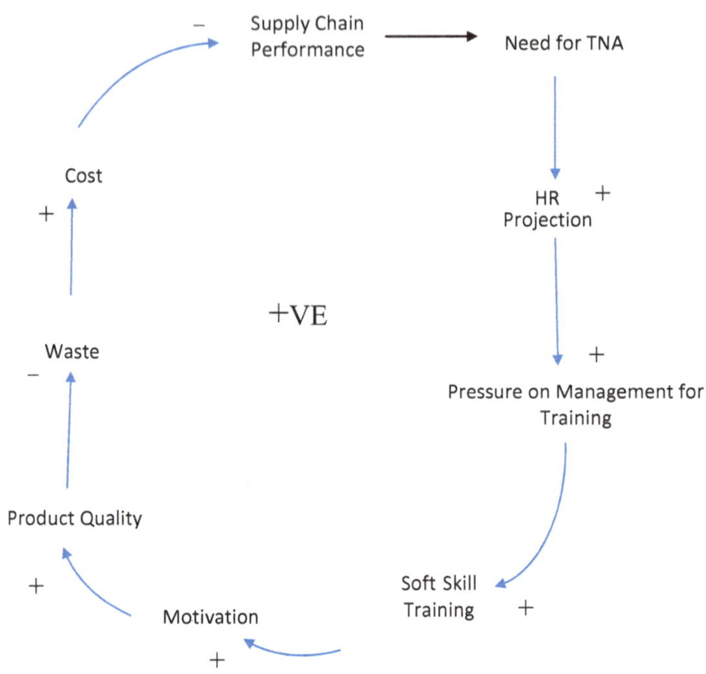

Fig. 6.3 Soft skills training loop is the result of management pressure to improve the quality of product and process; soft skills training changes the attitude and behavior of the employee which appear in the form of motivation (motivation is measured on the Likert scale from 1 to 5 indicating 1 low level of motivation and 5 highest level of motivation). It is a well-known fact that motivated employees find innovative ways to enhance the product quality that leads to reduction in waste. Consequently the cost of failures is reduced in the system. This is the reinforcing loop

The model is the fifth-order differential equation with associated flows (Forrester 1968). The order of the model depends upon the number of levels and the number of delays (Sushil 1993). As in this model, there is no delay function so the number of levels decides the order of the model. Various perception maps have been used in the model to translate the ground realities on the basis of information provided the company management. The range of the different variables in the graphical functions and its curvilinear pattern is discussed and approved by the concerted company staff. Polarity of the positive and negative feedback loops generates the behavioral patterns. The system consists of major positive and negative feedback loops (Lane 2007) (Figs. 6.3, 6.4, 6.5, and 6.6).

The Reinforcing Loops The model contains almost all positive feedback loops. More training hours will enhance the product quality due to motivation level and improve technical skill of the employee, but there are many other factors that also influence the product quality index which are not considered in the scope of model. Table 6.3 shows one example.

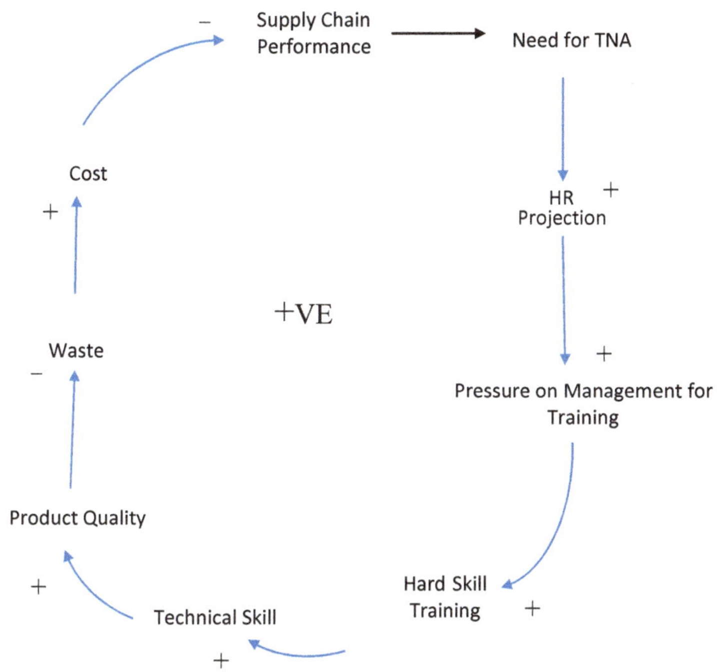

Fig. 6.4 Hard skills training loop improves the technical skill of the employee that ensures the improvements in the product quality; as a result defective units are reduced, lowering the cost of failures (internal and external). This is a positive loop for amplifying the increase in technical capability

Training need analysis variable is coupled with two exogenous variables step function and ramp function to generate the employees for the execution of trainings. Base run graphs indicate as the system goes on over a period of 130 months the waste level declines from 10% to 0%. The policies that are implemented so far seem to be effective. Allocation of the training budget Rs 1,000,000 on annual basis, 40% weight-age to soft skills training and multiplicative effect of on-job trainings and outside country training on hard skills training, widens the technical knowledge of the employees that brings improvement in product quality and process improvement. Decision about those 20 employees for soft skills training session and around 10 employees for hard skills training programs indicates that for in-house customized training workshops, management focus is to boost the motivation level of the employees generating team synergy and building positive attitude. Graphical functions mentioned in Fig. 6.35 for management pressure and HR projection and other table functions mentioned in Figs. 6.36, 6.37, 6.38, 6.39, 6.40, and 6.41 indicate the interesting behavior of the model due to nonlinearities that prevail in those perception maps.

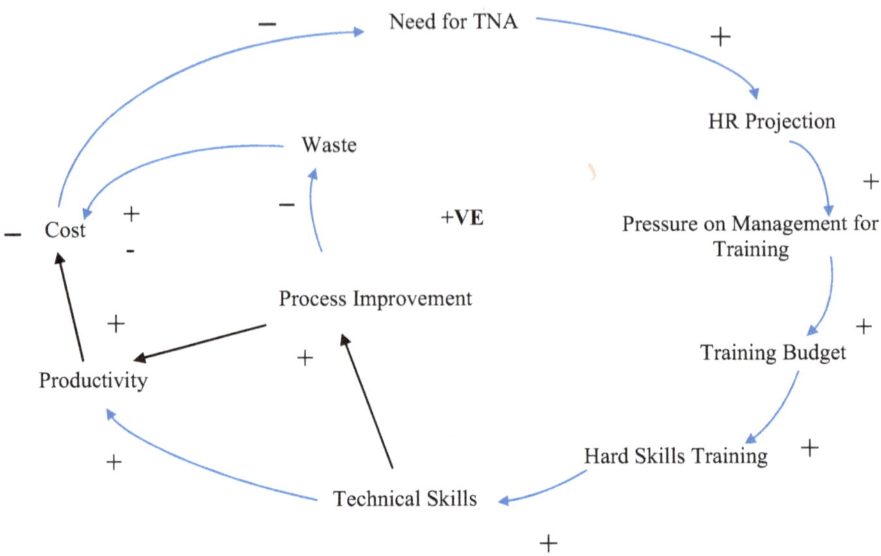

Fig. 6.5 Process improvement and productivity loop indicates that hard skills training increases the employee capability that contributes for the good for the process specification; consequently process quality is improved, and machine productivity has gone up reducing the cost of failures. This is also the reinforcing or positive loop

6.3 Policy Experiments

Model structure and interactions of dynamic variables generate multiple patterns that are understandable, interesting, and for some variables surprising. Dominance of the polarity and its shift from positive to negative and negative to positive add complexity; the mechanisms of change from one pattern to another can be searched through experimentation (Saeed 1991). Experimentation with the model not only allows us to understand the diversity of patterns, but it also helps us to unveil underlying structures for policy design. Various policies based on parametric changes are tested in the model to gain the insight and suggest some decisions to the management for enhanced supply chain performance and reduced waste level. Primarily, policies tested are based on sensitive parameters which can be the entry point for the policy interventions. For this type of scenario, it is assumed that the policy-maker is satisfied with the equations, and the only doubt remains as to what values the parameters or the policy levers should have (Duggan 2008). Secondly, policy experiments are performed on the structural changes having a base of mental intuition and perception about the problem solution. Human decisions are guided by profit and utility maximization (Sterman 1987). This is the equation perspective to optimization.

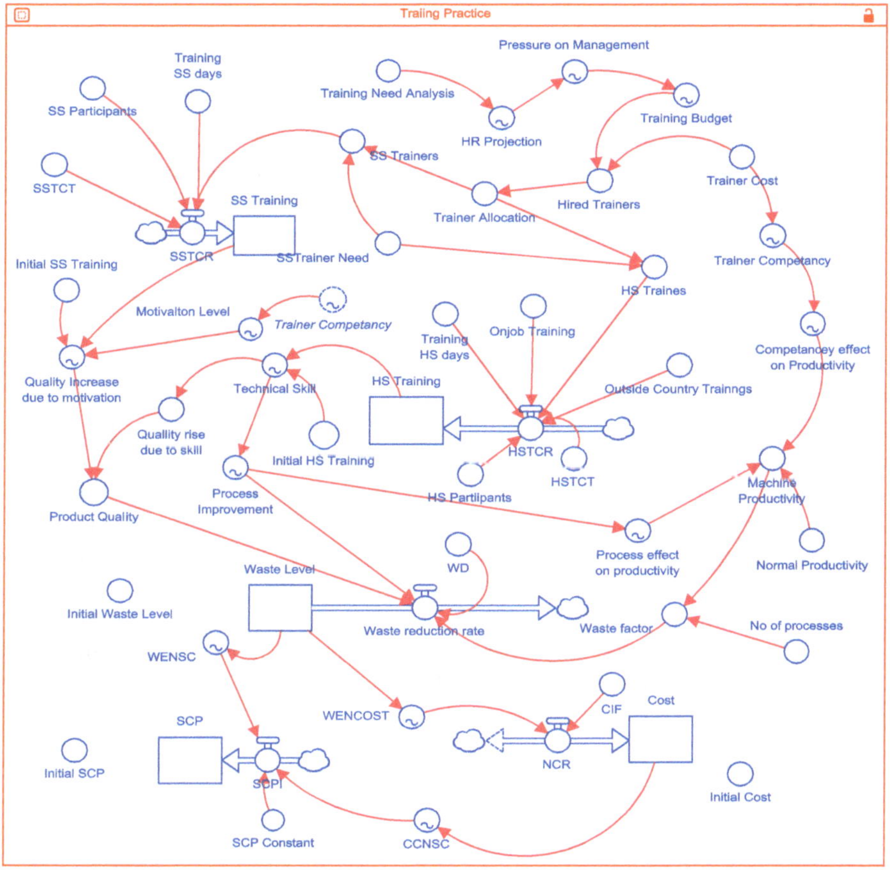

Fig. 6.6 Block diagram of training model

Table 6.3 Key Construct Variable and Sub Variables

Key parameter	Influencing parameters
Product quality index	Quality increase due to motivation
	Quality increase due to technical skill
	Quality increase due to good preventive and planned maintenance (not discussed)
	Quality increase due to life of equipment/machinery (not discussed)
	Quality increase due to quality of raw material (not discussed)

6.4 Policy Design on the Basis of Parametric Changes

Policy Run 1 Engaging the Low-Rate Trainers

Management has decided the training budget including all sort of training around one million Pak Rupees (USD 10,000 per year). The HR director has got the training schedules and curriculum vitae of the professional trainers from the various training institutes on yearly basis. The rate of the trainer indicates the competency of the trainer and its market repute. Survey guides the management a competent trainer charge around Pak Rupees hundred thousand (USD 1000 per trainer per day). Management has decided to hire the low-rate trainers and has reduced trainer rate from USD 1000 to USD 500 per trainer per day to USD 500. The pre-simulation prediction was that while lowering the rate, more training hours will be available and accordingly technical skill will increase to improve the product quality and enhance the process improvement. Consequently the waste level will reduce; cost of failures will decrease, and supply chain performance score (Beamon 1998) will increase. But the simulation results show, from Figs. 6.10, 6.11, 6.12, 6.13, 6.14, 6.15, 6.16, and 6.17 in Appendix E, that no doubt technical skills and soft skills training hours and hard skills training hours have acquired the highest number, but against the pre-simulation prediction, waste level has gone up, supply chain performance score more or less remains the same, and there is an interesting finding that instead of more soft skills training hours, the motivation level of the participants has gone down as well.

Policy Run 2 Focus on Cultural and Behavioral Issue of Organization (0.4 to 0.8)

Employee survey has been conducted in the organization. Survey results highlighted the attitude and behavioral issues of the management. The management has decided to increase the focus from hard skills trainings to soft skills trainings to motivate the employees, build the team synergy, and change the attitude and behavior of the employee. It was decided from the management that more focus will be on soft skills trainings rather than hard skills training and control parameter soft skills training need has been increased from 0.4 to 0.8 (assuming enhancing from 40% to 80%). Our pre-simulation understanding was that the stock of soft skills training hours will increase and stock of hard skills training hours will decrease, technical skill will reduce, but motivation of the employee will increase that will contribute for increased product quality and reduced level of waste; consequently the supply chain performance score will increase. Surprisingly their waste level has increased instead of the employee attitude and behavioral changes. Only the cost of failures has gone down slightly, whereas all other parameters are worsened. In Appendix E, Figs. 6.18, 6.19, 6.20, 6.21, 6.22, 6.23, 6.24, and 6.25 depict the complete story.

Policy Run 3 Policy on Enhancing the Technical Base by Introducing Three Decision Streams

1. First trainer cost is reduced from USD 1000 to USD 250 per trainer per day.
2. Second soft skills training need is reduced from 0.4 to 0.2 (40–20%).
3. Third hard skills training participants increased from 10 to 30 persons.

The results of policy run compared with the base run values have been shown in Figs. 6.26, 6.27, 6.28, 6.29, 6.30, 6.31, 6.32, 6.33, and 6.34 mentioned in Appendix E.

Introducing the reduced rate of trainer means to have more training hours that will increase the stock of soft skills training hours and hard skills training hours. Both enhancements in these stocks will lead to improvement in motivation level and technical skill of the employees. Lowering the soft skills training need from 0.4 to 0.2 highlights that the management focus is widening the technical base and improving the technical skill of the employees. The management message is very clear, and we have seen the results that on the basis of this management vision, the technical skill of the employee has gone up; there is a significant increase in product quality index, and waste level has gone down till 80 months.

6.5 Model Validation

Model is often seen as the external and explicit representation of the part of reality observed by the group of people who wish to use that model to understand, to change, to manage, to control, and to improve that part of reality (Pidd 2010). It is easy to build the model but difficult to make it acceptable to the community even if it depicts the real-life settings. Model validation using various tests is an essential step to enhance the confidence upon the simulated system dynamics model. Founder of System Dynamics Jay Forrester has jot down the list of tests for model validation (Forrester and Senge 1980; Sterman 2000). Forrester emphatically states that the validity of the model should be judged by its suitability, usefulness, and purpose of the model for the stakeholders, modeler, clients, and management of the company and policy researchers and boosts the image of modeler with regard to its usefulness (Sterman 2000; Qudrat-Ullah 2008; Pidd 2010; Forrester 1987; Forrester and Senge 1980; Qudrat-Ullah and Seong 2010). Model structure and interactions of dynamic variables generate multiple patterns that are understandable, interesting, and for some variables surprising. Dominance of the polarity and its shift changes the behavioral pattern.

6.5.1 Dimensional Consistency

Dimensional consistency in the model is verified and validated by making all equations dimensionally balanced, and each variable in the equations has unit of measure which represents the real-life system (Forrester and Senge 1980; Qudrat-Ullah 2008).

6.5.2 *Structure Verification*

Structure verification leads to determine whether the model structure is compatible, and it is a representation of the relevant case company descriptive knowledge. Is there any support from the literature to represent the variables, constructs, sub-model, and model structures (Qutrat-Ullah 2008, 2010)? It is carefully determined that model structure must be consistent with the XYZ Packaging Company system operating procedures (SOPs) and the variables chosen in the model and causal loop relationships also be rooted in literature (see the following Table 6.4).

Table 6.4 Structures. (Adapted from literature)

Variables/ structures/concepts	Sources	Remarks
Training need analysis	Sila and Ebranhimpour (2005), Ahire et al. (1996), and Kaynak(2003)	
Training	Sila and Ebranhimpour (2005), Ahire et al. (1996), Flynn et al. (1995), Adm et al. (1997), Kaynak(2003), and Tan et al.(1999),	
Trainer competency	Sila and Ebranhimpour (2005), Ahire et al. (1996), and Kaynak(2003)	
Motivation	Sila and Ebranhimpour (2005), Dow et al. (1999), and Kaynak(2003)	
Technical skill	Tan et al. (1999)	
Product quality	Sila and Ebranhimpour (2005), Ahire et al. (1996), Tan et al. (1999), and Flynn and Flynn (2005)	
Waste level	Sila and Ebranhimpour (2005), Anderson et al. (1995), Flynn et al. (1995), and Tan et al. (1999)	
Productivity	Sila and Ebranhimpour (2005), Anderson et al. (1995), and Flynn et al. (1995)	
Process improvement	Sila and Ebranhimpour (2005), Anderson et al. (1995), Flynn et al. (1995), and Flynn and Flynn (2005)	
Training budget	Sila and Ebranhimpour (2005), Anderson et al. (1995), Flynn et al. (1995), and Kaynak(2003)	
Trainer cost	Sila and Ebranhimpour (2005), Anderson et al. (1995), and Flynn et al. (1995)	
Supply chain performance	Sila and Ebranhimpour (2005), Anderson et al. (1995), Flynn et al. (1995), and Flynn and Flynn (2005)	
Pressure on management	Sila and Ebranhimpour (2005), Ahire et al. (1996), and Kaynak(2003)	
HR projection	Sila and Ebranhimpour (2005), Anderson et al. (1995), Flynn et al. (1995), and Kaynak(2003)	
Cost of quality	Sila and Ebranhimpour (2005), Anderson et al. (1995), Tan et al. (1999), and Kaynak (2003)	
On-job training	Sila and Ebranhimpour (2005), Anderson et al. (1995), and Flynn et al. (1995)	
Outside country training	Sila and Ebranhimpour (2005), Anderson et al. (1995), and Flynn et al. (1995)	

6.5.3 Extreme Condition Verification

This test checks that the behavior of the model structure for the selected variables must justify the extreme condition (Sterman 2000; Forrester and Senge 1980) and exhibit the logical behavior. In our case extreme condition test examines the conformance to basic physical laws, like basic assumption, that is, there is no HR projection against the TNA, pressure on the management should be zero, and accordingly the training budget allocation is also zero (Fig. 6.7) (Sterman 2000). There is no amount to hire the trainer for soft skills and hard skills that is why there is no increase and decrease in the training hours as shown in Fig. 6.8. As a result motivation level remains the same, product quality index is the same, and process improvement is also unchanged. Consequently the waste level has declining trend over the periods of time; this interesting phenomenon is due to the constant rate of reduction in the waste level; as a result the waste level has declining trend over the period of time justifying the extreme condition verification as shown in Fig. 6.9.

6.5.4 Parametric Verification

Parameter verification means comparing model parameters to the observation of the real-life system (Forrester and Senge 1980). Are the parametric values consistent with the relevant knowledge of the case company? Company documentation, archival materials, judgmental opinions on the basis of interviews and participant experience, and expert opinion of the top management ensure the parameter justification for the real system. Appendix B and Appendix C indicate list of variables with base run values and policy run values.

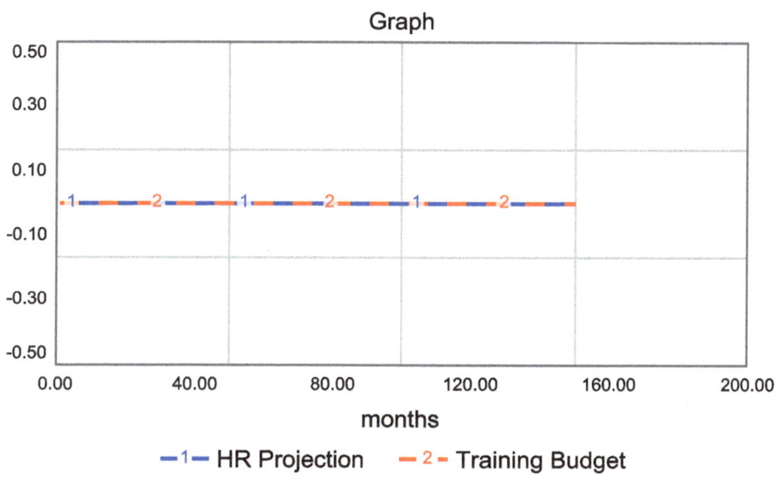

Fig. 6.7 Depicting HR projection and training budget

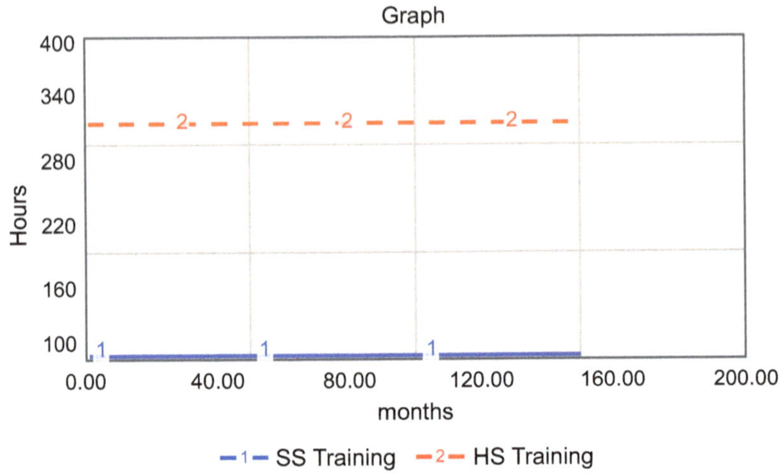

Fig. 6.8 Depicting SS trainings and HS trainings

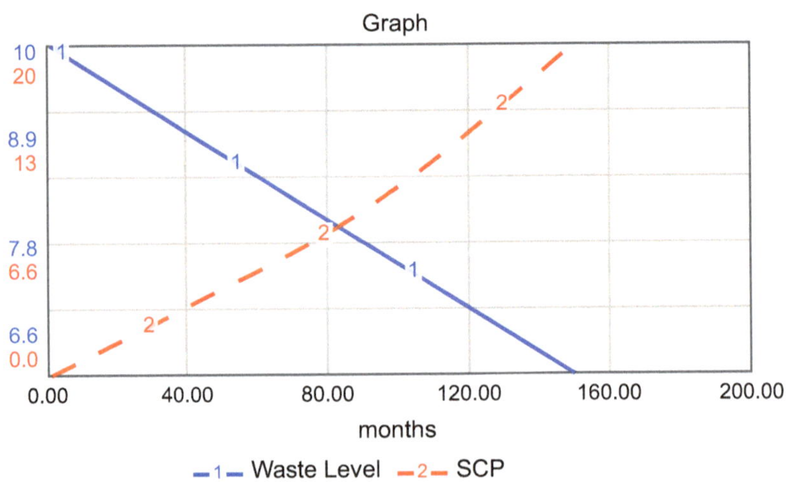

Fig. 6.9 Depicting waste level and SCP

6.5.5 Boundary Adequacy Test

This test evaluates the important concepts and structures for addressing the policy issues (Sterman 2000; Forrester and Senge 1980). It identifies endogenous and exogenous variables. In our model the list of variables are mentioned in Table 6.5.

Table 6.5 List of endogenous and exogenous variables

Endogenous	Exogenous
HR projection	Waste level
Pressure on management	Supply chain performance
Training budget	Cost of quality
Trainer cost	
Trainer competency	
Trainer allocation	
Motivation level	
Product quality index	
Productivity	
Process improvement	
Process effect on productivity	
Quality rise due to technical skill	
Quality rise due to motivation level	

6.5.6 Behavior Reproduction Test

This test is being carried out for the validation of the model behavior. Behavioral validity is to compare the model-generated behavior to the observed behavior (Sterman 2000; Qudrat-Ullah and Seong 2010) of the case company. Historical pattern of the various parameters of XYZ Packaging Company like number of training hours for soft skills and hard skills, Human Resource (HR) projections, waste level, and cost of quality qualifies this test.

6.6 Conclusion

Experimentation with the model suggests to different policy interventions to reduce the waste level, lower the cost of failures, increase the motivation level, and enhance the product quality index and supply chain performance. Policy design on the basis of single parameter may help to achieve certain goals, but improvement in overall all necessary variables is missing. The results of policy run 1 and policy run 2 indicate the same story. In policy run 3 when we make a policy based on three decision streams, the results are more favorable as compared to the previous policy run 1 and policy run 2. There is a need to design the policies on the bases of system structures for long run and sustainable results. Concept of feedback system structure for organizing the explicit and tacit knowledge about the system (Saeed 2008) explores the avenues that make the system better behaved over a period of time. Translation of dynamic hypothesis into a model provides insight to management about the causes of behavior. This chapter formulates the policies for the management of the XYZ Packaging. The policies are thought provoking and eye-opening because pre-simulation predictions of the management were contrary to the actual policies suggested by the model. These policies and their results enhanced the management perceptions that the issues which seem to be so simple are not simple and policies based on mere intuition may appear as the disposable initiatives in terms of sustainable improvement. This chapter leads to future direction to suggest more plausible policies on the basis of structural changes rather than parametric changes.

Appendix A

Table 6.6 Symbols for flow diagram

Description	Symbol	Associated equation type	Explanation
Level		L	Stock
Rate	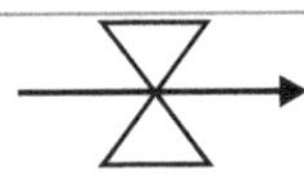	R	Flow
Auxiliary	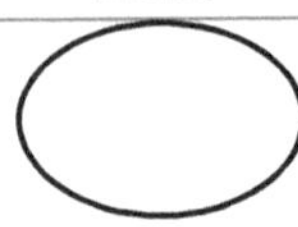	A	Convertor
Table function	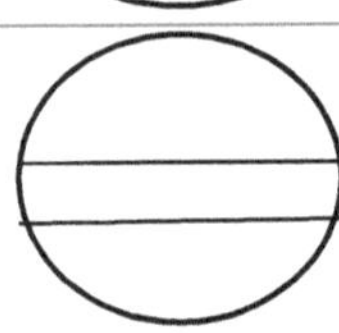	T	Perception map between *x* and *y* plane
Exogenous variable	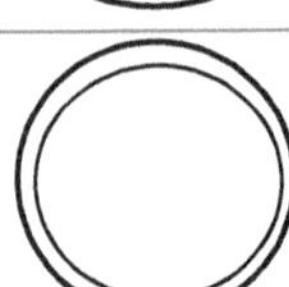	E	Occasionally affect the model behavior but not part of model
Constant		C	Constant which has unique value
Source or sink of material		Define	Out of boundary, defines the model scope
Material/ information flow			Use for the movement of material and information

Table 6.7 Details of trainings

Sr. No.	Hard skills trainings	Soft skills trainings
1	Inventory management	Attitude, team building, and communication
2	Effective production techniques	Motivating self and others
3	Quality management systems	Peak performance
4	Effective selling techniques	Human productivity
5	Statistical techniques	Quality teams
6	Energy management	Neuro-linguistic programming (NLP)
7	Condition-based monitoring maintenance	Effective managerial skills
8	TPM	Supercharged business creativity
9	Waste heat recovery boilers	Creative leadership and team building
10	HFO/LFO separators	Quality circles facilitators
11	Boiler inspection and testing	How to conduct performance evaluation
12	Finance for non-finance managers	Internal quality auditing
13	Assessors of quality awards	Creative leadership and team building
14	An overview of ERP-SAP	Leaders and legends
15	Machinery management system (MAMA)	Time management
16	Seven basic tools	Decision-making skills
17	HACCP awareness	Labor policies and laws
18	Microsoft excel advanced	Preparing documentations and manuals
19	Warehouse management	ISO 9001 for service sector
20	Lean manufacturing systems	5S housekeeping

Appendix B

List of variables

Variables	Description	UOM	Equation type	Parametric value
SSTCT	Soft skills training conduction time	Hours per trainer per month	C	8
HSTCT	Hard skills training conduction time	Hours per trainer per month	C	8
SS training	Soft skills training stock	Hours imparted	L	
HS training	Hard skills training stock	Hours imparted	L	
Initial SS training	Initial soft skills training	Hours imparted	C	104
Initial HS training	Initial hard skills training	Hours imparted	C	320
Trainer cost	Trainer cost	Rupees	C	100,000
Pressure on management	Pressure on management	Percentage	C	0–100
Training budget	Training budget	Rupees	C	1,000,000
SS training need	Soft skills training need	Dimensionless	C	0.4
No of training SS days	Number of training soft skills days	Number of training days per trainer	C	1

(continued)

Variables	Description	UOM	Equation type	Parametric value
No of training HS days	Number of training hard skills days	Number of training days per trainer	C	1
SS participants	Soft skills training participants	Number of persons in a training	C	20
HS participants	Hard skills training participants	Number of persons in a training	C	20
No of processes	Number of processes/ machines	Number	C	10
Product quality	Product quality	Percentage	A	
SSTCR	Soft skills conduction rate	Hours per month	R	
HSTCR	Hard skills conduction rate	Hours per month	R	
On-job training	On-job training	Hours	C	2
Outside country training	Outside country training	Hours	C	10
Initial waste level	Initial waste level	Percentage	C	10
Initial cost	Initial failure costs	Rupees	C	500,000
Initial SCP	Initial supply chain performance score	Number	C	10
Normal productivity	Normal machine productivity	Percentage per process per month	C	0.025
NCR	Nonconformance rate	Rupees per month	R	
SCP constant	Supply chain performance constant	Months	C	10
Trainer competency	Trainer competency	Dimensionless	T	1–5
Motivational level	Motivational level	Dimensionless	T	1–5
Quality Incr due to motivation	Quality increase due to motivation	Dimensionless	T	0–1
Process improvement	Process improvement	Dimensionless	T	0–1
Quality rise due to technical skill	Quality rise due to technical skill	Dimensionless	T	0–1
SCP	Supply chain performance score	Number	L	
SCPI	Supply chain performance score increase rate	Number per month	R	
Competency effect on productivity	Competency effect on productivity		A	
Process effect on productivity	Process effect on productivity		A	
WENSC	Waste effect on supply chain performance		T	

(continued)

Variables	Description	UOM	Equation type	Parametric value
WENCOST	Waste effect on cost of failures		T	
WD	Waste deduction factor	Months	C	180
Waste factor	Waste factor		A	
Waste reduction rate	Waste reduction rate		R	

Appendix C: Variable with Base Run and Policy Run Parametric Values

List of variables

Variables	Description	UOM	Base run parametric value	Policy run parametric value
SSTCT	Soft skills training conduction time	Hours per trainer per month	8	
HSTCT	Hard skills training conduction time	Hours per trainer per month	8	
SS training	Soft skills training stock	Hours imparted		
HS training	Hard skills training stock	Hours imparted		
Initial SS training	Initial soft skills training	Hours imparted	104	
Initial HS training	Initial hard skills training	Hours imparted	320	
Trainer cost	Trainer cost	Rupees	100,000	50,000
Pressure on management	Pressure on management	Percentage	0–100	
Training budget	Training budget	Rupees	1,000,000	
SS training need	Soft skills training need	Dimensionless	0.4	0.8, 0.2
No of training SS days	Number of training soft skills days	Number of training days per trainer	1	
No of training HS days	Number of training hard skills days	Number of training days per trainer	1	
SS participants	Soft skills training participants	Number of persons in a training	20	
HS participants	Hard skills training participants	Number of persons in a training	10	30

(continued)

Variables	Description	UOM	Base run parametric value	Policy run parametric value
No of processes	Number of processes/ machines	Number	10	
Product quality	Product quality	Percentage		
SSTCR	Soft skills conduction rate	Hours per month		
HSTCR	Hard skills conduction rate	Hours per month		
On-job training	On-job training	Hours	2	
Outside country training	Outside country training	Hours	10	
Initial waste level	Initial waste level	Percentage	10	
Initial cost	Initial failure costs	Rupees	500,000	
Initial SCP	Initial supply chain performance score	Number	10	
Normal productivity	Normal machine productivity	Percentage per process per month	0.025	
NCR	Nonconformance rate	Rupees per month		
SCP constant	Supply chain performance constant	Months	10	
Trainer competency	Trainer competency	Dimensionless	1–5	
Motivational level	Motivational level	Dimensionless	1–5	
Quality Incr due to motivation	Quality increase due to motivation	Dimensionless	0–1	
Process improvement	Process improvement	Dimensionless	0–1	
Quality rise due to technical skill	Quality rise due to technical skill	Dimensionless	0–1	
SCP	Supply chain performance score	Number		
SCPI	Supply chain performance score increase rate	Number per month		
Competency effect on productivity	Competency effect on productivity			
Process effect on productivity	Process effect on productivity			
WENSC	Waste effect on supply chain performance			
WENCOST	Waste effect on cost of failures			
WD	Waste deduction factor	Months	180	
Waste factor	Waste factor			
Waste reduction rate	Waste reduction rate			

Appendix D

Programming for System Dynamics Simulation Model on STELLA Software

Note: Trainings_SCM Model

```
Top-Level Model:
Cost(t) = Cost(t - dt) + (NCR) * dt
INIT Cost = Initial_Cost
INFLOWS:
NCR = -CIF*WENCOST
HS_Training(t) = HS_Training(t - dt) + (HSTCR) * dt
INIT HS_Training = Initial_HS_Training
INFLOWS:
HSTCR
=
(HSTCT*HS_Traines*HS_Partiipants*Training_HS_days)+Onjob_
Training+Outside_Country_Trainngs
SCP(t) = SCP(t - dt) + (SCPI) * dt
INIT SCP = Initial_SCP
INFLOWS:
SCPI = (WENSC+CCNSC)/SCP_Constant
SS_Training(t) = SS_Training(t - dt) + (SSTCR) * dt
INIT SS_Training = Initial_SS_Training
INFLOWS:
SSTCR = SSTCT*SS_Participants*Training_SS_days*SS_Trainers
Waste_Level(t) = Waste_Level(t - dt) + ( - Waste_reduction_rate)
* dt
INIT Waste_Level = Initial_Waste_Level
OUTFLOWS:
Waste_reduction_rate = ((Process_Improvement+Product_Qaulity)/
WD)+Waste_factor
CCNSC = GRAPH(Cost)
(0, 0.965), (52631.5789474, 0.871), (105263.157895, 0.698),
(157894.736842, 0.535), (210526.315789, 0.347), (263157.894737,
0.297), (315789.473684, 0.287), (368421.052632, 0.267),
(421052.631579, 0.252), (473684.210526, 0.223), (526315.789474,
0.203), (578947.368421, 0.188), (631578.947368, 0.168),
(684210.526316, 0.158), (736842.105263, 0.153), (789473.684211,
0.149), (842105.263158, 0.139), (894736.842105, 0.134),
(947368.421053, 0.119), (1000000, 0.099)
CIF = 0.015
```

```
Competancey_effect_on_Productivity = GRAPH(Trainer_Competancy)
(0.000, 0.318), (0.500, 0.346), (1.000, 0.389), (1.500, 0.436),
(2.000, 0.464), (2.500, 0.493), (3.000, 0.526), (3.500, 0.588),
(4.000, 0.635), (4.500, 0.678), (5.000, 0.701)
Hired_Trainers = Training_Budget/Trainer_Cost
HR_Projection = GRAPH(Training_Need_Analysis)
(0.00, 0.00), (4.54545454545, 0.89), (9.09090909091, 1.85),
(13.6363636364, 2.36), (18.1818181818, 4.72), (22.7272727273,
6.94), (27.2727272727, 13.59), (31.8181818182, 21.36),
(36.3636363636, 29.85), (40.9090909091, 36.89), (45.4545454545,
44.42), (50.00, 50.00)
HS_Partiipants = 10
HS_Traines = (1-SSTrainer_Need)*Trainer_Allocation
HSTCT = 8
Initial_Cost = 500000
Initial_HS_Training = 320
Initial_SCP = 0
Initial_SCP_S = 10
Initial_SS_Training = 104
Initial_Waste_Level = 10
Machine_Productivity = Normal_Productivity*Process_effect_on_
productivity*Competancey_effect_on_Productivity
Motivalton_Level = GRAPH(Trainer_Competancy)
(0.000, 0.000), (0.500, 0.470), (1.000, 0.767), (1.500, 1.163),
(2.000, 1.460), (2.500, 2.104), (3.000, 2.599), (3.500, 3.069),
(4.000, 3.564), (4.500, 3.911), (5.000, 5.000)
No_of_processes = 10
Normal_Productivity = 0.025
Onjob_Training = (2000*8*25*.01/2000)
Outside_Country_Trainngs = 10
Pressure_on_Management = GRAPH(HR_Projection)
(0.00, 0.0), (5.00, 5.9), (10.00, 10.4), (15.00, 16.8), (20.00,
25.7), (25.00, 32.2), (30.00, 41.1), (35.00, 48.0), (40.00,
59.4), (45.00, 70.8), (50.00, 88.6)
Process_effect_on_productivity = GRAPH(Process_Improvement)
(0.000, 0.100), (0.100, 0.156), (0.200, 0.190), (0.300, 0.256),
(0.400, 0.313), (0.500, 0.346), (0.600, 0.389), (0.700, 0.417),
(0.800, 0.460), (0.900, 0.517), (1.000, 0.592)
Process_Improvement = GRAPH(Technical_Skill)
(0.000, 0.043), (0.100, 0.066), (0.200, 0.071), (0.300, 0.095),
(0.400, 0.123), (0.500, 0.156), (0.600, 0.194), (0.700, 0.251),
(0.800, 0.332), (0.900, 0.393), (1.000, 0.417)
Product_Qaulity = (Quality_Increase_due_to_motivation*Quallity_
rise_due_to_skill)*100
```

```
Quality_Increase_due_to_motivation = GRAPH(Motivalton_Level*(SS_
Training/(Initial_SS_Training*100)))
(0.000, 0.045), (0.500, 0.129), (1.000, 0.246), (1.500, 0.346),
(2.000, 0.436), (2.500, 0.512), (3.000, 0.635), (3.500, 0.678),
(4.000, 0.810), (4.500, 0.882), (5.000, 0.950)
Quallity_rise_due_to_skill = Technical_Skill
SCP_Constant = 10
SS_Participants = 20
SS_Trainers = Trainer_Allocation*SSTrainer_Need
SSTCT = 8
SSTrainer_Need = 0.4
Technical_Skill = GRAPH(HS_Training/(Initial_HS_Training*10))
(0.00, 0.005), (1.00, 0.074), (2.00, 0.261), (3.00, 0.370),
(4.00, 0.488), (5.00, 0.559), (6.00, 0.664), (7.00, 0.744),
(8.00, 0.829), (9.00, 0.926), (10.00, 1.000)
Trainer_Allocation = Hired_Trainers
Trainer_Competancy = GRAPH(Trainer_Cost)
(0, 0.124), (7142.85714286, 0.119), (14285.7142857, 0.178),
(21428.5714286, 0.233), (28571.4285714, 0.421), (35714.2857143,
0.668), (42857.1428571, 1.015), (50000, 1.312), (57142.8571429,
1.609), (64285.7142857, 1.931), (71428.5714286, 2.302),
(78571.4285714, 2.649), (85714.2857143, 2.995), (92857.1428571,
4.431), (100000, 5.000)
Trainer_Cost = 100000
Training_Budget = GRAPH(Pressure_on_Management)
(0.0, 0), (5.26315789474, 10000), (10.5263157895, 50000),
(15.7894736842, 100000), (21.0526315789, 140000), (26.3157894737,
210000), (31.5789473684, 390000), (36.8421052632, 480000),
(42.1052631579, 540000), (47.3684210526, 600000), (52.6315789474,
650000), (57.8947368421, 720000), (63.1578947368, 780000),
(68.4210526316, 820000), (73.6842105263, 840000), (78.9473684211,
850000), (84.2105263158, 870000), (89.4736842105, 890000),
(94.7368421053, 910000), (100.0, 920000)
Training_HS_days = 1
Training_Need_Analysis = 10+STEP(10, 10)+RAMP(0.15, 10)
Training_SS_days = 1
Waste_factor = No_of_processes*Machine_Productivity
WD = 180
WENCOST = GRAPH(Waste_Level)
(0.00, 32000), (2.00, 54000), (4.00, 106000), (6.00, 145000),
(8.00, 237000), (10.00, 303000), (12.00, 353000), (14.00,
399000), (16.00, 433000), (18.00, 473000), (20.00, 493000)
WENSC = GRAPH(Waste_Level)
(0.0, 0.936), (4.7619047619, 0.861), (9.52380952381, 0.767),
```

```
(14.2857142857, 0.584), (19.0476190476, 0.391), (23.8095238095,
0.337), (28.5714285714, 0.302), (33.3333333333, 0.287),
(38.0952380952, 0.267), (42.8571428571, 0.243), (47.619047619,
0.223), (52.380952381, 0.208), (57.1428571429, 0.198),
(61.9047619048, 0.188), (66.6666666667, 0.183), (71.4285714286,
0.178), (76.1904761905, 0.178), (80.9523809524, 0.168),
(85.7142857143, 0.168), (90.4761904762, 0.168), (95.2380952381,
0.163), (100.0, 0.163)
{ The model has 53 (53) variables (array expansion in parens).
In 1 Modules with 1 Sectors.
Stocks: 5 (5) Flows: 5 (5) Converters: 43 (43)
Constants: 21 (21) Equations: 27 (27) Graphicals: 13 (13)}
```

Appendix E

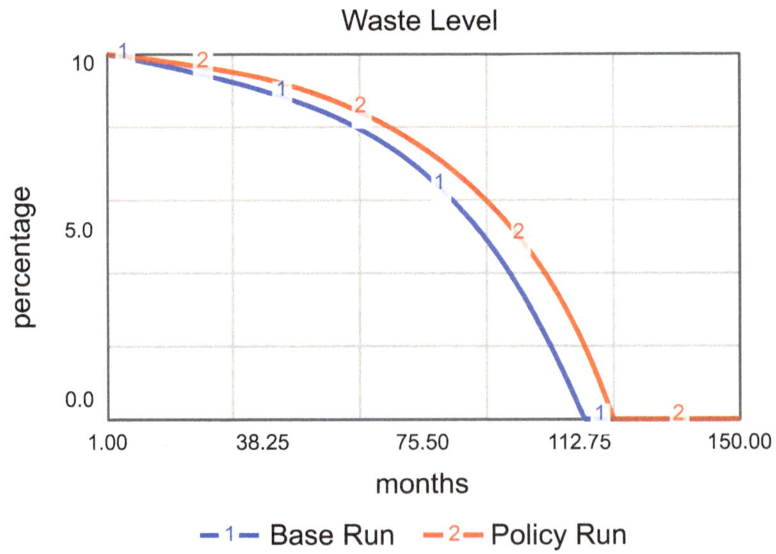

Fig. 6.10 *Base run and policy run 1* Engaging low-rate trainer from USD 1000 to USD 500 per day

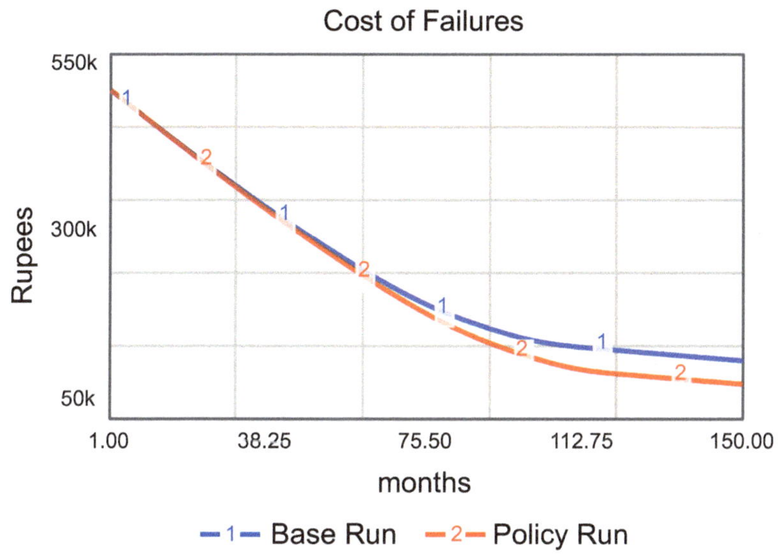

Fig. 6.11 *Base run and policy run 1* Engaging low-rate trainer from USD 1000 to USD 500 per day

Fig. 6.12 *Base run and policy run 1* Engaging low-rate trainer from USD 1000 to USD 500 per day

Fig. 6.13 *Base run and policy run 1* Engaging low-rate trainer from USD 1000 to USD 500 per day

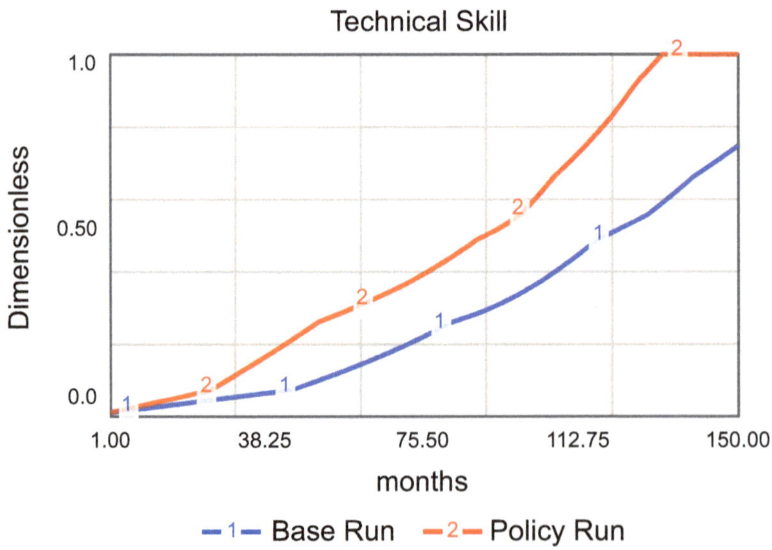

Fig. 6.14 *Base run and policy run 1* Engaging low-rate trainer from USD 1000 to USD 500 per day

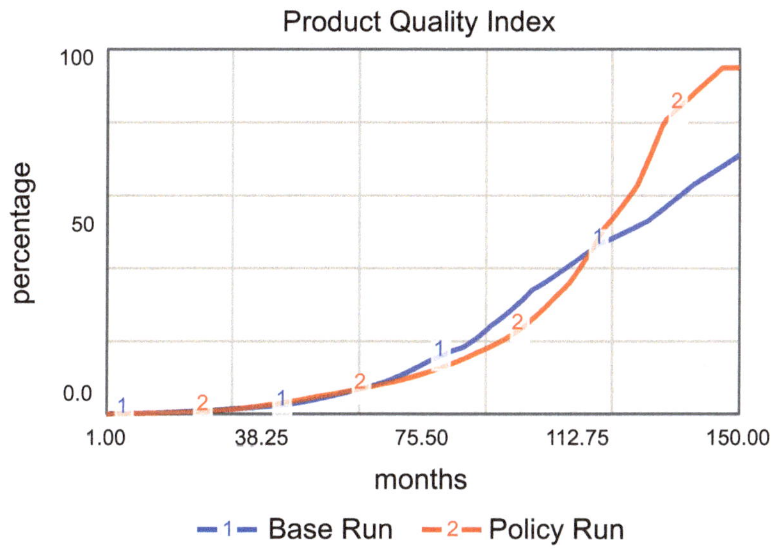

Fig. 6.15 *Base run and policy run 1* Engaging low-rate trainer from USD 1000 to USD 500 per day

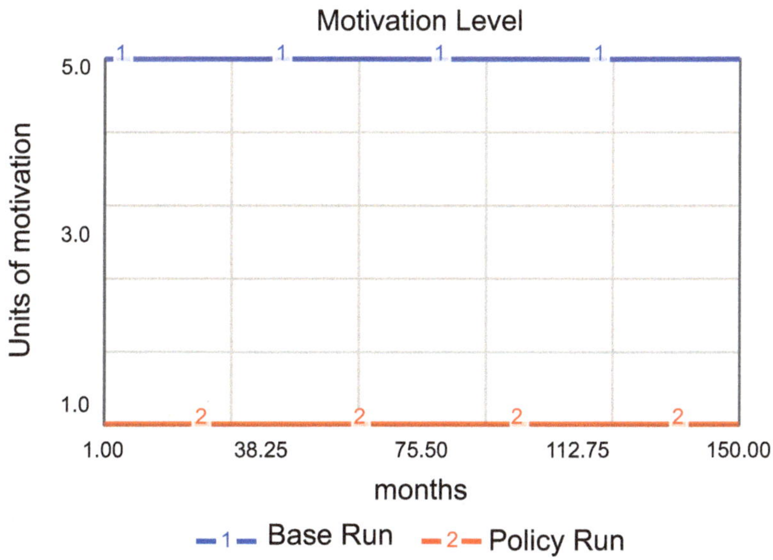

Fig. 6.16 *Base run and policy run 1* Engaging low-rate trainer from USD 1000 to USD 500 per day

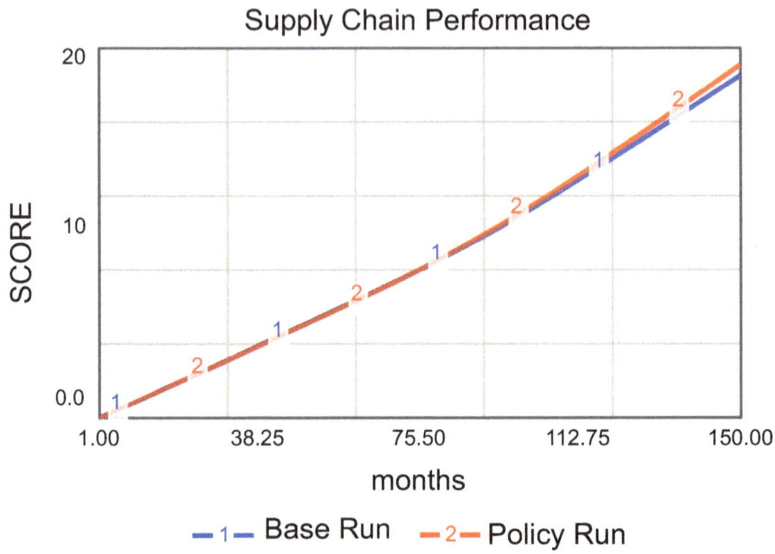

Fig. 6.17 *Base run and policy run 1* Engaging low-rate trainer from USD 1000 to USD 500 per day

Fig. 6.18 *Base run and policy run 2* Focus on cultural and behavioral issue of organization (soft skills training need increased from 0.4 to 0.8)

Fig. 6.19 *Base run and policy run 2* Focus on cultural and behavioral issue of organization (soft skills training need increased from 0.4 to 0.8)

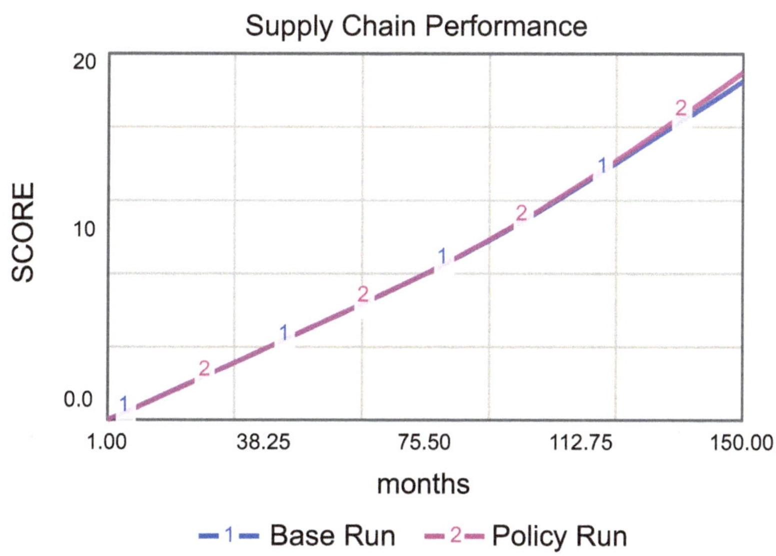

Fig. 6.20 *Base run and policy run 2* Focus on cultural and behavioral issue of organization (soft skills training need increased from 0.4 to 0.8)

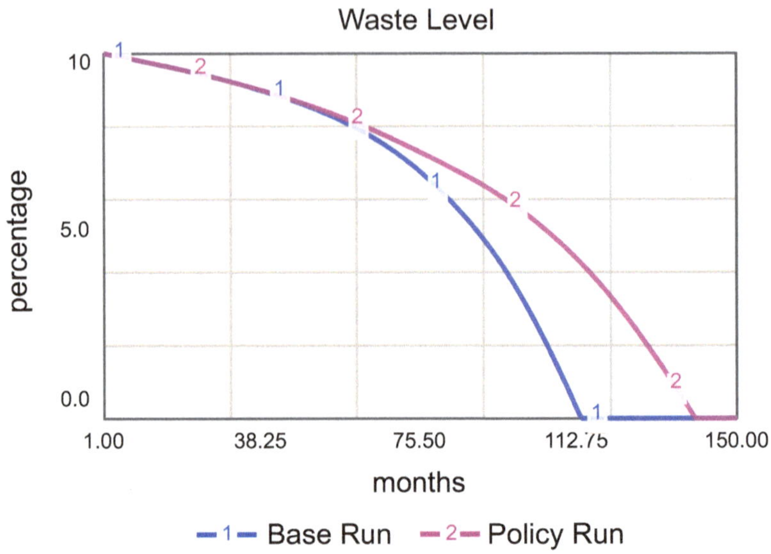

Fig. 6.21 *Base run and policy run 2* Focus on cultural and behavioral issue of organization (soft skills training need increased from 0.4 to 0.8)

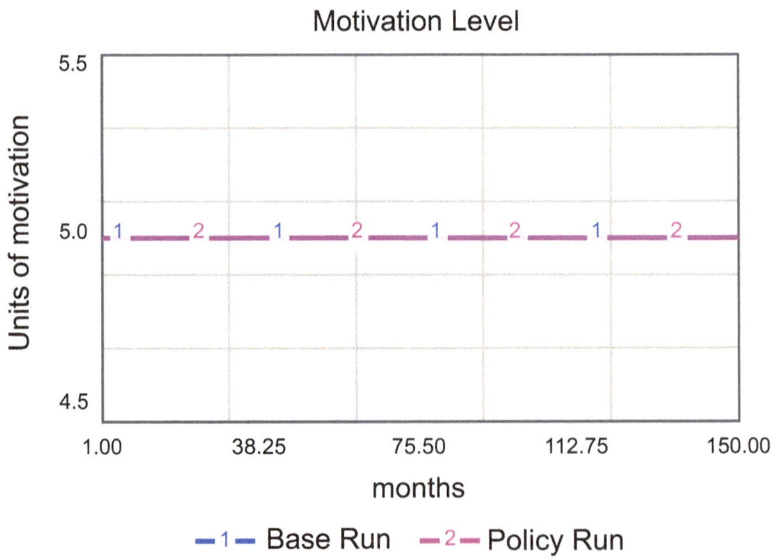

Fig. 6.22 *Base run and policy run 2* Focus on cultural and behavioral issue of organization (soft skills training need increased from 0.4 to 0.8)

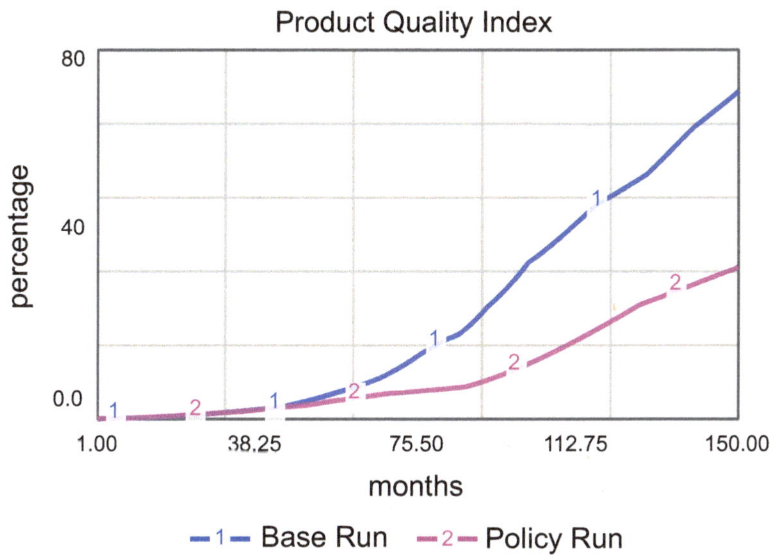

Fig. 6.23 *Base run and policy run 2* Focus on cultural and behavioral issue of organization (soft skills training need increased from 0.4 to 0.8)

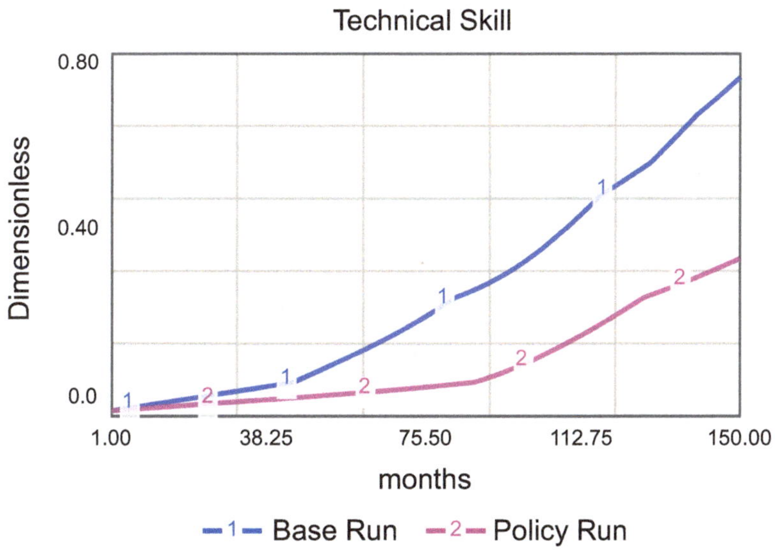

Fig. 6.24 *Base run and policy run 2* Focus on cultural and behavioral issue of organization (soft skills training need increased from 0.4 to 0.8)

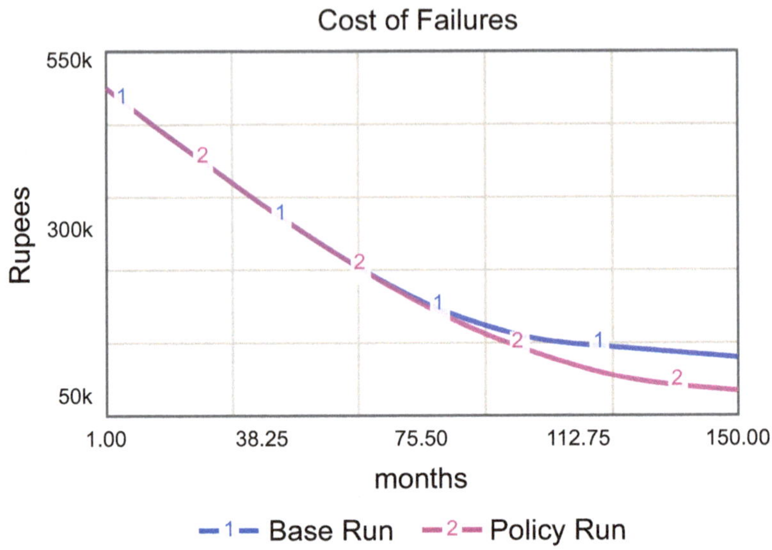

Fig. 6.25 *Base run and policy run 2* Focus on cultural and behavioral issue of organization (soft skills training need increased from 0.4 to 0.8)

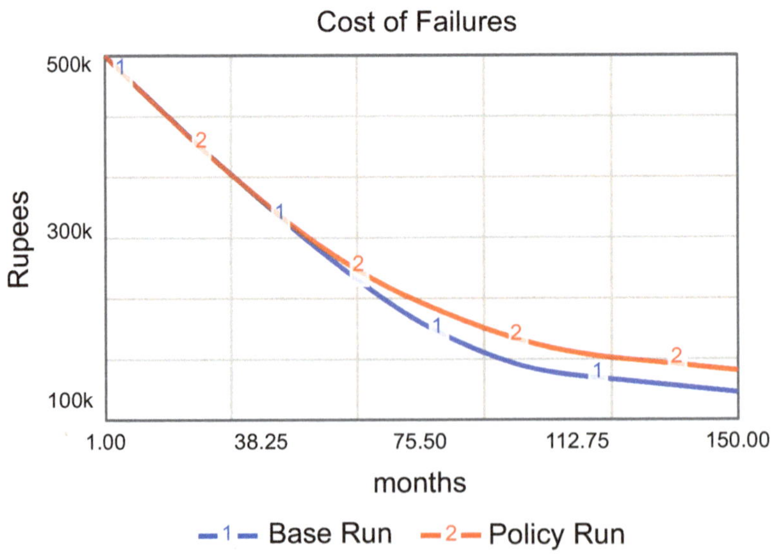

Fig. 6.26 *Base run and policy run 3* Policy on enhancing the technical base of employees by experimenting with the three parameters. First trainer cost is reduced from USD 1000 to USD 250 per trainer per day, second soft skills training need is reduced from 0.4 to 0.2 (40–20%), and third is hard skills training participants increased from 10 to 30 persons; see the results of those policies

Fig. 6.27 *Base run and policy run 3* Policy on enhancing the technical base of employees by experimenting with the three parameters. First trainer cost is reduced from USD 1000 to USD 250 per trainer per day, second soft skills training need is reduced from 0.4 to 0.2 (40–20%), and third hard skills training participants increased from 10 to 30 persons; see the results of those policies

Fig. 6.28 *Base run and policy run 3* Policy on enhancing the technical base of employees by experimenting with the three parameters. First trainer cost is reduced from USD 1000 to USD 250 per trainer per day, second soft skills training need is reduced from 0.4 to 0.2 (40–20%), and third hard skills training participants increased from 10 to 30 persons; see the results of those policies

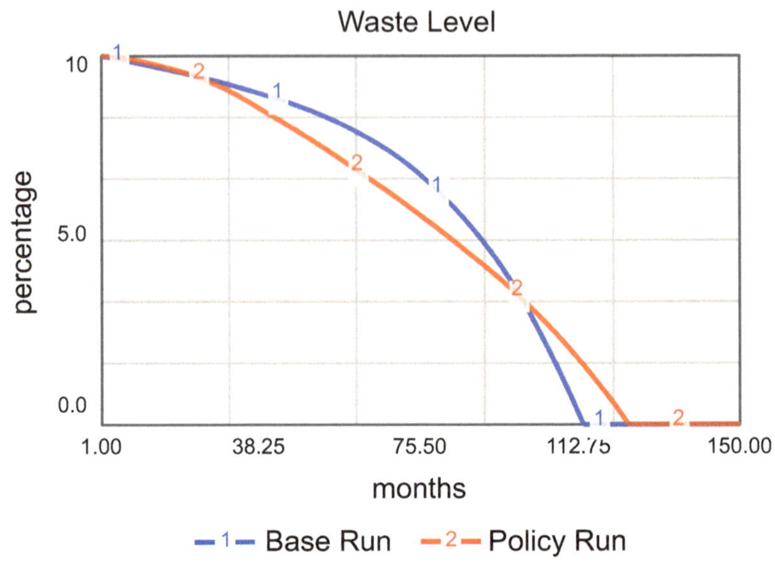

Fig. 6.29 *Base run and policy run 3* Policy on enhancing the technical base of employees by experimenting with the three parameters. First trainer cost is reduced from USD 1000 to USD 250 per trainer per day, second soft skills training need is reduced from 0.4 to 0.2 (40–20%), and third hard skills training participants increased from 10 to 30 persons; see the results of those policies

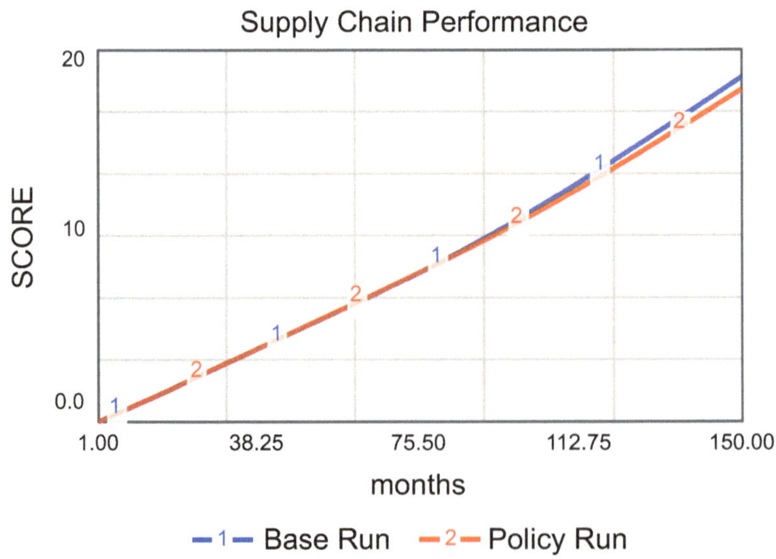

Fig. 6.30 *Base run and policy run 3* Policy on enhancing the technical base of employees by experimenting with the three parameters. First trainer cost is reduced from USD 1000 to USD 250 per trainer per day, second soft skills training need is reduced from 0.4 to 0.2 (40–20%), and third hard skills training participants increased from 10 to 30 persons; see the results of those policies

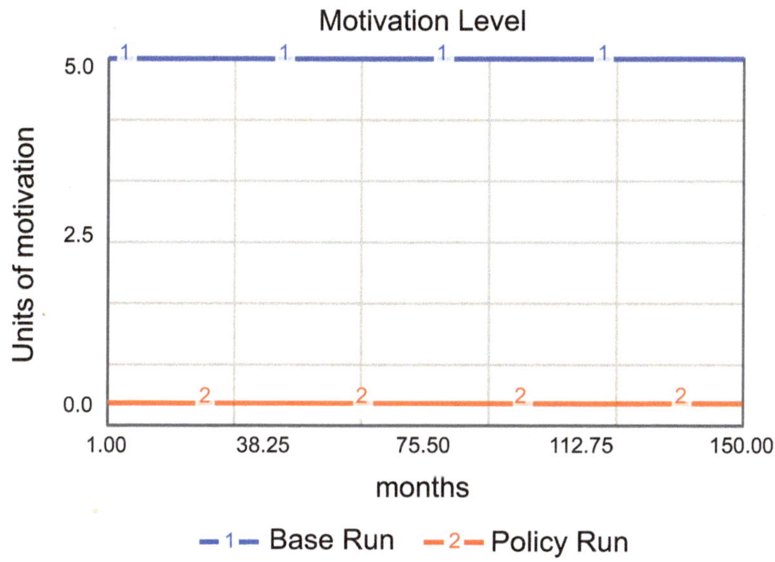

Fig. 6.31 *Base run and policy run 3* Policy on enhancing the technical base of employees by experimenting with the three parameters. First trainer cost is reduced from USD 1000 to USD 250 per trainer per day, second soft skills training need is reduced from 0.4 to 0.2 (40–20%), and third hard skills training participants increased from 10 to 30 persons; see the results of those policies

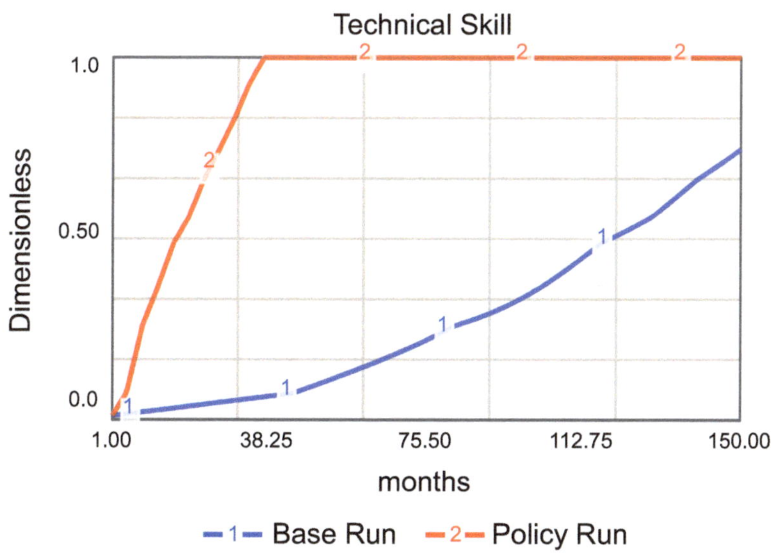

Fig. 6.32 *Base run and policy run 3* Policy on enhancing the technical base of employees by experimenting with the three parameters. First trainer cost is reduced from USD 1000 to USD 250 per trainer per day, second soft skills training need is reduced from 0.4 to 0.2 (40–20%), and third hard skills training participants increased from 10 to 30 persons; see the results of those policies

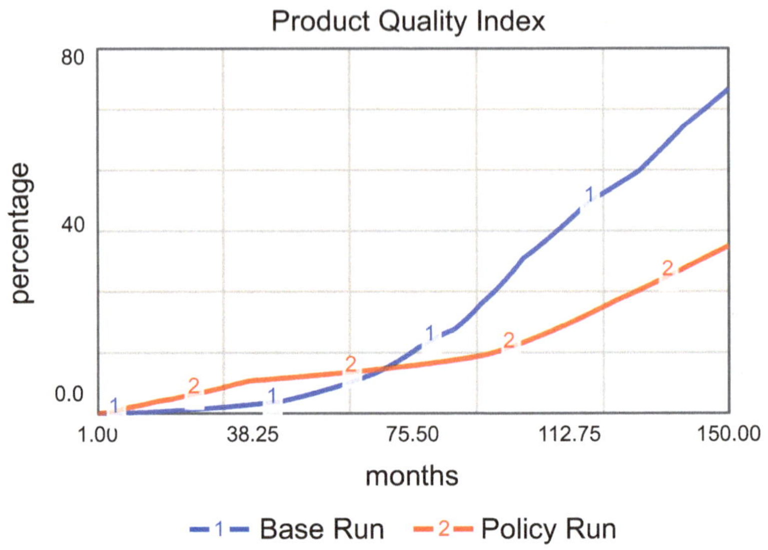

Fig. 6.33 *Base run and policy run 3* Policy on enhancing the technical base of employees by experimenting with the three parameters. First trainer cost is reduced from USD 1000 to USD 250 per trainer per day, second soft skills training need is reduced from 0.4 to 0.2 (40–20%), and third hard skills training participants increased from 10 to 30 persons; see the results of those policies

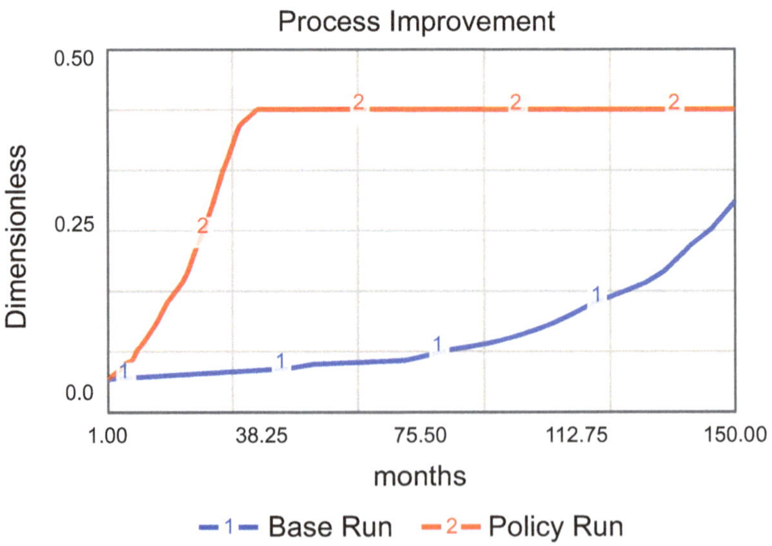

Fig. 6.34 *Base run and policy run 3* Policy on the enhancing the technical base of employees by experimenting with the three parameters. First trainer cost is reduced from USD 1000 to USD 250 per trainer per day, second soft skills training need is reduced from 0.4 to 0.2 (40–20%), and third hard skills training participants increased from 10 to 30 persons; see the results of those policies

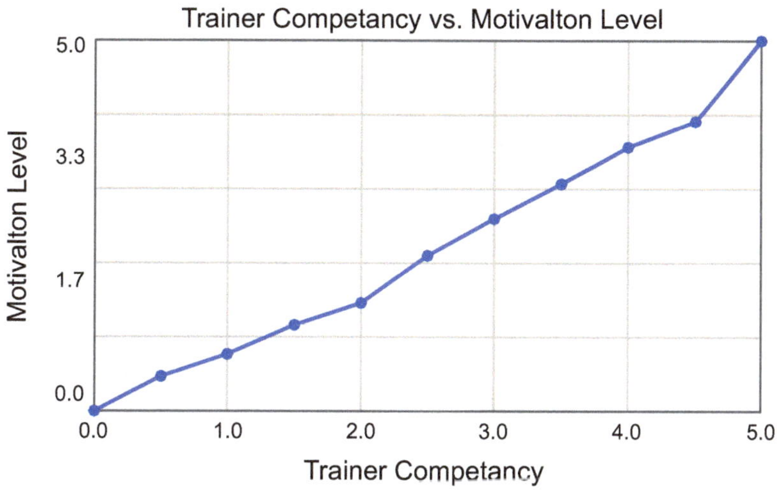

Fig. 6.35 Graphical function between motivation level and trainer competency

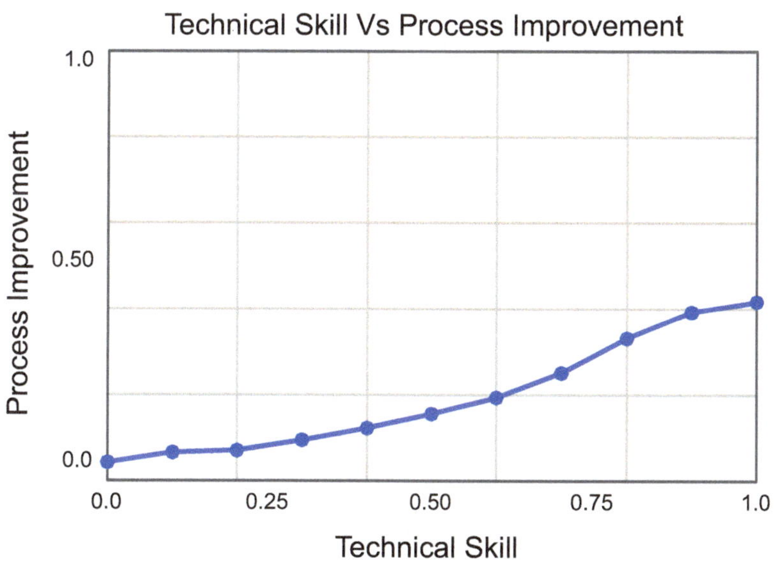

Fig. 6.36 Graphical function between process improvement and technical skill

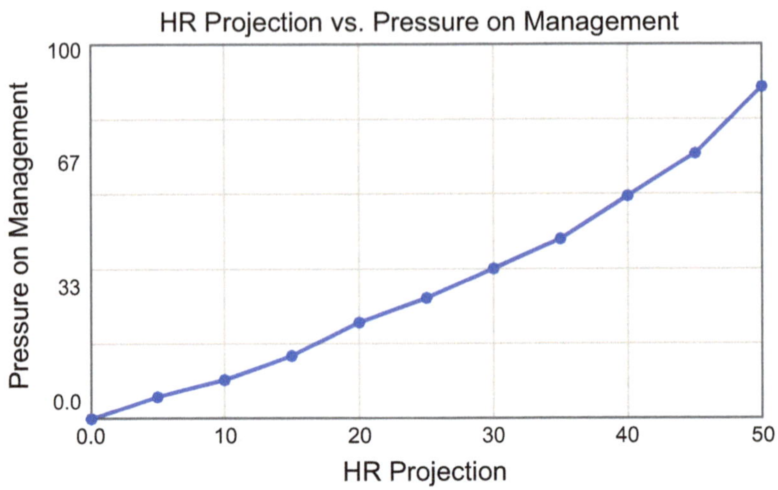

Fig. 6.37 Graphical function between pressure on management and HR projection

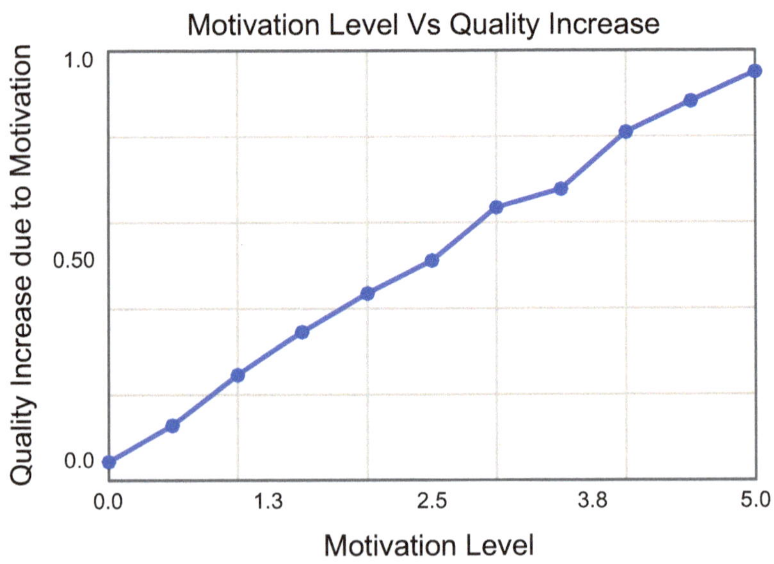

Fig. 6.38 Graphical function between quality increase due to motivation and motivation level

Fig. 6.39 Graphical function between cost of failures and waste level

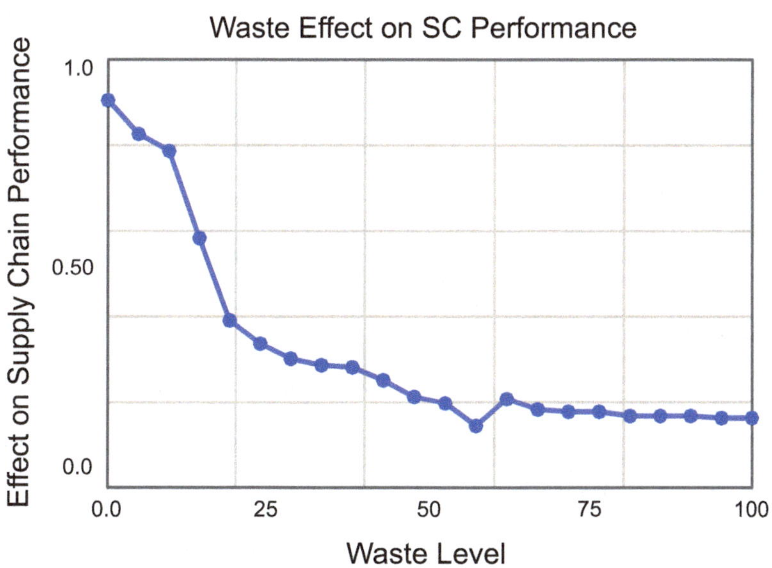

Fig. 6.40 Graphical function between effect on supply chain performance and waste level

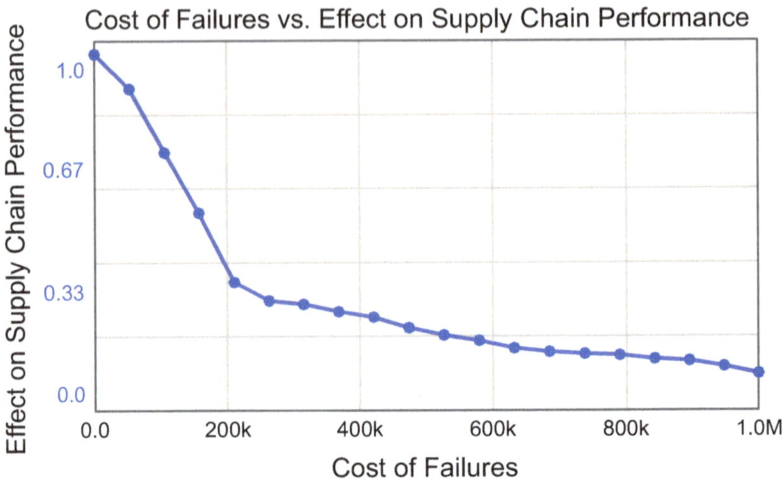

Fig. 6.41 Graphical function between effect on supply chain performance and cost of failures

References

Adm, et al. (1997). An International study of quality improvement approach and firm performance. *International Journal of Operations and Production Management, 17*(9), 842–873.

Ahire, S. L., Golhar, D. Y., & Waller, M. A. (1996). Development and validation of TQM implementation constructs. *Decision Sciences, 27*(1), 23–56.

Anderson, et al. (1995). A path analytic model of theory of quality management underlying the Deming management method: preliminary empirical findings. *Decision Sciences, 26*(5), 637–658.

Beamon, B. M. (1998). Supply Chain Design and Analysis: Models and Methods. *International Journal of Production Economics, 55*(3), 281–294.

Dow, D., Samson, D., & Ford, S. (1999). Exploding the myth: do all quality management practices contribute to superior quality performance. *Production and Operations Management, 8*(1), 1–27.

Duggan, J. (2008). Equation based policy optimization for agent-oriented system dynamics models. *System Dynamics Review, 24*(1), 97–118.

Flynn, B. B., & Flynn, E. J. (2005). Synergies between supply chain management and quality management: emerging implications. *International Journal of Production Research, 43*(16), 3421–3436.

Flynn, et al. (1995). The impact of quality management practices on performance and competitive advantages. *Decision Sciences, 26*(5), 659–691.

Forrester, J. W. (1961). *Industrial dynamics*. MIT Press.

Forrester, J. W. (1968). *Principles of systems*. MIT Press.

Forrester, J. W. (1987). 14 Obvious Truths. *System Dynamics Review, 3*(2), 156–159.

Forrester, J. W., & Senge, P. M. (1980). *Tests for Building Confidence in System Dynamics Models, TIMS Studies in Management Sciences* (Vol. 14, pp. 209–228). North Holland Publishing Company.

Kaynak, H. (2003). The relationship between total quality practices and their effects on firm performance. *Journal of Operations Management, 21*(4), 405–435.

Lane, C. D. (2007). The power of the bond between cause and effect: Jay Wright Forrester and the field of system dynamics. *System Dynamics Review, 23*(2–3), 95–118.

Meadows, D. L., et al. (1974). *Dynamics of growth in finite world.* MIT Press.
Morgan, D. L. (2007). Paradigms lost and pragmatism regained methodologies implications of combining qualitative and quantitative methods. *Journal of Mixed Method Research, 1*(1), 48–76.
Oliva, R., et al. (2006). Loop eigenvalue elasticity analysis: three case studies. *System Dynamics Review, 22*(2), 141–162.
Pidd, M. (2010). Why modelling and model use matter. *Journal of Operations Research Society, 61*, 14–24.
Qudrat-Ullah, H. (2008). Behaviour validity of a simulation model for sustainable development. *International Journal of Management and Decision Making, 9*(2), 129–140.
Qudrat-Ullah, H., & Seong, B. S. (2010). How to do structural validity of a system dynamics type simulation model: The case of an energy policy model. *Energy Policy, 38*, 2216–2234 Elsevier Journal.
Richardson, G. P. (1986). Problems with Causal-loop Diagrams. *System Dynamics Review, 2*(2 Summer), 158–170.
Richmond, B. (1993). Systems Thinking: critical thinking skills for the 1990 and beyond. *Systems Dynamics Review, 9*(2), 113–133.
Saeed, K. (1991). Slicing a complex problem for system dynamics modelling. AIT, IE&M#275.
Saeed, K. (2008). *Towards sustainable development: Essays on system dynamics of National Policy*. Lahore: Progressive Publishers.
Sila, I., & Ebranhimpour, M. (2005). Critical linkages among TQM factors and business results. *International Journal of Operations & Production Management, 25*(11), 1123–1155.
Sterman, J. D. (1987). Testing Behavioural Simulation Models by Direct Experiment. *Management Science, 3*(12), 1572–1592.
Sterman, J. D. (2000). Business dynamics systems thinking and modeling for a complex world. published by Irwin McGraw-Hill, USA.
Sterman, J. D. (2007). Exploring the next great frontiers: system dynamics at fifty. *System Dynamics Review, 23*(2–3), 89–93.
Sushil. (1993). System dynamics a practical approach for managerial problems, published by H.S.Popali for Wiley Eastern Limited, New Delhi.
Tan, et al. (1999). Supply chain management: an empirical study of its impact on performance. *International Journal of Operations and Production Management, 19*(10), 1034–1052.
Towill, D. R. (1996). Industrial dynamics modelling of supply chains. *International Journal of Physical Distribution & Logistics Management, 26*(2), 23–42.
Yusuf, I., & Azhr, T. M. (2017). Dynamics of costs of quality: The case of Expresspac Pvt Ltd., 2nd International Conference on Dynamic Innovation held at MUST University in Kuala Lumpur, Malaysia 2017.
Yusuf, I., & Ramish, A. (2011). Modelling the factory dynamics: The case of CCRS, International Conference on Business held at UCP, Lahore 2011.
Zagonel, A. A. (2002). Model conceptualization in group model building, proceedings of 2002 international system dynamics conference, Palermo, Italy.

Chapter 7
The Application of System Dynamics for a Sustainable Procurement Operation

Sherif Barrad, Raul Valverde, and Stéphane Gagnon

7.1 Introduction

System dynamics is an approach to modeling complex systems using feedback loops to explain relationships between variables and to reflect their nonlinear interdependencies through time, along with their underlying driving forces (Sterman 2000). This chapter has the objective of reviewing system dynamics and the supply chain and discussing sourcing and procurement functions in relationship to cost savings via sustainable and green cost-reduction strategies and initiatives.

The chapter raises awareness on the various factors enabling or impeding organizations in reaching a sustainable procurement operation. For example, a key issue in sustainable procurement is procurement's capacity to select the most appropriate roster of suppliers and negotiate the best possible deals due to their inability to engage in negotiations with the entire supply base on a regular basis (Winter and Knemeyer 2013; Varsei et al. 2014). If a procurement function has several thousands of suppliers and must renew 500+ contracts on an annual basis, they may not have the bandwidth to engage in the end-to-end sourcing cycle for contracts coming up to renewal. Another factor discussed is how procurement is constrained by business and corporate strategies and the negative and reinforcing system behavior that ultimately leads to value leakage, hence the incapability to reach sustainable value creation.

S. Barrad
Massachusetts Institute of Technology, Cambridge, MA, USA
e-mail: sbarrad@sloan.mit.edu

R. Valverde (✉)
Concordia University, Montreal, QC, Canada
e-mail: raul.valverde@concordia.ca

S. Gagnon
Université du Québec en Outaouais, Gatineau, QC, Canada
e-mail: stephane.gagnon@uqo.ca

H. Qudrat-Ullah (ed.), *Innovative Solutions for Sustainable Supply Chains*, Understanding Complex Systems, https://doi.org/10.1007/978-3-319-94322-0_7

We conclude this chapter with an interpretation of our systems dynamics model, emphasizing the challenge of fine-tuning such a system in real-life operations. We also infer as to what extent cognitive computing, especially through the merging of structured and unstructured data, may help automate system balancing and sustainability objectives as it relates to contract management and optimization.

7.2 Literature Review

7.2.1 System Dynamics

System dynamics (Sterman 2000) is a method aimed at identifying and understanding behaviors between variables in a system in order to understand how the world works as a whole. These behaviors are typically part of a larger system and can be analyzed not only as a subsystem but also as an actor within a larger system.

Causal loop diagrams are diagrams that depict relationships between variables. These causal loops can be positive (reinforcing) or negative (balancing). The strength of these relationships can also vary over time. A positive relationship is typically annotated with a "+" sign and arises when an increase in one variable causes an increase in another variable. These positive causal relationships are also known as reinforcing loops as one behavior is reinforcing another. In contrast, a negative relationship (also referred to as a balancing loop) occurs when an increase in one variable creates a decrease on another variable. A negative relationship is typically annotated with a "-" sign and can also vary over time. Arrows are drawn in a circular manner indicating the causes and effect leading to a feedback loop which consists of a closed sequence of cause and effects (Tulinayo et al. 2012).

Applying this concept, from a qualitative aspect, within a procurement operation, an increase in the amount of "spend under management" otherwise known as the spend being managed by the "procurement" function (i.e., purchasing/sourcing experts) typically leads to an increase in the likelihood that savings will be captured following expert intervention. Procurement includes a company's requisitioning, purchasing, transportation, warehousing, and inbound receiving process (Stephens and Valverde 2013); this can be explained by the sourcing expert's ability to analyze the category of spend in question and develop an intelligent and fact-based sourcing strategy ahead of supplier negotiations. By better understanding the internal organization history of spend with a key supplier, assessing the supply market for competitive offers, decomposing and understanding the cost structure to clearly understand which cost elements can be negotiated, procurement can enter supplier negotiations with a dominant position. This position can enable them to negotiate better pricing by committing to larger volumes, harmonizing product lists, concentrating/consolidating purchases for volume discounts, etc. Assessing the supply market for competitive offers can enable sourcing experts to be informed of their bargaining position and leverage this during supplier negotiations. For example, procurement

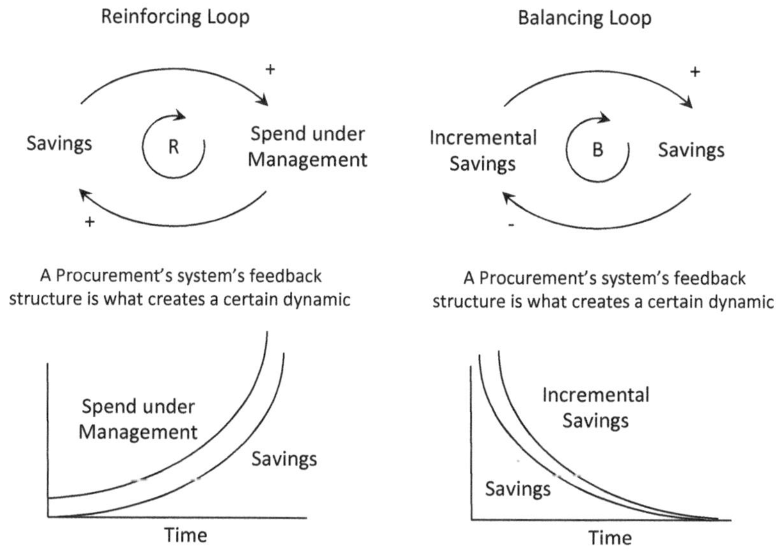

Fig. 7.1 Reinforcing and balancing loops in a procurement system

can uncover that the business offered to a certain supplier represents over 50% of the supplier's total sales volume and as such command lower costs or better value.

Another dynamic within the same procurement system emphasizes the positive relationship of increased spend under management as a result of increased savings. This can be explained by suggesting that the more value (i.e., savings, better purchasing options, etc.) procurement is able to deliver, the increased likeliness that the function will become more solicited from the various internal functions. To further illustrate this dynamic, perhaps a simple and straightforward example, when a business unit (i.e., marketing, operations, manufacturing, etc.) is able to gain more from suppliers with the same budget, this allows them to free up budget to purchase other necessities once constrained by their current budget. Improving supplier value (i.e., negotiating aggressive cost reductions, improved service levels, etc.) allows for more options. Therefore, the procurement function can deliver value by expanding options for the function, and this in turn is a rewarding process as it generates more dependence on the function. The dynamic results in an increase in the amount of spend under management by procurement. The figure below (Fig. 7.1) attempts to demonstrate the positive relationship between savings and spend under management as well as the balancing relationship between savings and incremental savings.

It is important to note that the positive system dynamic between "spend under management" and "savings" cannot go on infinitely. There will be constraints once the system has reached full capacity. This is where the concept of the "balancing loops" comes into play (see right-hand side of Fig. 7.1). In a balancing loop, we have attempted to demonstrate that the increase in incremental savings will eventually taper off. The savings a procurement function can generate is not infinite.

In other words, you can only negotiate suppliers for savings up until a certain point. After that point, incremental savings begin declining to a point where no more incremental savings can be achieved as the supplier has been fully "squeezed" and can no longer operate sustainably. Hence, incremental savings begin declining once the entire supplier based has been fully optimized/negotiated.

Applying the same concept, from a quantitative aspect, as the requests come into the procurement department, there are delays involved, and these requests also have an impact on capacities (staff). Depending on what the request might be (i.e., negotiate pricing with an already selected supplier as opposed to sourcing a vendor all the way to operationalizing a contract), the average length of the delay must be considered. The outputs around the average delay time must also be assessed.

In terms of loop polarity, when the polarity is positive, we can refer to it as a gain. The gain strength will depend on the correlation between variables. If a gain is evaluated at 0.5, this means that an increase in the initial variable creates a gain in the correlated variable by half the amount. These loop gains can be calculated.

This can be interpreted using mathematics in the following way. Considering all else is equal, if variable A increases (or decreases), then Y increases (or decreases) above (below) what it would have been (Sterman 2000):

$$\frac{\partial Y}{\partial X} > 0$$

and in the case of accumulations,

$$Y = \int_{t0}^{t} (X + \ldots) ds + Y_{t0}$$

$$Y_{t0} = \text{initial value of } Y \text{ at initial time } t0$$

If we take, for example, a loop gain in the dynamics of a sustainable procurement dynamics, we could assume that if the number of requisitions that are funneled through procurement increase, then savings would also increase given procurement's intervention in the end-to-end sourcing event.

On the flip side, and when all else is equal, if X increases (decreases), then Y decreases (increases) below (above) what it would have been:

$$\frac{\partial Y}{\partial X} < 0$$

and in the case of accumulations:

$$Y = \int_{t0}^{t} (-X + \ldots) ds + Y_{t0}$$

7.2.2 Supply Chain Management

Supply chain management consists of all parties involved, directly or indirectly, in fulfilling a customer requests (Chopra and Meindl 2009). The supply chain deals with the flow of material, information, and money. SCM can also be seen as the oversight of materials, information, and finances as they move in a process from supplier to manufacturer to wholesaler to retailer to consumer (Robertson et al. 2002). The management of the supply chain is emerging as an important strategic factor (Valverde and Saadé 2015) that is critical for industries with global operations (Valverde et al. 2016).

7.2.3 Sustainable Supply Chains

The sustainable supply chain extends beyond the traditional supply chain by integrating both environmental and social issues (Varsei 2016). In the traditional supply chain, there has always been the notion of classical trade-offs such as the one between production, inventory, and customer service levels. That is, increasing production capacity and warehousing costs as a trade-off in order to meet 99% customer service levels. The variables, when integrated in sustainable supply chains, have their own constraints and trade-offs. In a sustainable supply chain, the challenge is no different when time comes to manage between the three dimensions of sustainability (Winter and Knemeyer 2013; Varsei et al. 2014).

7.2.4 Procurement

Procurement is a key function within the supply chain and is now a significant area of cost that is garnering greater attention from both practitioners and scholars (Husted and Reinecke 2009). The notion of procurement and/or strategic sourcing as it relates to expenditures is often neglected by the boardroom (Booth 2010). Procurement requires collaboration as it extends outside the organization and across the supply chain by interacting with suppliers (Avédissian et al. 2015). It is not unusual, in a high-capital-intensive environment (i.e., manufacturing), most organizations spend at least half their costs with suppliers – often for business-critical goods and services. An effective procurement program that masters the system dynamics of a sustainable procurement operation can deliver direct impact to the P&L.

Table 7.1 Factors inhibiting good contract management processes

Contract management process	Assessing requirements	Drafting a contract	Managing costs and supplier value	Fulfilling orders
Challenges	Delays in providing information	Contract terms inappropriate and applied at the incorrect moment	Procurement procedures bypassed in order accelerate the buying process sacrificing higher acquisition costs	Cumbersome authorization process
	Communication across departments/ business units	No culture of contractual certainty	Limited access to historical pricing information	Terms and conditions disregarded by suppliers
	Lack of standardization	Legal process often bypassed	No incentive to control costs	Inadequate monitoring of the renewal of agreements
		Suppliers often draw up contract and specifies terms and conditions		

7.2.5 *Contract Management*

Contract management within the space of procurement deals with the effective negotiation and management of contracts. It is also referred to as the optimal implementation of contractual clauses (Schuhmann and Eichhorn 2015). Contracts are typically split into two key areas: the first being the statement of work (SOW) where detailed description of products and/or services are described and the second being the commercial side of the contract which refers to the monetary/financial aspect of contracts. In any procurement organization, the activity of managing contracts involves a series of processes and tasks, and several challenges exist within the contract management practice (Table 7.1).

7.2.6 *Analytics and Business Intelligence*

Some of the most sophisticated firms are deeply connected to the data as they truly see the value of data in supporting both strategic and operational decision-making. These firms build spend cubes to increase the flow of information and visibility across the entire supply chain (Viswanathan and Sadlovska 2010). The purpose of business analytics, especially when it comes to procurement and sourcing, is to facilitate the firm's ability to analyze trends and to predict future purchasing patterns for increased leverage through negotiations. In fact, Turban et al. (2011) refer to it as a broad range of technologies, analytical techniques, and methodologies combined to support decision-making. Souza (2014) suggests that supply chain analytics focuses on the use of information and analytical tools to make better decisions regarding material flow in the supply chain. Rafati and Poels (2015) suggest that the organization requires two critical competencies to enable fact-based decision-making: data management, which includes data architecture, extraction, transformation, movement, storage, integration, and governance, and data analytics which includes modeling (statistical, contextual, quantitative, predictive, cognitive, etc.), deep computing, simulation, data mining, and optimization. The ultimate objective is to find information that one would otherwise not be able to find while running daily operations. Some examples include, but are not limited to, the percentage (%) of spend not being purchased under existing contracts. This information is extremely useful as simple policy enforcement can generate significant savings for the organization.

Analytics techniques can be categorized into three types: descriptive, predictive, and prescriptive (Souza 2014). The first, descriptive analytics, gathers enormous amount of data to describe the current state. For example, in sourcing, you can determine what is your current spend year to date and validate results against the forecast to make sure you don't go over budget. Predictive, on the other hand, gathers past data and manipulates this dataset, using methods such as exponential smoothing and regression analysis, to predict what will happen in the future. An example would be forecasting where a firm would assume that demand for office supplies is seasonal with a peak around back-to-school. Finally, prescriptive analytics emphasizes mathematical optimization, drawing from both descriptive and predictive analytics to model the future. According to Souza (2014), there is limited use of analytics for strategic sourcing. One of the most common techniques in sourcing includes multi-criteria decision-making (MCDM) where a series of criteria are established and weighed to contribute to a final score. In their work (Barrad and Valverde 2015), suppliers were assessed in terms of risk by using key stock information and thus generating a risk score. They suggested that in order to improve the accuracy of the model, an MCDM would have to be considered, whereby firms are scored on a series of criteria such as industry, geopolitical, geography, liens, judgments, proceedings, etc.

7.3 Problem Definition

System dynamics, within the sourcing division of a supply chain function, is one of the most important challenges in today's global supply chain. There exists great opportunity for the sourcing and procurement department to capitalize on their potential to generate cost savings through sustainable cost-reduction initiatives. Today, this is not always the case in every organization, and there are several reasons for this. First, procurement does not always have the capacity to enter into negotiations with the entire supply base on an annual basis – some organizations deal with several thousands of suppliers with thousands of renewals on a monthly basis – they just don't have the resources to manage contracts expiries. Second, and most importantly, they do not have the capability to strategically negotiate with suppliers. This causes a negative and reinforcing system dynamic that ultimately leads to value leakage.

Procurement organizations are sometimes prioritizing tactical sourcing events which typically consist of reviewing contract clauses and trying to negotiate price without understanding internal spend and purchasing patterns, cost structures, or even understanding industry dynamics (i.e., Porter's five forces). This makes it extremely difficult to identify and secure savings – contracts are renewed year over year with little negotiation and almost no savings as a key output. The root cause for this is that employees have not been trained in strategic sourcing, and the result is that millions of dollars are being left on the table each year.

Today, many organizations do not leverage their size to purchase goods and services. This creates many sustainability issues across the entire supply chain. Instead, all business units operate independently, and purchasing is decentralized. Most products and services are purchased through independent non-optimized contracts. This increases the costs of doing business, and these costs are reflected in the prices paid by customers.

To appropriately leverage its size, "procurement" must overcome one key challenge which is to position procurement, within the rest of the organization, as a center for value creation. Today, this is not the case, and this creates a significant problem as it reduces the amount of spend that is passed through procurement for value creation (decentralized procurement model). Several impacts are associated to this such as poorly negotiated contracts, fragmented spend which minimizes volume discounts, etc.

Procurement must increase the value by creating strategic alliances and by improving the relationship between suppliers. By successfully doing this, procurement can increase internal credibility by providing expertise. Once it does, this will have a direct positive impact on the % of spend going through procurement. This will in turn grow the department.

7.3.1 Key Variables Driving Growth in Procurement

1. % of potential spend funneled through procurement (spend handled by procurement ($)/total spend ($))
2. Confirmed savings ($ and %)
3. Productivity ($ cost/$ revenue)
4. Growth

7.3.2 *Reference Modes*

7.3.2.1 Percentage of Total Spend Funneled Through Procurement

In many organizations, less than 100% of the spend are funneled through procurement (i.e., spend under management). This may be because procurement is still not yet viewed as a "value creation" department. Moreover, the organization does not yet see the links between cost reduction → improved efficiency → growth (increased market share). The dynamic suggests that the more spend goes through procurement, the higher the likelihood of negotiating optimized contracts and reducing overall supplier-related costs.

Another important notion to consider when assessing the amount of total spend funneled through procurement is the concept of delays.

Delays result from deltas (difference) between inflows and outflows. If the inflow is greater than the outflow, delays will occur. This can be described mathematically as follows:

Outflow = DELAYn (Inflow;D):

Total Requisitions Being rocessed = $\sum_{i=0}^{n} \textit{Requisitions in process.}$

Requisitions Being Processed$_i$ = INTEGRAL (NetInflow$_i$, BeingProcessed^i (0))

Being Processed$_i$ (0) = Inflow * D/n

$$\text{Net Inflow rate}_i = \begin{cases} \text{Exit Rate Stage}_{i-1} - \text{Exit Rate Stage}_i & \text{for } I \in (2,\ldots n-1) \\ \text{Inflow} - \text{Exit Rate Stage}_1 & \text{for } i=1 \end{cases}$$

$$\text{Exit Rate Stage}_{n-1} - \text{Outflow} \qquad \text{for } i = n$$

Exit Rate Stage$_i$ = Requisitions being processed/(D/n) for $I \in (1,\ldots, n-1)$

Outflow = Requisitions being processed$_n$ /(D/n)

7.3.2.2 Confirmed Savings ($ and %)

The objective is to concentrate to overall volumes at top suppliers for volume discounts (i.e., shifting the smaller and fragmented volumes of spend from smaller suppliers and allocating it to the larger ones in return for volume discounts). A key objective could be to reduce total spend by supplier through improved demand management (specification harmonization, relaxing service-level agreements (SLAs), demand reduction, etc.) and through aggressive supplier discounts via negotiations. The objective would be to hit a specific percentage point (%) in savings.

7.3.2.3 Productivity

Organizations typically measure productivity as a cost to revenue ratio. For example, it may cost a bank $0.57 earnings per share (EPS) to generate $1 of revenue. The objective could be to reduce this to $0.53. Cost efficiencies would need to be

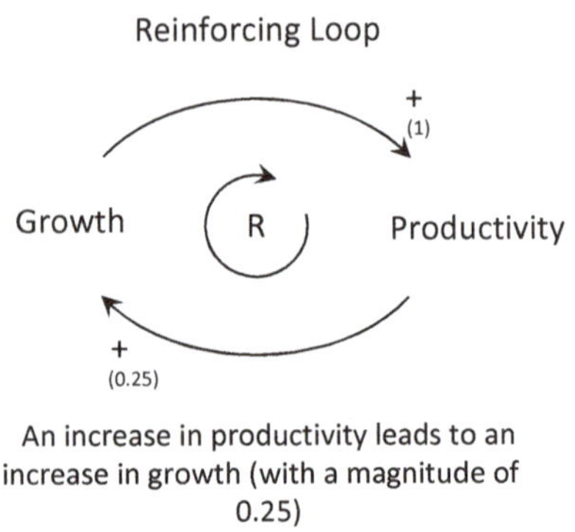

Fig. 7.2 Reinforcing relationship between productivity and growth

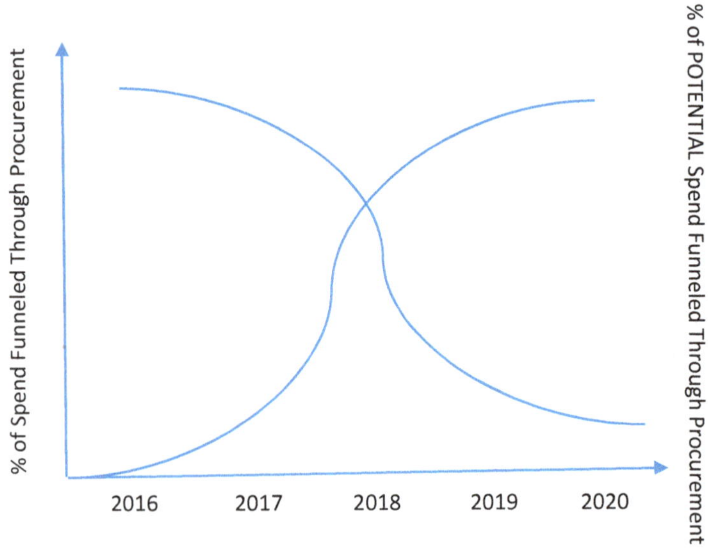

Fig. 7.3 Shifting to a centralized procurement model

take place in order for the bank to hit productivity targets (or the $0.04 EPS reduction). For example, depending on the financial statements, the $0.53 could be reached once the bank hits $100 M of savings over a certain period of time. In other words, $25 M of savings translates into $0.01 of productivity improvement (or $100MM in savings for $0.04 increase in EPS). If we relate productivity to growth, another assumption would be to increase productivity by $0.01 and to secure 0.25% of growth. This could also be interpreted as a positive (reinforcing loop with a strength of 25%). See Figs. 7.2 and 7.3.

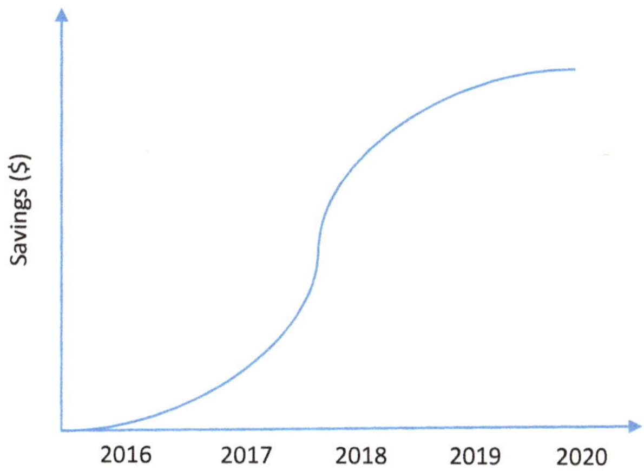

Fig. 7.4 Reinforcing relationship between spend under management and savings

7.3.2.4 Growth

The department's growth depends on the efforts invested in marketing their service as a key internal service and their ability to execute once they're solicited by internal business units. If they are able to generate demand for their services and are able to deliver once challenged, they will become more and more solicited and hence will become a key business unit within the bank, hence fueling the growth of the department.

7.3.2.5 Time Horizon

Five years is the typical time required for a new and sustainable system dynamic to stabilize and reach anticipated benefits. At full maturity, the organization should be able to generate annual recurring savings in a sustainable way. This should contribute to the departments growth as the more spend is funneled through procurement, more resources will be required to continue creating value.

Five years is also the amount of time required to transform the culture of the organization where internal business units start leveraging procurement as a center of excellence (CoE) and where procurement starts focusing on strategic activities rather than tactical purchasing processes (e.g., issuing RFIs, RFPs, scoring suppliers, etc.).

Channel more spend through procurement to eventually eliminate the decentralized procurement model.

Channeling spend has a positive impact on savings – the more spend is funneled through procurement (spend under management), the greater the possibility of applying sourcing expertise to generate savings (Fig. 7.4).

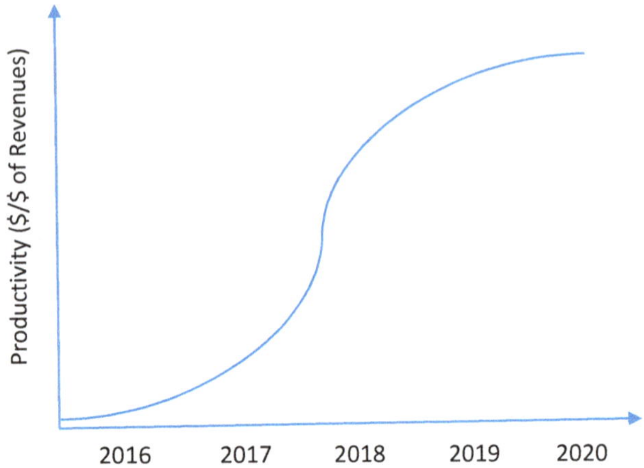

Fig. 7.5 Reinforcing relationship between productivity and growth

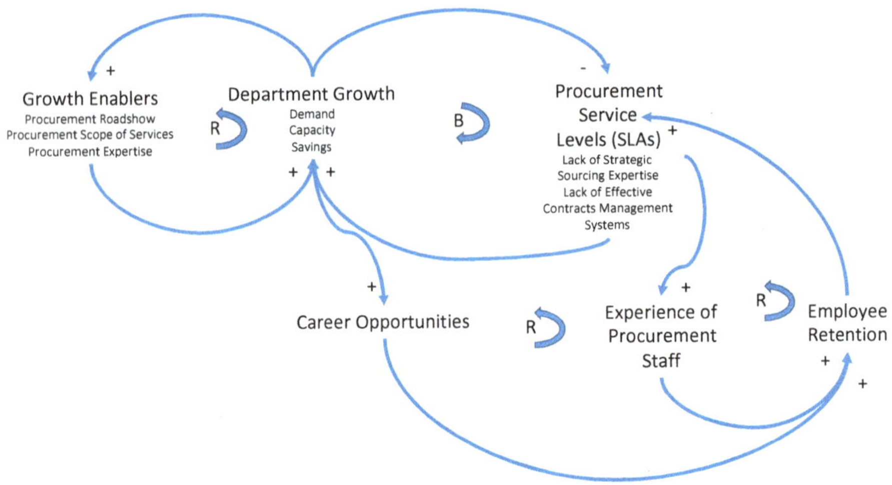

Fig. 7.6 Determinants of procurement attractiveness

Savings generated from a procurement program have a positive direct contribution to the efficiency program's $100 M objective and therefore would increase productivity to reach the $0.53 objective (Fig. 7.5).

7.4 Causal Diagrams and Determinants of Procurement Attractiveness

In the system dynamic figure below, there are a total of three reinforcing loops (R) and one balancing loop (B). These loops are further explained in the next paragraph (Fig. 7.6).

1. **R1:** As procurement engages in doing a road show and creating awareness in terms of their potential contribution, the more solicited they will become. Also, the more procurement expands its scope of services (by covering more categories of spend), the more they'll become valued in the rest of the organization. This will create departmental growth. This growth will become procurement's fuel to grow even larger and serve more internal clients.
2. **R2:** As the department grows, it will become more attractive given the strategic work they do (shifting away from pure tactical sourcing). This will create interest from employees operating in other business units who will now consider joining procurement. This will in turn promote employee retention. Employees that stick around will enjoy satisfaction from the work they do, and through their satisfaction, procurement secures good service levels which will increase the participation in sourcing deals.
3. **R3:** Delivering great service levels to internal clients will in turn promote a great environment where procurement staff will be satisfied with the work they perform not to mention the recognition they will receive for their work. This will induce employee retention, and service levels will continue to rise.
4. **B1**: As procurement grows and, as their scope of services expands, service-level agreements (SLAs) (between procurement and internal business functions) will eventually take a hit. Once it does, some internal customers (especially the ones that have been impacted by negative experiences) may decide to go their own way and purchase independently going forward (going back to a partially decentralized procurement operating model).

The most important predictor of success in the procurement department lies in its ability to shift their focus to strategic activities as opposed to tactical activities such as getting contracts signed and entered into a system. By becoming more strategic, the procurement department can generate millions in savings for the organization through strategic supplier negotiations. There is clearly a system dynamic that exists. This dynamic must be understood and managed in order maximize supplier value and minimize contract leakage.

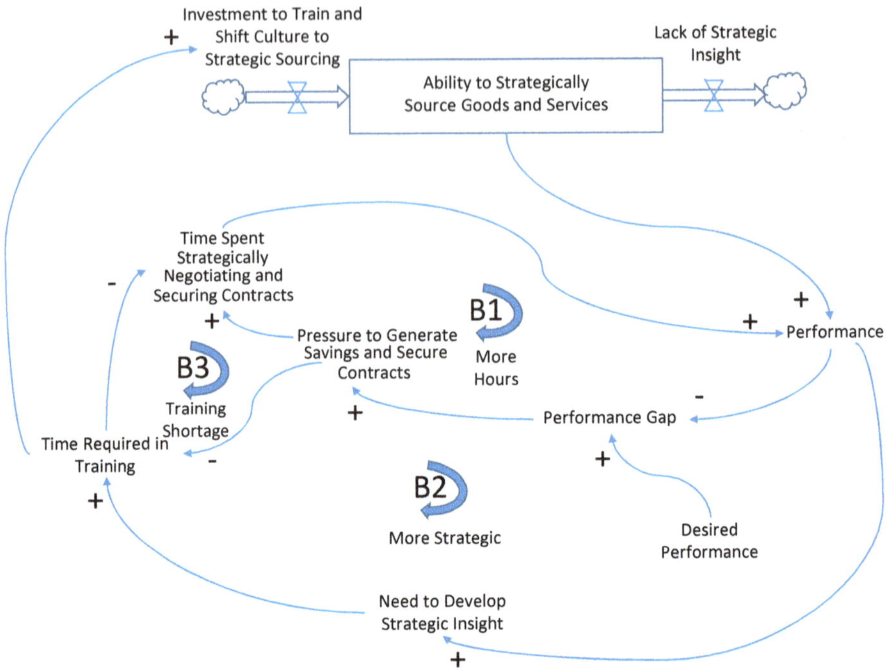

Fig. 7.7 The capabilities trap

7.5 Procurement Dynamics and the Capabilities Trap

The figure on the next page attempts to describe the system dynamics behind the capabilities trap in procurement and outlines the trade-off a procurement organization must make in order to shift its culture from a tactical-based operation toward a strategic mindset. Capabilities trap is a term coined by Sterman (2000). When a company has a set of resources committed to a product and it doesn't meet a target quality, there is a tendency to redouble efforts to meet that target, but by doing this, the company becomes stuck in a "work harder loop" to solve a problem that the company might not have the capability to fix. System dynamics theory provides an interesting lens to view this challenge known as the capabilities trap (Sterman 2000). The figure below uses the concept of capabilities trap proposed by Sterman (2000) applied to the procurement process (Fig. 7.7).

B1 The lack of training leads to continuously having to review and optimize supplier contracts over shorter periods of time (e.g., they negotiated a good price but forgot to negotiate volume discounts or extended payment terms). This multiplies the efforts required to complete a sourcing event. That is, by the time theses clauses are renegotiated, dollars have been left on the table. The staff spends too much time working on tactical activities which takes away the time they have to take a step back and think about categories and suppliers more strategically.

If staff can approach opportunities more strategically, they would spend less time doing day-to-day activities and would generate much more value to the organization. If the staff is not trained adequately to perform strategic sourcing, they cannot effectively deliver value, their capabilities decreases, and this has a direct impact on performance (creating a larger gap between actual and desired performance). This creates more work as contracts aren't optimized and are in continuous need to be rereviewed.

B2 Because performance is limited and decreases, the organization starts realizing the opportunity costs of poor sourcing decisions. It then needs to invests in consultants to come and train staff on strategic sourcing. This requires the staff to leave their daily routine and to go through the training in sourcing (which also requires them to attend workshop and produce certain analyses). Once the staff understands the value of approaching suppliers strategically, performance then increases.

B3 When performance decreases, this creates an even larger gap in performance expectations. This can potentially slow down operations because contracts aren't being managed appropriately and suppliers may end up causing severe supply disruptions. To avoid this, we would require the staff to let go of training in order to deal with emergencies (which is actually the case at the moment). As staff handles emergencies, less time is devoted to training and developing strategies, and this only delays the strategic value they could have created through capability development.

The net savings the procurement organization would be able to generate on an annual basis would represent very little of the total spend because of the lack for strategic insight to analyze opportunity from a strategic lens.

7.6 Application of System Dynamics for a Sustainable Supply Chain

To address the capabilities trap, as part of the entire system dynamics, several options exist. The first option is to hire consulting firms specialized in sourcing and procurement to perform a high-level diagnostic of the procurement function which would also include an evaluation of the team's procurement maturity and capabilities.

Following this diagnostic, the organization would kick off an exercise (which would also end up becoming an annual procurement process) called "strategic sourcing," a sequential methodology to systematically identify savings. This initiative would cost the organization up-front but would secure a prioritized pipeline of opportunities.

In this activity, the external partner would solicit procurement staff to participate in the activity which aims at developing their sourcing capabilities (typically category managers who know enough about the category to be able to brainstorm on pragmatic strategies).

A typical organization may have between 5 and 20 spend categories split between business and IT. The organization would start with either one and request that each of the category managers be dedicated to the exercise at 50% of their time for a period of 12 weeks. Each category manager would have to build a "category profile" which includes all the category info including strategies (e.g., spend analysis, cost structure analysis, total cost of acquisition analysis, major players, industry trends, and, finally, potential procurement levers to generate savings, i.e., demand management, supplier consolidation, specification harmonization, joint process improvement, etc.).

The external partner would then meet with each of the category managers once a week to review progress over a 60-min period. During the last 4 weeks, they would set up four 3-h workshops (one per week) to brainstorm on strategies and package the final pipeline of opportunities for each category. The incumbent would make sure the organization has gone through the entire strategic sourcing cycle to truly understand how to identify savings and initiatives and ultimately manage a category strategically.

In order to finance this effort, all category managers would have to identify at least a certain amount of "Quick-Hits" initiatives in each of their categories for an ROI ratio of at least 2:1. "Quick-Hits" would consist of initiatives that are easily implementable within 3 months and that have already been approved by management for a "go-ahead."

The notion of cognitive could also be integrated in order to leverage this effort. Take, for example, technologies such as IBM's Supplier IQ which attempts to merge structured data (contract clauses, spend data, supplier performance) along with unstructured data (market information, price benchmarking, news events, financial and stock information, etc.) in order to identify potential risks in contracts and allow procurement to focus on efforts on managing high-risk suppliers. Another example is the Barrad and Valverde (2015); the articles propose a web services architecture that takes public financial information in order to estimate risk of bankruptcy of suppliers for procurement transactions.

7.7 Conclusion

Achieving a sustainable procurement operation is no easy task. There are several intricacies involved in managing this complex process within the entire supply chain operation. Procurement has typically been seen as a cost center with little influence on value creation. This results in many inefficiencies such as decentralized procurement operating models, fragmented spend, maverick buying, etc. In order to truly understand the dynamics involved in managing a complex procurement operation, several of the key variables (spend under management, savings, efficiency, productivity, growth, etc.) were analyzed as key variables within a system as a whole. We attempted to describe, using causal diagrams, the causality between variables from both a qualitative and quantitative standpoint. We discussed the capabilities trap and the need to isolate strategy from operations by dedicating resources which are entirely focused on strategic procurement coupled with industry experts (external support) to help baseline spend and procurement processes to set the tone and deliver a road map for improvement.

References

Avédissian, A., Valverde, R., & Barrad, S. (2015). An extension proposition for the agent-based language modeling ontology for the representation of supply chain integrated business processes. *WSEAS Transactions on Business and Economics, 12*, 211–228.

Barrad, S., & Valverde, R. (2015). Mitigating supply chain risks by evaluating supplier bankruptcy probabilities through web services and the Black-Scholes-Merton model. *WSEAS Transactions on Information Science and Applications, 12*, 186–193 ISSN 2224–3402.

Booth, C. (2010). *Strategic procurement: organizing suppliers and supply chains for competitive advantage* (p. 2010). London, Philadelphia: Kogan.

Chopra, S., & Meindl, P. (2009). *Supply chain management: Strategy, planning, and operation* (4th ed.). Saddle River: Prentice Hall.

Husted, C., & Reinecke, N. (2009). Improving public sector purchasing. *McKinsey Quarterly, 4*(2), 18–25.

Rafati, L., & Poels, G. (2015, February). Towards model-based strategic sourcing. In *International workshop on global sourcing of information technology and business processes* (pp. 29–51). Cham: Springer.

Robertson, B., Fagerhaug, T., Randmoel, S., Schuldmaier, J., & Prenninger, J. (2002). Benchmarking supply chain management: Finding best practices. *Journal of Business & Industrial Marketing, 14*(5/6), 378–389.

Schuhmann, R., & Eichhorn, B. (2015). From contract management to contractual management. *European Review of Contract Law, 11*(1), 1–21. https://doi.org/10.1515/ercl-2015-0001.

Souza, C. G. (2014). Supply chain analytics. *Business Horizons, 57*(5), 595–605.

Stephens, J., & Valverde, R. (2013). Security of e-procurement transactions in supply chain reengineering. *Computer and Information Science, 6*(3), 1.

Sterman, J. (2000). *Business dynamics: systems thinking and modeling for a complex world* (p. c2000). Boston: Irwin/McGraw-Hill.

Tulinayo, F., van Bommel, P., & Proper, H. (2012). From a system dynamics causal loop diagram to an object-role model: a stepwise approach. *Journal of Digital Information Management, 10*(3), 174–186.

Turban, E., Sharda, R., Delen, D., & King, D. (2011). *Business intelligence: A managerial approach*. Upper Saddle River: Prentice Hall.

Valverde, R., & Saadé, R. G. (2015). The effect of E-supply chain management systems in the North American electronic manufacturing services industry. *Journal of Theoretical and Applied Electronic Commerce Research, 10*(1), 79–98.

Valverde, R., Saadé, R. G., & Barrad, S. (2016). Empirical investigation of e-supply chain management experience in North American electronic manufacturing services. In *The development of e-commerce management for the book industry*. Hershey: Igi global.

Varsei, M. (2016). Sustainable supply chain management: A brief literature review. *Journal of Developing Areas, 50*(6), 411–419. https://doi.org/10.1353/jda.2016.0129.

Varsei, M., Soosay, C., Fahimnia, B., & Sarkis, J. (2014). Framing sustainability performance of supply chains with multidimensional indicators. *Supply Chain Management: An International Journal, 19*(3), 242–257.

Viswanathan, N., & Sadlovska, V. (2010). *Supply chain intelligence: Adopt role-based operational.* Boston: Aberdeen Group.

Winter, M., & Knemeyer, A. M. (2013). Exploring the integration of sustainability and supply chain management: current state and opportunities for future inquiry. *International Journal of Physical Distribution & Logistics Management, 43*(1), 18–38.

Part III
The Innovative Solutions for Industry-Specific Supply Chains

Chapter 8
Alternative Energy Policy for Mitigating the Asynchrony of the Wind-Power Industry's Supply Chain in Brazil

Milton M. Herrera, Isaac Dyner, and Federico Cosenz

8.1 Introduction

One of the main concerns of the Brazilian electricity market is its high dependency on hydroelectricity. The impacts generated by changing climate conditions and the sedimentation of dams have strongly affected hydroelectricity generation, causing a reduction of storage levels in dams (De Lucena et al. 2009). In recent decades, Brazil initiated its commitment to wind power in order to meet the challenges of sustainable energy development. Though renewable energy policies are achieving improvements in the power sector, the problem of time delays in the construction of new transmission stations affects the rapid expansion of wind power.

In 2013, Brazil faced problems regarding the lack of sufficient transmission lines, which affected generation by the wind farms. Consequently, the country was forced to use diesel and natural thermoelectric power generation. The cost of generation shot up, and this obliged the system to compensate the wind-power companies that had completed their projects by 2012 but had no means to dispatch their

M. M. Herrera (✉)
Universidad Jorge Tadeo Lozano, Bogotá, Colombia

Department of European Studies and International Integration (DEMS), Università Degli Studi di Palermo, Palermo, Italy
e-mail: miltonm.herrerar@utadeo.edu.co

I. Dyner
Faculty of Natural Science and Engineering, Universidad Jorge Tadeo Lozano, Bogotá, Colombia
e-mail: isaac.dynerr@utadeo.edu.co

F. Cosenz
Department of European Studies and International Integration (DEMS), Università Degli Studi di Palermo, Palermo, Italy
e-mail: federico.cosenz@unipa.it

H. Qudrat-Ullah (ed.), *Innovative Solutions for Sustainable Supply Chains*, Understanding Complex Systems, https://doi.org/10.1007/978-3-319-94322-0_8

power. Though Brazil possesses an extensive network of long-distance transmission lines, the current capacities of transmission facilities are not sufficient to guarantee energy supply, especially in isolated regions. Therefore, this situation compromises the rapid expansion of wind farms in the northern region of Brazil.

Germany and China are making significant progress in developing their transmission infrastructure, thanks to their favorable legal and regulatory frameworks, which positively promote their wind-power supply chains (Yuan et al. 2014; de Melo et al. 2016). Conversely, Brazil during the last two decades has been facing two critical concerns relating to the transformation of its energy sector: accumulated time delays in the expansion of transmission lines and rapid expansion of wind power. This situation has generated an asynchronous growth between the transmission industry and the wind industry that affects the sustainability of the supply chain. The concern is thus the misalignment between the industry policy of wind-power development and energy policy for transmission expansion.

The Brazilian power sector is the largest in Latin America, with 158 GW (ANEEL 2016). While hydropower represents 61% of the total capacity, with 96.8 GW, wind power embodies 6.2% of the total and solar only 0.0145%. However, this profile may change significantly, as the current energy policy considers climate variation issues, including lower flows into hydroelectric reservoirs. Aligned with this, since 2004, Brazil has promoted the development of the wind-power industry through its Incentive Programme for Alternative Sources (PROINFA), resulting in increases of wind-power participation in the electricity matrix. This initiative, based on the European feed-in policies, principally promotes the diversification of power generation sources and the reduction of greenhouse gas emissions. In this sense, Brazil and South Korea are emerging as producers of wind technology, evidenced by their increased numbers of wind energy manufacturers (Foley et al. 2015).

Supply chain management is an area of growing attention due to its impact on the industry (Georgiadis and Besiou 2008). The structure of the energy supply chain is characterized by situations of dynamic complexity when information is not available among all the players involved in the different industrial stages. In this context, econometric methods and linear programming are not adequate to deal with the complexity of the relationships between the players involved in the wind-power supply chain (Dyner and Larsen 2001; Kunsch and Friesewinkel 2014; Qudrat-Ullah 2015; Franco et al. 2015). Therefore, system dynamics modeling is a more appropriate approach as it can capture much of the complexity of supply chains.

Numerous studies have examined the various perspectives regarding the expansion of wind power in Brazil (Pereira et al. 2013; Silva et al. 2013; Juárez et al. 2014; Brannstrom et al. 2017). However, few have analyzed the effects of the expansion of renewable energy on the sustainability of the wind-power supply chain. In this sense, Wee et al. (2012) assessed renewable energy growth from the supply chain perspective. This study provides managerial insights into renewable energy use and makes proposals for overcoming the barriers to its development. Others studies have also identified the relationships and concerns of stakeholders along energy supply chains associated with wind industry growth (Yuan et al. 2014; Wüstemeyer et al. 2015). Despite the fact that some studies show the concerns of stakeholders in the wind-power supply chain, the asynchrony between the transmission capacity and the wind industry has not been taken into account. In particular, the wind industry growth effects in Brazil have not yet been sufficiently analyzed, from an energy supply chain perspective.

Given the complexities of asynchronies, this research uses simulation modeling to gain better understanding of the dynamics of wind-power supply chain performance and its effect on the energy sector. This modeling approach is also helpful for assessing alternative effective energy policy that may mitigate the asynchrony resulting from insufficient transmission infrastructure for the wind-power supply chain. Thus, this approach better gauges the consequences of the energy policies that might affect the performance of the supply chain.

This chapter is divided into five parts. The introduction is followed by a section that reviews the development of the wind-power supply chain and its asynchrony in Brazil. Section 8.3 describes the model-based methodology used for studying this issue. Section 8.4 discusses simulation results, which explain the asynchrony of the wind-power supply chain and its effects on performance and sustainability. Further, a model-based alternative policy is proposed that shows the synchrony from the supply chain perspective. Finally, the conclusions section establishes the benefits of coordinated policies on the supply chain performance and identifies future research.

8.2 Asynchronous Policy in the Wind-Power Supply Chain Industry

In many countries, the rapid penetration of renewable energies has been driven by several factors, primarily those related to environmental and energy policy. Furthermore, as renewable energies are intermittent, they require important infrastructure and technology support for their synchronization with other supply sources. Despite the several studies aimed at estimating Brazil's renewable energy potential (Geller et al. 2004; Pereira et al. 2012; de Jong et al. 2015; Corrêa Da Silva et al. 2016), the synchronization problems between the Brazilian policies on transmission infrastructure and renewables generation will affect the sustainability and performance of the wind-power supply chain.

The Brazilian energy sector has been highly dependent on hydropower, which has motivated the diversification of its electricity matrix. In response to the dependency on hydropower, Brazil is promoting other renewable energies to hedge hydropower risks through the PROINFA program (Corrêa Da Silva et al. 2016).

8.2.1 Development of the Wind-Power Supply Chain in Brazil

Hydropower has dominated the Brazilian electricity market for years, and this has made the sector vulnerable to droughts and consequently to high electricity prices. As wind power is getting cheaper, it is becoming an interesting option for complementing hydropower during dry seasons (Brannstrom et al. 2017; de Melo et al. 2016; Juárez et al. 2014; Silva et al. 2016; Aslani et al. 2014). Reports on the Brazilian Northeast are highly optimistic for energy generation because of the periods of high wind potential (Brannstrom et al. 2017). According to ANEEL (2016), the total installed wind-power capacity changed between 2005 and 2016, from 0.027 GW to 9.8 GW. The North and Northeast regions have experienced a rapid growth in wind farms (Tiba et al. 2010; de Jong et al. 2015).

In Brazil, industry promotion policies play an important role in achieving economic viability for wind-power projects (Kissel and Krauter 2006). Indeed, the wind-power industry has had an annual growth rate of about 15% since 2010. Brazil started by importing wind turbines, but this has changed rapidly as the policy has been to promote the domestic supply chain of the wind industry. In this direction, the wind turbine industry is experiencing rapid growth (Simas and Pacca 2014). Despite the efforts of the Brazilian government, the performance and sustainability of the supply chain have been affected by the lack of synchronization between wind industry expansion and development of the transmission infrastructure.

Brazil faces barriers associated with delays in the delivery of new transmission lines postponing the operation of wind farm generation (de Melo et al. 2016). This situation is the result of a backlog of construction delays in recently auctioned transmission lines, for connecting wind projects, across the country. Transmission infrastructure has not kept up with the rapid growth in capacity, leaving wind farms with inappropriate infrastructure for power transmission to demand centers in the south and east (Foley et al. 2015). Even though the North and Northeast regions are quite suitable for the generation of wind power, environmental licensing issues have been a matter of concern in the construction of transmission infrastructure. Additionally, land price has been a constraint on the expansion of transmission in these regions. Thus, the great difficulties for transmitting wind power from generation towers in the Brazilian North to demand centers elsewhere involve a lack of synchrony between energy policies and the supply chain development.

8.2.2 Asynchrony Between Policies and Performance of the Wind-Power Supply Chain

Development of the wind-power industry is driven by the initiative of the federal government. In this way, Brazil has demonstrated its ability to adopt and implement energy policies through electricity-sector incentives (Geller et al. 2004). These Brazilian energy policies have a significant impact on both emission reductions and the diffusion of clean technologies. Despite this, the wind-power supply chain performance was strongly out of synchrony with the transmission infrastructure. Thus, the asynchronous implementation of energy policies involves a conflict of interest associated with the targets of each stakeholder in the supply chain and electricity supply. In terms of transmission network investment, the stakeholders have different interests;

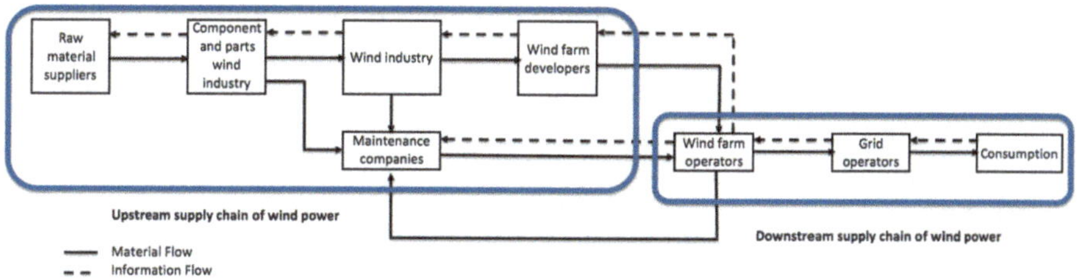

Fig. 8.1 Structure of the wind-power supply chain

in other words, they will have different investment perspectives and strategies along with the wind-power supply chain, including the network expansion (Pudjianto et al. 2016). This motivates studying the asynchrony between energy policies and its effects on the wind-power supply chain in terms of a performance measure and security of energy supply.

Regarding development of the wind-power supply chain in Brazil, de Melo et al. (2016) mention that the critical issues are the legal and regulatory framework, the institutions created to support the energy system, and the long-term energy planning. Because of high electricity market uncertainty and rapid change in the wind industry, the energy planning process enables a better understanding of the supply chain. This implies the need for synchronization of the planning process among the energy system stakeholders. In this sense, there are several model-based planning methodologies that are useful for approaching this issue (Dyner and Larsen 2001).

Study of the wind-power supply chain involves understanding the execution of a sequential chain of operations and flows, as shown in Fig. 8.1. This mean that the dynamic behavior of the supply chain is associated with the complex relationships among the raw material suppliers, components, wind turbine manufacturers, wind farm developers, grid operators, and service suppliers (Prostean et al. 2014; Yuan et al. 2014). The wind-power supply chain can be divided into two parts, upstream and downstream (Yuan et al. 2014). The first part involves suppliers, industry, and developers. The second part is related to the wind-power generation, electricity grid, and consumption. Some authors have studied the performance of the wind-power supply chain and its implications, but little attention has been paid to the energy policy implications (Wee et al. 2012; Wüstemeyer et al. 2015), which is the concern of this chapter.

8.2.3 *Asynchrony and Its Implications for the Sustainability of the Wind-Power Supply Chain*

Sustainable supply chain management theory highlights the complex issue associated with stakeholders' relations and energy policy implications (Linton et al. 2007; Wüstemeyer et al. 2015). The interaction between the stakeholders and policies can occur either synchronously or asynchronously (Sahay et al. 2014). While synchrony in the supply chain is achieved by appropriate information exchange resulting from coordination between policy and planning, asynchrony occurs when coordination is not in place. In this context, clean energy policy may lead to poor power industry outcomes (Lund 2009) without careful planning of the entire supply chain.

In the case of Brazil, policy has led to a rapid expansion of the Brazilian wind-power industry, but the development of transmission infrastructure has not kept pace. In this context, Brazil faces delays in the delivery of new transmission lines that generate increased grid congestion (de Melo et al. 2016). To minimize the transmission problem, the government has limited auctioning wind-power projects and, consequently, has slowed the rate of increase of wind generation. This situation limits the performance of wind industry operations due to the reduction of financial resources used, which affects the finances of the wind-power supply chain.

Expansion of the wind generation industry involves varied interests that are not aligned with transmission development, producing an asynchrony in the wind-power supply chain. The consequence is that the asynchrony generated by different investment decisions of the various stakeholders might affect the electricity price to customers and compromise security of supply. Therefore, the continuing expansion of the wind-power supply chain is affected.

8.3 Methodology

This chapter addresses the problem of insufficient electricity transmission infrastructure resulting from the rapid expansion of wind power that affects the North and Northeast of Brazil. This section proposes a model-based approach for assessing alternative sustainable policy that deals with asynchrony in the development of the wind-power supply chain.

8.3.1 The System Dynamics Approach

A considerable amount of literature has supported renewable energy policy based on general equilibrium modeling and agent-based simulation (Ioakimidis et al. 2012; Bale et al. 2015; Cabalu et al. 2015; Troost et al. 2015; Abrell and Rausch 2016; Karplus et al. 2016). Abundant literature also reports renewable energy policy assessment based on system dynamics (Ford 1997; Ford 1999; Dyner and Larsen 2001; Zuluaga and Dyner 2007; Ochoa et al. 2013; Qudrat-Ullah 2014). The research reported here found no system dynamics literature that supports renewable policy from a supply chain perspective, the purpose of this chapter.

Broadly speaking, the system dynamics approach has been used to help understand supply chain behavior for over 40 years (Sterman 2000; Asif et al. 2012; Rendon-Sagardi et al. 2014; Tian et al. 2014; Orjuela-Castro et al. 2017). The system dynamics modeling approach facilitates the analysis of problems with complex causality and time delays. More recently, system dynamics modeling of the supply chain has been used to assess transport policies and capacity planning (Vlachos et al. 2007; Orjuela et al. 2015). Though many researchers have addressed the expansion of renewables and the challenges involved in the sustainability of energy systems (Qudrat-Ullah 2013; Franco et al. 2015; Jimenez et al. 2016; Qudrat-Ullah 2016; Castaneda et al. 2017), these researchers have not, however, addressed asynchrony problems from the wind-power supply chain perspective.

The wind-power supply chain consists of many stakeholders with different roles and targets (Wee et al. 2012). System dynamics modeling is useful for simulating and understanding the integration in energy supply chains of distinct roles and targets, in emerging markets (Rendon-Sagardi et al. 2014). The first published system dynamics work related to supply chain management established a link between the downstream flow of material and upstream flow of information (Forrester 1961; Sterman 2000). In the wind-power supply chain, downstream and upstream flows are key factors for improving the supply chain performance. These flows are the main concern of both the supply chain's dynamics and the supply chain's stakeholders.

The system dynamics approach, as incorporated here, establishes first a dynamic hypothesis of the system that involves the problem situation. This is followed by the simulation model structure, which combines the main relationships in a stock and flow diagram through a set of differential equations. Finally, the simulation model is validated to ensure that it behaves as expected and its behavior coincides with the dynamic hypothesis.

8.3.2 *Dynamic Hypothesis*

This section presents a general framework that depicts the asynchrony of the wind-power supply chain, which supports the dynamic hypothesis. The main aspects defining the dynamics within the system under study are described next:

- *Promotion and development of the Brazilian wind-power supply chain*: Since 2002, the Incentive Programme for Alternative Sources of Energy – PROINFA – Aimed to increase the electric energy produced from renewable energy, which includes wind projects. Consequently, the Brazilian government has been making efforts for promoting scientific and technological development by investing in research (Juárez et al. 2014). From 2002 to 2016, the wind industry has been reporting rapid growth all along the supply chain. In this sense, the new wind equipment suppliers have experienced significant growth through the financial incentives for the development of a national wind industry supported by the Brazilian government.
- *The impact of inadequate development on the transmission lines and towers*: Despite the best wind conditions being found in the Northeast region, Brazil suffers an electricity supply problem due to the lack of transmission infrastructure

serving the North and Northeast. According to de Jong et al. (2016), the problem transmission lines will need to be expanded and upgraded to accommodate the expected increase in wind-power penetration in the Northeast region. Otherwise, the currently inadequate transmission infrastructure in the Northeast region can lead to significant increases in electricity prices.

- *Asynchrony between energy policies and supply chain performance*: Despite the increase in renewable energy in Brazil, the lack of political support for facing Brazil's current economic crisis is one point that can prevent the development of the wind-power supply chain (Silva et al. 2013). Although a set of policies was adopted to expand the share of renewable energy, the projects for transmission line construction have been delayed in the north of Brazil. The transmission construction delays have a significant impact on the wind-power supply chain development, which involves the synchrony problem. This problem directly affects the performance of the wind-power supply chain as well as the Brazilian electricity market. Government and researchers have not properly addressed this asynchrony. Therefore, this study presents results on the effects of asynchrony on the supply chain performance.

A dynamic hypothesis is a theory of structure and relationship policies (i.e., decision-making) that affect the observed behavior (Oliva 2003). To assess the asynchrony of the wind-power supply chain, this study uses a dynamic hypothesis. The causal loop diagram represents the dynamic hypothesis that describes the asynchrony between the growth of the wind industry and the lack of transmission capacity, as shown in Fig. 8.2. The balancing loop (B1) shows the capacity margin, which depends on the difference between electricity demand and installed capacity. Therefore, the capacity margin affects electricity price, and this influences electricity demand, creating the balancing loop B1. The interaction of these variables represents the functioning of the electricity market as well as the relationships between

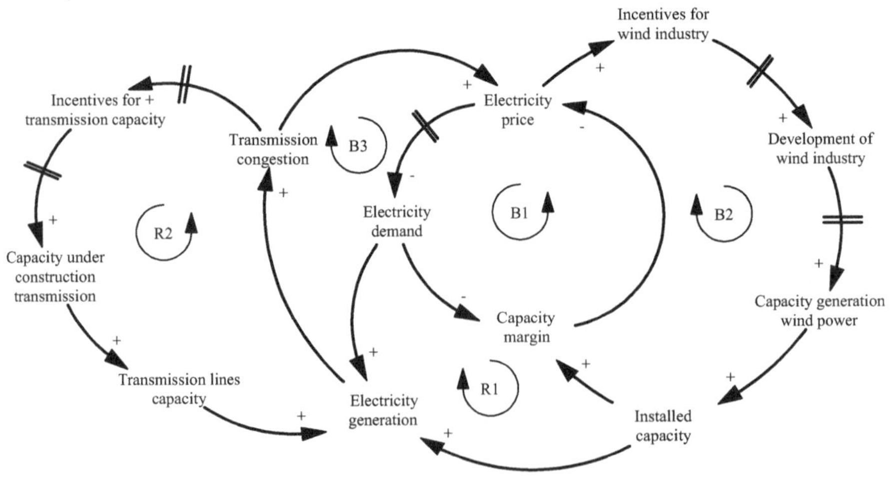

Fig. 8.2 Dynamic hypotheses for the asynchrony of the wind-power supply chain

supply capacity and demand. Therefore, the capacity margin depends on the total installed capacity and electricity demand. A large capacity margin induces lower electricity prices; however, higher prices incentivize the wind industry. The response of price to investment incentives involves long delays that affect development of the wind industry. Therefore, this will lead to an increase in the total installed capacity. In this way, the balancing loop B2 describes electricity price effects on investment incentives for the wind industry, which in the medium term influences the capacity margin. Likewise, this loop shows the supply chain development effect promoted by Brazilian government policy. In this case, the creation of additional wind-power capacity affects the total installed capacity, which determines the offer curve for electricity demand. This feedback structure represents the penetration of wind power as a consequence of the energy policies adopted by Brazil.

The third balancing loop (B3) describes the dynamics between electricity generation and transmission congestion as well as their effects on electricity price and demand. The transmission congestion depends on the electricity generation. Therefore, the lack of transmission infrastructure causes a decrease in the electricity supply, which could result in high electricity price. Because of this, the electricity price will be affected by the inadequate expansion of transmission capacity as well as by development of the wind industry. Feedback loop R1 encourages wind industry expansion, which affect transmission capacity.

Because of the increase in total installed capacity, electricity generation will have an effect on the transmission congestion. Consequently, this is promoting the incentives for transmission lines, which affect its capacity under construction, creating the reinforcing loop R2. This feedback structure is the result of a time delay that occurs between the investment decision and the needed transmission line capacity. Therefore, the loop (R2) limits the effectiveness of wind-power supply chain adoption due to the high cost of electricity transmission.

As the result of huge increases in energy demand, Brazil is projecting important economic growth as well as low-cost energy generation. However, the lack of financial resources and construction delays for transmission infrastructure will act as barriers to reaching this objective. In this sense, the lack of transmission infrastructure may be critical to Brazilian energy sustainability. This problem may cause a delay to widespread use of renewable energy sources and affect the performance of the wind-power supply chain. Consequently, conversion to wind power may become expensive due to the shortage of financial resources for the transmission system.

8.3.3 Simulation Model

The simulation approach used here allows evaluation of the dynamic characteristics of a regional energy industry system. The simulation model was built using detailed descriptions of the players in the wind-power supply chain, such as suppliers, the wind industry itself, and assemblers of wind towers. Likewise, the electricity demand, installed capacity for the wind power, as well as transmission lines were

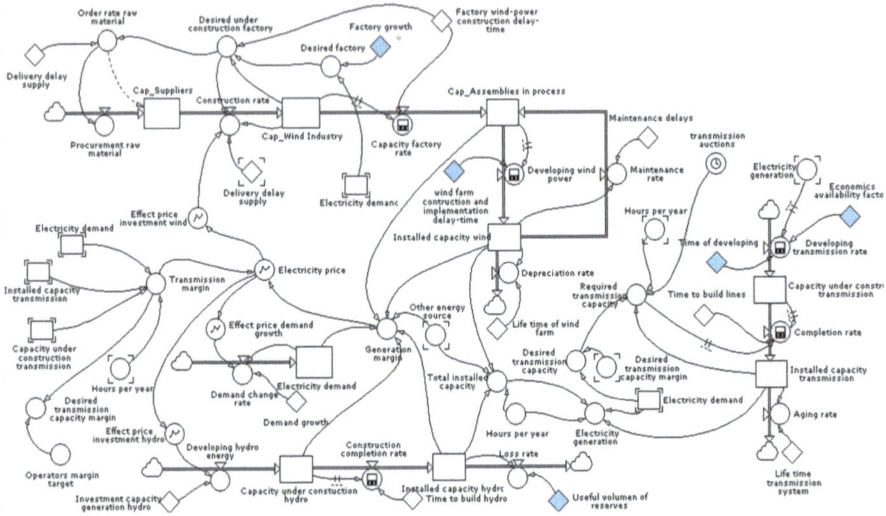

Fig. 8.3 Stock and flow diagram of wind-power supply chain in Brazil

also taken into account. In addition to this, other Brazilian energy sources, such as fossil fuels, nuclear, importation of power, and solar photovoltaic are considered. Therefore, the stock and flow diagram represents the Brazilian energy system structure, as shown in Fig. 8.3.

The model design was based on structures proposed by Cardenas et al. (2016), Franco et al. (2015), and Sterman (2000). However, an extension of this model includes a structure for the wind-power supply chain that allows analysis of the asynchrony of Brazilian energy policy. The model structure considers three basic subsystems, as follows: (1) the wind-power supply chain, (2) the electricity market, and (3) the dynamics of transmission capacity.

Wind-power supply chain The structure of the supply chain includes the capacities of suppliers, the wind industry (manufacturers), assemblers, and installed capacity of wind power. Other system-related factors identified in the wind-power supply chain are as follows: procurement of raw materials, delivery delay of supply, capacity factory rate, and maintenance and disposal rates.

The wind-power supply chain in Brazil is characterized by significant uncertainties associated with logistics, environmental permits, and time lost during infrastructure construction. These uncertainties affect the development of total installed capacity and wind industry capacity. In this sense, the delay function is used to capture the delayed availability of capacities in the supply chain for the simulation model. For instance, the *capacity factory rate* (CF) and *developing wind power* (DWP) capture this delay time and are determined by delaying the values of the *capacity of wind industry* (CWI_w) and *capacity of assemblers* (CA), respectively. In addition to this, the *installed capacity of wind power* (IC_w) has a useful life cycle, which regulates the depreciation and maintenance rates. Therefore, a closed-loop supply chain modeling for the maintenance service of wind-power towers was included. The mathematical equations relating to the main drivers of the wind-power supply chain are the following:

$$\mathrm{IC}_w(t) = \mathrm{IC}_w(t - dt) + \int_{t=o}^{T} [\mathrm{DWP}(s) - \text{Depreciation rate}(s) + \text{Maintenance rate}(s)]\, ds \; [\mathrm{GW}] \tag{8.1}$$

$$\mathrm{CF} = \mathrm{CWI}_w(1 - \mathrm{fwcdt}) \quad \left[\frac{\mathrm{GW}}{year}\right] \tag{8.2}$$

$$\mathrm{DWP} = \mathrm{CA}(1 - \mathrm{wfcdt}) \quad \left[\frac{\mathrm{GW}}{\mathrm{year}}\right] \tag{8.3}$$

where:

fwcdt: factory wind-power construction delay time
wfcdt: wind farm construction and implementation delay time

Electricity market The price of electricity is the crucial factor to be considered for investment decisions regarding the different generation technologies. In this sense, reserve margins determine the electricity price, which has an effect on investment in new wind-power capacity and electricity demand (balancing loops B1, B2, and B3). The transmission congestion and generation margin are taken into account while computing the electricity price.

Dynamics of transmission capacity Considering that transmission lines are essential components of energy supply chain structure, decision mechanisms must be designed so that they promote adequate expansion of renewable energy. The transmission lines' capacity under construction depends on two main aspects: *economics availability factors* (eaf) and *time of developing* (tdt). These variables affect the "flow" of developing transmission capacity. The *developing transmission rate* (dtr) captures the delay time involved. This is determined by delaying the value of the needed capacity for transmitting *electricity generation* (CTEG_T), as shown in Eq. (8.5). Given that wind power has had an increase, transmission lines must be built to carry the resulting electricity. Therefore, time is required to build the transmission lines and towers that are needed as a result of the auctions delaying wind-power projects in Brazil. The amount of *installed capacity of transmission lines* (IC_T) is calculated as shown in Eq. (8.6).

$$\mathrm{CUC}_T(t) = \mathrm{CUC}_T(t - dt) + \int_{t=o}^{T} [\mathrm{dtr}(s) - \mathrm{tlc}(s)]\, ds \; [\mathrm{GW}] \tag{8.4}$$

$$\text{dtr} = \left[(\text{CTEG})(1 - \text{tdt})\right] * \text{eaf} \quad \left[\frac{\text{GW}}{\text{year}}\right] \tag{8.5}$$

$$\text{IC}_T(t) = \text{tsc}(t - dt) + \int_{t=o}^{T} \left[\text{tlc}(s) - \text{ar}(s)\right] \text{ds} \; [\text{GW}] \tag{8.6}$$

where:

tlc: transmission lines completion rate
ar: aging rate
tsc: installed capacity of transmission

In each period, the transmission system establishes whether the desired capacity is triggered (Eq. 8.7). This is based on the difference between expected transmission margin (projected demand vs. expected capacity) and the transmission operator's margin target. The required transmission capacity (RTC) is calculated as indicated by Eq. (8.8):

$$\text{DTC}(t) = \text{epd} * (1 + \text{dtcm}) \; [\text{GWh}] \tag{8.7}$$

where:

DTC: desired transmission capacity
epd: electricity peak demand
dtcm: desired transmission capacity margin

$$\text{RTC}(t) = \text{DTC} - \text{IC}_T \quad [\text{GWh}] \tag{8.8}$$

The modeling assumptions and parameter values for the simulation model are as follows:

- Time horizon of the simulation model ranged between 2016 and 2050, to capture the time delay in transmission construction and the effects of the asynchrony of the wind-power supply chain.
- Initial data on the installed capacity of each technology corresponds to the year 2016, according to ANEEL (2016). In 2016, the installed system capacity in Brazil was 157.86 GW, which was supplied by hydroelectric plants (61%), wind power (6%), fossil fuels (18%), nuclear (1%), importation of power (5%), and biomass (9%) (ANEEL 2016). In this sense, other generation technologies are considered within the model.
- From 2001 to 2016, Brazil experienced an average 3% per annum increase in electricity demand (Pereira et al. 2011; ANEEL 2016). However, between 2005 and 2014, the annual growth averaged 4.1% (de Jong et al. 2016).
- Transmission system average life is 20 years (Pudjianto et al. 2016).

- The structure of the wind-power supply chain as represented in the model is supported by previous studies (Varella 2013; ABDI 2014).
- Forecasting of the wind power and electricity demand used in the validation model was based on Fiestas and GWEC (2011), ANEEL (2016), Ministério de Minas e Energia (2007), and WWF-Brazil (2015).
- The simulation model evaluated the asynchrony of the wind-power supply chain in Northeast region. In this region of Brazil, the total installed capacity of wind power could reach 16.40 GW by 2020 (de Jong et al. 2016).

8.3.4 Model Validation

Validation is a continuous process of testing, building confidence in the model (Sterman 2000; Ahmad et al. 2016). This section conducts several model validation tests, including dimensional consistency and behavior reproduction tests (Sterman 2000; Luna-Reyes and Andersen 2003; Qudrat-Ullah and Seong 2010). According to Oliva (2003), statistically based approaches have been used to make the parameter estimation process more rigorous. In order to evaluate the historical fit of the simulation model, the mean absolute percentage error statistic (MAPE) was used, as shown in Table 8.1.

To validate the accuracy of the proposed model, the results were compared to real outputs and forecasts for the years 2005–2019, in terms of the installed capacity of wind power and Brazilian electricity demand. Reviewing the variables with MAPE, the error rate is below 6%; R^2 equals 0.97, which presented a small bias; and the residuals do not show a significant trend. The results obtained from the validation analysis are shown in Fig. 8.4. Thus, the simulation has the ability to capture trends and replicate historical data, as shown in the results. In the model calibration process, the results examined differences between simulated output and historical data. Then, we adjusted the model parameters to correct the discrepancy and simulated again.

The purpose of the behavior validation process is to compare how the model reproduces the Brazilian experience of the real system, with regard to electricity supply and demand, in the wind-power supply chain. Thus, these findings suggest that the model does a good job of tracking the observed data.

Table 8.1 Error analysis of the simulation model in the validation process

Variable	MAPE	R^2	Standard deviation (simulation)	Standard deviation (actual)
Installed capacity of wind power	0.031	0.99	5.75	6.73
Electricity demand	0.061	0.98	83.62	78.73
Total installed capacity	0.037	0.98	123.29	124.99

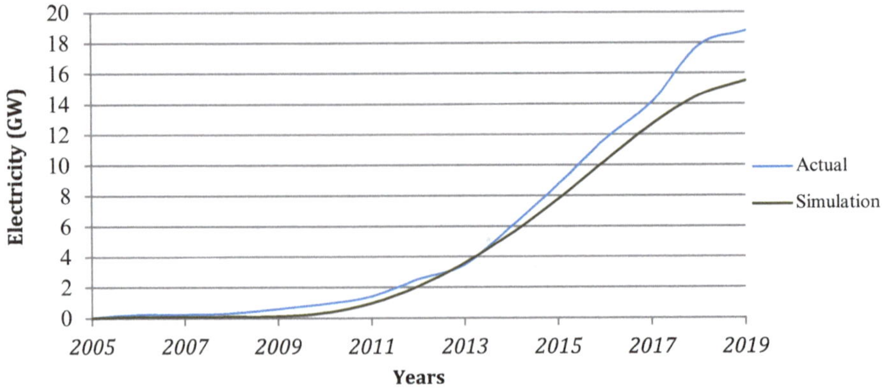

(a) Installed capacity of wind power

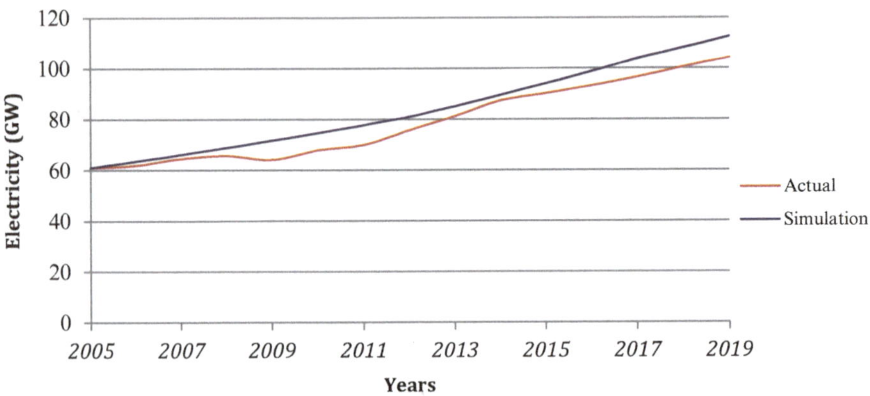

(b) Electricity demand

Fig. 8.4 Validation process, calibration, and analysis of the simulation model

8.4 Simulation Results

This section includes simulation results that show the asynchrony between the development of wind-power farms and transmission facilities. Subsequently, alternative energy policies for mitigating the asynchrony and improving the performance of the wind-power supply chain are assessed.

8.4.1 Synchrony in the Wind-Power Supply Chain and Its Implications for the Industry Performance

The simulation model assesses the performance of the wind-power supply chain of Brazil as affected by the asynchrony of energy policies, as shown in Fig. 8.5a. The capacity development of transmission facilities and wind farms was assessed

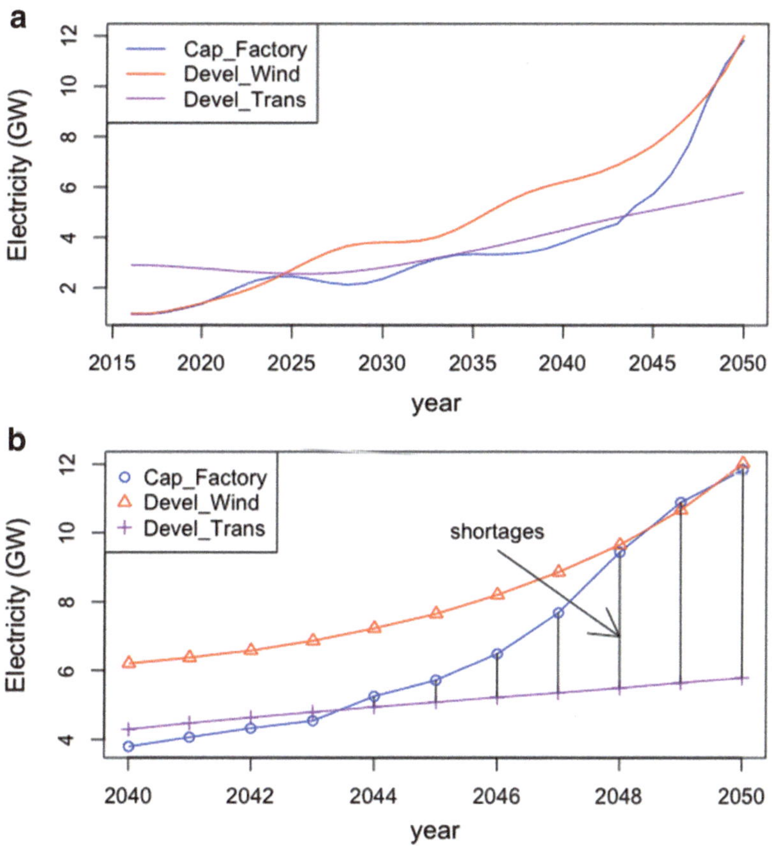

Fig. 8.5 Impact of asynchronous energy policies over the time period 2016–2050 on the wind-power supply chain, with a 5% growth in wind industry electricity generation (**a**) Dynamics of the development of transmission infrastructure (Devel_Trans) and wind-power rate (Devel_Wind) with a 5% increase in investment for the wind industry and 2.5% for transmission line development (**b**) Asynchrony between the capacity development of transmission infrastructure and wind power associated with the growth in the wind industry over the time period 2040–2050 as from a blowup of Fig. 8.5a

by conducting two simulation tests. Simulations test the hypothesis that if no synchronization is in place, investments in transmission capacity may prove to be inadequate in the face of rapid growth of wind farms, because of the time lags inherent in building the required infrastructure. Figure 8.5b shows shortages that may occur in transmission lines as the result of a 5% increase for the wind industry power electricity generation over the time period 2040–2050. This shortage may be the result of both long construction time delays for new transmission lines and the inadequacy of investment incentives associated with delays in auctions. Additionally, the delivery times for parts and components for wind farm construction produce an asynchrony in the operations running along the supply chain. Therefore, even though there is an initial steady growth in the wind industry, the

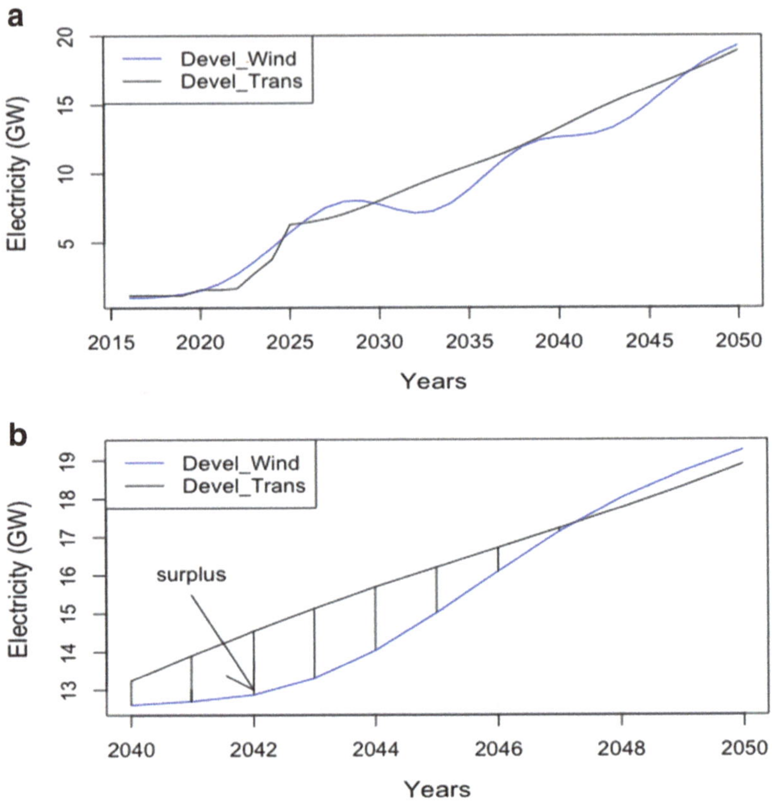

Fig. 8.6 Impact of synchronous energy policies over the time period 2016–2050 on the wind-power supply chain, with a 10% increase in wind industry electricity generation (**a**) Dynamics of the development of transmission infrastructure (Devel_Trans) and wind-power rate (Devel_Wind) with a 10% increase in the investment of wind industry and 6% for the developing transmission lines (**b**) Synchrony between the capacity development of transmission infrastructure and wind power associated with the growth in wind industry over the time period 2040–2050 as from a blowup of Fig. 8.6a

developing transmission lines do not keep pace with its growth, consequently affecting the activity of wind farms.

Thus, the wind-power supply chain in Brazil needs to address the issue of coordinating energy policy along the supply chain, including the appropriate incentives for transmission infrastructure (considering the unavoidable delays). A synchronized energy policy might be attained with the help of better coordination and conjoint planning among suppliers, industry, and the transmission operator in terms of investment, delivery times, and the scheduling of times for wind farm maintenance. The way in which this might be achieved is discussed next.

Three coordination policies are examined: (a) promotion of coordinated investment policies for the transmission and manufacturing sectors; (b) promotion of reductions in delivery time for developing wind farms, construction of windmills (wind-power manufacturing), and the manufacture of windmill components; and (c) reduction of time taken for wind farm maintenance.

Coordinating investment policy Figure 8.6a shows surpluses in transmission lines as investment increases at a rate of 6% per annum and as the manufacturing input into the wind industry increases its throughput at a rate of 10% per annum. Naturally, when increasing investments in transmissions at higher rates, synchronous coordination might be attained. In this exercise, it was accomplished by increasing the transmission incentives by 40% of the original value (2.5%). Interestingly, a significant increase in the developing transmission lines results in synchrony along the entire supply chain (up to electricity delivery), as illustrated in Fig. 8.6b. During the period 2020–2050, the developing transmission lines exhibit a pronounced rise attributable to investment incentives for the construction of transmission capacity that guarantees the security of energy supply as well as the synchrony between the transmission and wind-power development.

Although the wind industry grows in this scenario, assembly of wind farms fell behind as the result of delivery delays from suppliers of parts for towers, which delayed the supply chain downstream. The effects of reductions in delivery time associated with industry capacity are evaluated next.

Promoting reductions in delivery time To attain the appropriate wind-power capacity, it is necessary to consider the delivery delay between industry developers and building wind farms. Delivery time is the main driver for the expansion of installed capacity in the supply chain, which causes the operational resources of both the industry and developers to increase or decrease. Equation (8.7) calculates the *accumulated capacity average* (ACA) used for the industry and developers of the wind-power supply chain, during the simulation periods.

$$\text{ACA} = \frac{\frac{I_0 + I_T}{2} + \sum_{t=0}^{T} I_t}{n} \quad [\text{GW}] \tag{8.7}$$

where:

I_0: initial installed capacity
I_T: final installed capacity for the period T
I_t: installed capacity for each period t
n: number of periods of simulation

The total generation capacity in place, for each year, is established according to Eq. (8.8) and depends on the wind industry, installed capacity of wind farms, the capacity production of developers, and delivery time.

$$\mathrm{GCPR} = \frac{\frac{\mathrm{ACA}_{\mathrm{IND}} + \mathrm{ACA}_{\mathrm{WP}}}{2} + \left[\mathrm{ACA}_{\mathrm{IND}} + \mathrm{ACA}_{\mathrm{D}} + \mathrm{ACA}_{\mathrm{WP}}\right]}{\mathrm{Dt}} \quad [\mathrm{GW}] \tag{8.8}$$

where:

ACA_{IND}: average capacity of wind industry
ACA_{WP}: average installed capacity of wind farms
ACA_{D}: average capacity of developers
Dt: delivery time
GCPR: total generation capacity in place, for each year

The results of the installation rate in the wind-power supply chain, according to the delivery delays between industry and developers, are shown in Table 8.2. Results show increases in generation capacity in place as deliveries present reduction in time, supporting the expansion of the transmission system. Thus, the reduction of delivery time will help to coordinate investment policies in transmission and in developing wind farms. In this sense, the simulations suggest that it should be possible to attain policy synchronization.

Promoting reductions in maintenance time The same type of closed-loop stock and flow structure used for the wind-power supply chain model is also used for representing the supply chain for the maintenance of wind turbines and towers. In this structure, the maintenance time plays an important role regarding the performance of wind farm developers as well as the ongoing development of wind farms. Also, the maintenance rate means the product of the number and power rating of wind turbines that are maintained in a year. Two scenarios have been analyzed with the objective of assessing the impact of a reduction in maintenance time along the supply chain. The first scenario showed the maintenance time of the current state of wind turbines and towers. The second scenario exhibits the effect of a 4% reduction in maintenance time observable when simulating the synchrony of an investment policy that's coordinated with reductions in delays of maintenance of wind farms. As shown in Fig. 8.7, the effect of reducing maintenance time has a significant impact on the developing wind-power infrastructure. In the other words, the developing wind power is subject to a decrease or increase in the maintenance times for

Table 8.2 Simulation for different delivery times between industry and developers of the Brazilian wind farms vs. coordinating investment policy for transmission capacity

Delivery time	2 years	3 years	5 years
Generation capacity in place (GW)	4.85	2.39	1.20
	Uncoordinated investment policy	Coordinated investment policy	
Transmission capacity in place (GW)	3.78	5.53	

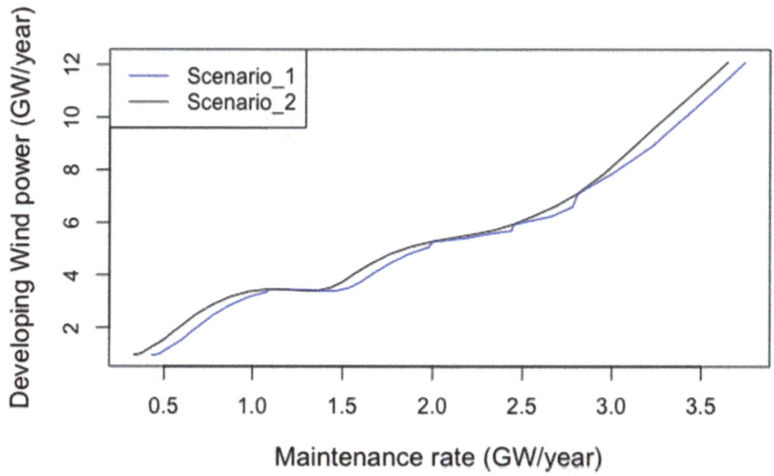

Fig. 8.7 Effects of maintenance rate on the developing wind power. Scenario_1, current state; and Scenario_2, with a 4% decrease in the maintenance time

the wind turbines. Results indicate that a maintenance strategy associated with energy policies will have a significant impact on the development of wind farms.

Simulation results show the need for policy that synchronizes the development of wind capacity and the transmission facilities. Even under a scenario with significant increases of wind power in Brazil, synchronization of the wind-power supply chain is achieved by investment policy coordination, promotion of reductions in delivery times, and reductions in the maintenance time of wind farms. Indeed, synchrony in the wind-power supply chain appears to be achievable with a 6% increase in the investment for developing transmission lines and a steady improvement in the construction times for the wind farms. Simulations of synchronized policy provide significant insights for both sustainable supply chain management and for energy policies, in the context of strategic planning of resources.

8.5 Conclusions

This study uses system dynamics modeling for understanding how to synchronize wind-power expansion and the corresponding transmission facilities to guarantee power supply. The chapter proposes a synchronized alternative policy for mitigating the asynchrony between wind-power expansions when there is insufficient transmission infrastructure. The modeling results indicate that despite current wind industry growth in Brazil, the Northeast and North regions will be facing a strong reduction in the energy supply as the result of time delays in the construction of transmission lines. As the best wind conditions can be found in the Brazilian Northeast, and as the transmission infrastructure is not well developed (Kissel and

Krauter 2006), Brazil will benefit from the design of appropriate energy policies and must consider both the expansion of wind-power generation and transmission capacity development. In this sense, the investment costs in wind power should decrease with expansion of transmission systems, making wind power more attractive for other regions in the near future. This chapter shows that synchrony in the supply chain may be achieved with coordinated investment policy together with reductions in delivery times and the maintenance time of wind farms.

Despite Brazil having a major potential for low-cost energy production, a lack of transmission lines should be a concern for the sustainable development of the wind industry. In this sense, the research considers that the investment incentives must be applied throughout the wind-power supply chain, using integrated and coordinated policies. However, the required conditions for this to happen are not yet well established. When both industrial growth policies and energy incentive policies are applied simultaneously, they can have a great impact on both industrial growth and the diffusion of renewables. Then, this situation of synchronous policies guarantees sustainability among stakeholders in the supply chain.

Brazil will benefit from coordinated investment policies in the wind-power supply chain as well as from good management of the supply of components and of wind farm maintenance. These are important elements for meeting targets in terms of energy security (Corrêa Da Silva et al. 2016). A similar approach can be used to study the effects of asynchronous policies associated with the electricity price. Therefore, an analysis of the asynchrony of the electricity market, in terms of the impacts on prices, should be an issue for future research.

References

ABDI. (2014). Mapeamento da cadeia produtiva da indústria eólica no Brasil. *Agência Bras Desenvolv Ind*, 1–177.

Abrell, B. J., & Rausch, S. (2016). Cross-country electricity trade, renewable energy and European transmission infrastructure policy. *Journal of Environmental Economics and Management*. https://doi.org/10.1016/j.jeem.2016.04.001.

Ahmad, S., Mat Tahar, R., Muhammad-Sukki, F., et al. (2016). Application of system dynamics approach in electricity sector modelling: A review. *Renewable and Sustainable Energy Reviews, 56*, 29–37. https://doi.org/10.1016/j.rser.2015.11.034.

ANEEL. (2016). Matriz de Energia Eléctrica. In: Capacid. Operación. www2.aneel.gov.br/aplicacoes/capacidadebrasil/OperacaoCapacidadeBrasil.cfm.

Asif, F. M., Bianchi, C., Rashid, A., & Nicolescu, C. M. (2012). Performance analysis of the closed loop supply chain. *Journal of Remanufacturing, 2*, 4. https://doi.org/10.1186/2210-4690-2-4.

Aslani, A., Helo, P., & Naaranoja, M. (2014). Role of renewable energy policies in energy dependency in Finland: System dynamics approach. *Applied Energy, 113*, 758–765. https://doi.org/10.1016/j.apenergy.2013.08.015.

Bale, C. S. E., Varga, L., & Foxon, T. J. (2015). Energy and complexity: New ways forward. *Applied Energy, 138*, 150–159. https://doi.org/10.1016/j.apenergy.2014.10.057.

Brannstrom, C., Gorayeb, A., de Sousa, M. J., et al. (2017). Is Brazilian wind power development sustainable? Insights from a review of conflicts in Ceará state. *Renewable and Sustainable Energy Reviews, 67*, 62–71. https://doi.org/10.1016/j.rser.2016.08.047.

Cabalu, H., Koshy, P., Corong, E., et al. (2015). Modelling the impact of energy policies on the Philippine economy: Carbon tax, energy efficiency, and changes in the energy mix. *Economic Analysis and Policy, 48*, 222–237. https://doi.org/10.1016/j.eap.2015.11.014.

Cardenas, L. M., Franco, C. J., & Dyner, I. (2016). Assessing emissions-mitigation energy policy under integrated supply and demand analysis: The Colombian case. *Journal of Cleaner Production, 112*, 3759–3773. https://doi.org/10.1016/j.jclepro.2015.08.089.

Castaneda, M., Franco, C. J., & Dyner, I. (2017). Evaluating the effect of technology transformation on the electricity utility industry. *Renewable and Sustainable Energy Reviews, 80*, 341–351. https://doi.org/10.1016/j.rser.2017.05.179.

Corrêa Da Silva, R., De Marchi, N. I., & Silva Seifert, S. (2016). Electricity supply security and the future role of renewable energy sources in Brazil. *Renewable and Sustainable Energy Reviews, 59*, 328–341. https://doi.org/10.1016/j.rser.2016.01.001.

Da Silva, N. F., Rosa, L. P., Freitas, M. A. V., & Pereira, M. G. (2013). Wind energy in Brazil: From the power sector's expansion crisis model to the favorable environment. *Renewable and Sustainable Energy Reviews, 22*, 686–697. https://doi.org/10.1016/j.rser.2012.12.054.

de Jong, P., Kiperstok, A., & Torres, E. A. (2015). Economic and environmental analysis of electricity generation technologies in Brazil. *Renewable and Sustainable Energy Reviews, 52*, 725–739. https://doi.org/10.1016/j.rser.2015.06.064.

de Jong, P., Kiperstok, A., Sánchez, A. S., et al. (2016). Integrating large scale wind power into the electricity grid in the Northeast of Brazil. *Energy, 100*, 401–415. https://doi.org/10.1016/j.energy.2015.12.026.

De Lucena, A. F. P., Szklo, A. S., Schaeffer, R., et al. (2009). The vulnerability of renewable energy to climate change in Brazil. *Energy Policy, 37*, 879–889. https://doi.org/10.1016/j.enpol.2008.10.029.

de Melo, C. A., Jannuzzi, G. d. M., & Bajay, S. V. (2016). Nonconventional renewable energy governance in Brazil: Lessons to learn from the German experience. *Renewable and Sustainable Energy Reviews, 61*, 222–234. https://doi.org/10.1016/j.rser.2016.03.054.

Dyner, I., & Larsen, E. R. (2001). From planning to strategy in the electricity industry. *Energy Policy, 29*, 1145–1154. https://doi.org/10.1016/S0301-4215(01)00040-4.

Fiestas, R., & GWEC. (2011). Analysis of the regulatory framework for wind power generation in Brazil, 1–46.

Foley, T., Thornton, K., Hinrichs-Rahlwes, R., et al. (2015) Renewables 2015 global status report.

Ford, A. (1997). System dynamics and the electric power industry. *System Dynamics Review, 13*, 57–85. https://doi.org/10.1002/(SICI)1099-1727(199721)13:1<57::AID-SDR117>3.0.CO;2-B.

Ford, A. (1999). Cycles in competitive electricity markets: A simulation study of the western United States. *Energy Policy, 27*, 637–658. https://doi.org/10.1016/S0301-4215(99)00050-6.

Forrester, J. W. (1961). Industrial dynamics. MIT Press, Cambridge, MA.

Franco, C. J., Castaneda, M., & Dyner, I. (2015). Simulating the new British electricity-market reform. *European Journal of Operational Research, 245*, 273–285. https://doi.org/10.1016/j.ejor.2015.02.040.

Geller, H., Schaeffer, R., Szklo, A., & Tolmasquim, M. (2004). Policies for advancing energy efficiency and renewable energy use in Brazil. *Energy Policy, 32*, 1437–1450. https://doi.org/10.1016/S0301-4215(03)00122-8.

Georgiadis, P., & Besiou, M. (2008). Sustainability in electrical and electronic equipment closed-loop supply chains: A system dynamics approach. *Journal of Cleaner Production, 16*, 1665–1678. https://doi.org/10.1016/j.jclepro.2008.04.019.

Ioakimidis, C., Koukouzas, N., Chatzimichali, A., et al. (2012). Energy policy scenarios of CCS implementation in the Greek electricity sector. *Energy Procedia, 23*, 354–359. https://doi.org/10.1016/j.egypro.2012.06.025.

Jimenez, M., Franco, C. J., & Dyner, I. (2016). Diffusion of renewable energy technologies: The need for policy in Colombia. *Energy, 111*, 818–829. https://doi.org/10.1016/j.energy.2016.06.051.

Juárez, A. A., Araújo, A. M., Rohatgi, J. S., & De Oliveira Filho, O. D. Q. (2014). Development of the wind power in Brazil: Political, social and technical issues. *Renewable and Sustainable Energy Reviews, 39*, 828–834. https://doi.org/10.1016/j.rser.2014.07.086.

Karplus, V. J., Rausch, S., & Zhang, D. (2016). Energy caps: Alternative climate policy instruments for China? *Energy Economics*, 422–431. https://doi.org/10.1016/j.eneco.2016.03.019.

Kissel, J. M., & Krauter, S. C. W. (2006). Adaptations of renewable energy policies to unstable macroeconomic situations-case study: Wind power in Brazil. *Energy Policy, 34*, 3591–3598. https://doi.org/10.1016/j.enpol.2005.07.013.

Kunsch, P. L., & Friesewinkel, J. (2014). Nuclear energy policy in Belgium after Fukushima. *Energy Policy, 66*, 462–474. https://doi.org/10.1016/j.enpol.2013.11.035.

Linton, J. D., Klassen, R., & Jayaraman, V. (2007). Sustainable supply chains: An introduction. *Journal of Operations Management, 25*, 1075–1082. https://doi.org/10.1016/j.jom.2007.01.012.

Luna-Reyes, L. F., & Andersen, D. L. (2003). Collecting and analyzing qualitative data for system dynamics: Methods and models. *System Dynamics Review, 19*, 271–296. https://doi.org/10.1002/sdr.280.

Lund, P. D. (2009). Effects of energy policies on industry expansion in renewable energy. *Renewable Energy, 34*, 53–64. https://doi.org/10.1016/j.renene.2008.03.018.

Ministerio de Minas e Energia E. (2007). *Plano Nacional de Energia*, 2030.

Ochoa, C., Dyner, I., & Franco, C. J. (2013). Simulating power integration in Latin America to assess challenges, opportunities, and threats. *Energy Policy, 61*, 267–273. https://doi.org/10.1016/j.enpol.2013.07.029.

Oliva, R. (2003). Model calibration as a testing strategy for system dynamics models. *European Journal of Operational Research, 151*, 552–568. https://doi.org/10.1016/S0377-2217(02)00622-7.

Orjuela, J. A., Herrera, M. M., & Casilimas, W. (2015). Impact analysis of transport capacity and food safety in Bogota. In *Workshop Engineering Application* (pp. 7–13).

Orjuela-Castro, J., Herrera-Ramirez, M., & Adarme-Jaimes, W. (2017). Warehousing and transportation logistics of mango in Colombia : A system dynamics model. *Revista Facultad de Ingeniería, 26*, 71–85 https://doi.org/10.19053/01211129.

Pereira, A. O., Pereira, A. S., La Rovere, E. L., et al. (2011). Strategies to promote renewable energy in Brazil. *Renewable and Sustainable Energy Reviews, 15*, 681–688. https://doi.org/10.1016/j.rser.2010.09.027.

Pereira, M. G., Camacho, C. F., Freitas, M. A. V., & Da Silva, N. F. (2012). The renewable energy market in Brazil: Current status and potential. *Renewable and Sustainable Energy Reviews, 16*, 3786–3802. https://doi.org/10.1016/j.rser.2012.03.024.

Pereira, A. O., Cunha Da Costa, R., Costa, C. D. V., et al. (2013). Perspectives for the expansion of new renewable energy sources in Brazil. *Renewable and Sustainable Energy Reviews, 23*, 49–59. https://doi.org/10.1016/j.rser.2013.02.020.

Prostean, G., Badea, A., Vasar, C., & Octavian, P. (2014). Risk variables in wind power supply chain. *Procedia - Social and Behavioral Sciences, 124*, 124–132. https://doi.org/10.1016/j.sbspro.2014.02.468.

Pudjianto, D., Castro, M., Strbac, G., et al. (2016). Asymmetric impacts of European transmission network development towards 2050: Stakeholder assessment based on IRENE-40 scenarios. *Energy Economics, 53*, 261–269. https://doi.org/10.1016/j.eneco.2014.05.003.

Qudrat-Ullah, H. (2013). Understanding the dynamics of electricity generation capacity in Canada: A system dynamics approach. *Energy, 59*, 285–294. https://doi.org/10.1016/j.energy.2013.07.029.

Qudrat-Ullah, H. (2014). Green power in Ontario: A dynamic model-based analysis. *Energy, 77*, 859–870. https://doi.org/10.1016/j.energy.2014.09.072.

Qudrat-Ullah, H. (2015). Modelling and simulation in service of energy policy. *Energy Procedia, 75*, 2819–2825. https://doi.org/10.1016/j.egypro.2015.07.558.

Qudrat-Ullah, H. (2016). *The physics of stocks and flows of energy systems applications in energy policy*. London: Springer.

Qudrat-Ullah, H., & Seong, B. S. (2010). How to do structural validity of a system dynamics type simulation model: The case of an energy policy model. *Energy Policy, 38*, 2216–2224. https://doi.org/10.1016/j.enpol.2009.12.009.

Rendon-Sagardi, M. A., Sanchez-Ramirez, C., Cortes-Robles, G., et al. (2014). Dynamic analysis of feasibility in ethanol supply chain for biofuel production in Mexico. *Applied Energy, 123*, 358–367. https://doi.org/10.1016/j.apenergy.2014.01.023.

Sahay, N., Ierapetritou, M., & Wassick, J. (2014). Synchronous and asynchronous decision making strategies in supply chains. *Computers and Chemical Engineering, 71*, 116–129. https://doi.org/10.1016/j.compchemeng.2014.07.005.

Silva, A. R., Pimenta, F. M., Assireu, A. T., & Spyrides, M. H. C. (2016). Complementarity of Brazil's hydro and offshore wind power. *Renewable and Sustainable Energy Reviews, 56*, 413–427. https://doi.org/10.1016/j.rser.2015.11.045.

Simas, M., & Pacca, S. (2014). Assessing employment in renewable energy technologies: A case study for wind power in Brazil. *Renewable and Sustainable Energy Reviews, 31*, 83–90. https://doi.org/10.1016/j.rser.2013.11.046.

Sterman, J. D. (2000). *Business dynamics: Systems thinking and modeling for a complex world*. Boston: McGraw-Hill.

Tian, Y., Govindan, K., & Zhu, Q. (2014). A system dynamics model based on evolutionary game theory for green supply chain management diffusion among Chinese manufacturers. *Journal of Cleaner Production, 80*, 96–105. https://doi.org/10.1016/j.jclepro.2014.05.076.

Tiba, C., Candeias, A. L. B., Fraidenraich, N., et al. (2010). A GIS-based decision support tool for renewable energy management and planning in semi-arid rural environments of northeast of Brazil. *Renewable Energy, 35*, 2921–2932. https://doi.org/10.1016/j.renene.2010.05.009.

Troost, C., Walter, T., & Berger, T. (2015). Climate, energy and environmental policies in agriculture: Simulating likely farmer responses in Southwest Germany. *Land Use Policy, 46*, 50–64. https://doi.org/10.1016/j.landusepol.2015.01.028.

Varella, H. (2013). Mediçao de desemprenho na cadeia de energia eólica: proposta de um conjunto de indicadores de desempenho.

Vlachos, D., Georgiadis, P., & Iakovou, E. (2007). A system dynamics model for dynamic capacity planning of remanufacturing in closed-loop supply chains. *Computers and Operations Research, 34*, 367–394. https://doi.org/10.1016/j.cor.2005.03.005.

Wee, H.-M., Yang, W.-H., Chou, C.-W., & Padilan, M. V. (2012). Renewable energy supply chains, performance, application barriers, and strategies for further development. *Renewable and Sustainable Energy Reviews, 16*, 5451–5465. https://doi.org/10.1016/j.rser.2012.06.006.

Wüstemeyer, C., Madlener, R., & Bunn, D. W. (2015). A stakeholder analysis of divergent supply-chain trends for the European onshore and offshore wind installations. *Energy Policy, 80*, 36–44. https://doi.org/10.1016/j.enpol.2015.01.017.

WWF-Brasil – Fundo Mundial para a Natureza. (2015). Desafios e oportunidades para a energia solar fotovoltaica no Brasil : recomendações para políticas públicas, 44.

Yuan, J., Sun, S., Shen, J., et al. (2014). Wind power supply chain in China. *Renewable and Sustainable Energy Reviews, 39*, 356–369. https://doi.org/10.1016/j.rser.2014.07.014.

Zuluaga, M. M., & Dyner, I. (2007). Incentives for renewable energy in reformed Latin-American electricity markets: The Colombian case. *Journal of Cleaner Production, 15*, 153–162. https://doi.org/10.1016/j.jclepro.2005.12.014.

Chapter 9
Resilience as Basis for Sustainability: Shortages in Production Supply Chains for Essential Consumer Goods

Stefan N. Grosser and Lize Duminy

9.1 Introduction

Certain goods such as water, basic food items, and pharmaceuticals are required for health and well-being or to preserve life. Continuous availability of such products is crucial. These goods are termed "essential consumer goods," defined as physical items that are required to maintain health or sustain a person's essential physical functions (Van Boom and Garde 2016). This study investigates the concept of sustainable supply chain management applied to production supply chains of essential consumer goods.

Supply chain management has long used the triple-bottom-line dimensions of organizational sustainability (Elkington 1998, 2004). The triple-bottom line is used to integrate environmental and social aspects with economic ambitions (Dubey et al. 2017). The three pillars of the triple-bottom line are economic development, environmental protection, and social development (Elkington 1998, 2004). Social sustainability is the least defined and least prominent of the three sustainability pillars (Hutchins and Sutherland 2008).

The least published sustainability pillar, social sustainability, is however the most prominent sustainability dimension when evaluating supply chains manufacturing essential consumer goods. A supply shortage in essential consumer goods could threaten physical well-being or life itself (Van Boom and Garde 2016) meaning continuous, quality supply is in the interest of social sustainability of society. On a global level, as far back as 1985, this fact was acknowledged by the United Nations General Assembly voting unanimously in favor of adopting general guidelines for consumer protection, explicitly including the need for reliable, quality supply of

S. N. Grosser (✉) · L. Duminy
Bern University of Applied Sciences, School of Engineering, Biel, Switzerland
e-mail: Stefan.groesser@bfh.ch

H. Qudrat-Ullah (ed.), *Innovative Solutions for Sustainable Supply Chains*, Understanding Complex Systems, https://doi.org/10.1007/978-3-319-94322-0_9

essential consumer goods including food, water, and pharmaceuticals (Harland 1987). This topic therefore provides a unique academic contribution to a topic of practical importance.

The most recent survey conducted by the European Association of Hospital Pharmacists (EAHP) identifies antimicrobial agents and oncological medications to be the products that experience the most frequent shortages (EAHP 2014). Between the two types of medicines, cancer medicines shortages are particularly detrimental to society since a shortage in oncological medications has the most substantial negative impact on individual patient care (Pauwels et al. 2014). Of all oncological medications, affordable, generic injectable oncological medications most often experience shortages (Economist Intelligence Unit 2017; Cherny et al. 2016; Pauwels et al. 2015).

Supply shortages of inexpensive, generic, injectable oncological medications have a tangible negative impact on social and economic sustainability. Cancer patient outcomes, including survival rates, are significantly influenced by failure to provide therapeutic treatment or delay in the dispensing of chemotherapy (Rider et al. 2013). Additionally, a single product shortage has far reaching effects since it is common for the same product to be used through different protocols for the treatment of a variety of different cancers (Pauwels et al. 2014). Moreover, oncological medications shortages have negative financial repercussions on healthcare funding bodies, providers, and patients (Fox et al. 2014; Hayes et al. 2014; Havrilesky et al. 2012). The primary source of increased costs for the healthcare system was staff time mitigating shortages, while payers and insurers are most affected by the increased cost of alternative treatments (Hayes et al. 2014; EAHP 2014; Becker et al. 2013).

There are no known quantitative simulation studies on this topic due to a lack of available data. Some recent strides have been made in the publication of oncological medications shortage causes in Europe (Pauwels et al. 2014). However, Bogaert et al. (2015) and Pauwels et al. (2014) emphasize the scarcity of quantitative data preventing policy-makers from assessing the real state of medicines shortages or making estimations of which countries are most affected (Economist Intelligence Unit 2017). Available data is not conclusive enough for any meaningful quantitative study requiring numerical validation. This means that there is a need for thorough analysis where the methodology is not solely dependent on data.

This study proposes the first causal model showing the underlying supply structure of the European essential consumer goods supply chain. The model provides structural insight in cases where data is unavailable or unreliable through relying on structural validation. The sustainability pillar that most resonates with the inherent nature of essential consumer goods is social sustainability. The key performance indicators developed and used in this study therefore center around social sustainability. Additionally, our study looks at financial sustainability in terms of additional costs to the public healthcare system as well as manufacturers' lost income. The objectives of this study are the following:

1. Determine the most common causes of supply shortages relevant to the supply chain structure of generic injectable oncological medications in Europe.

2. Offer a structural overview of oncological medicines supply chains.
3. Develop a model that can simulate the causes of supply chain disruptions of the oncology medicine supply chain to investigate impacts from a causal perspective.
4. Identify appropriate social and economic sustainability performance indicators for robust policy analysis.

This paper choses system dynamics modeling (Sterman 2000) to model and analyze the supply chain. System dynamics is a time-dependent simulation tool that captures key system variables and their interactions as explicitly defined differential equations. Model behavior is driven by feedback. It is particularly apposite that our study uses a system dynamics approach since the technique has its roots in the field of supply chain management (Forrester 1958, 1961). Of note in this field is the concept of demand amplification or the bullwhip effect (Wangphanich et al. 2010; Lin et al. 2014a, b; Disney and Towill 2003). More recently, the concept of remanufacturing has gain considerable attention with many papers on closed-loop supply chains in consumer electronics and appliances (Georgiadis and Athanasiou 2013; Georgiadis and Besiou 2008; Vlachos et al. 2007; Bhattacharjee and Cruz 2015; Lehr et al. 2013). Even though remanufacturing is concerned primarily with sustainability, it is not applicable to the essential consumer goods production supply chain. Other noteworthy papers in supply chain management include a risk assessment of the fast fashion apparel supply chain (Mehrjoo and Pasek 2016) as well as capacity planning of a short lifecycle product supply chain (Kamath and Roy 2007).

The rest of this study consists of four parts. The first section contains a literature review of oncology medicines shortages in the European Union. This is followed by a model description. Next, thorough qualitative model is validated and the results presented. Finally, implications, limitations, and directions for future research are offered.

9.2 Literature Review

The absence of medical shortages is a key element of the social sustainability of essential consumer goods' supply chains. A precise definition of medicines shortages is therefore the basis to any qualitative or quantitative study on this topic. De Weerdt et al. (2015) identified 56 different definitions of medicines shortages. The World Health Organization (WHO) and the European Medicines Agency (EMA) have made substantial progress toward congruency in the definition of medicines shortages (WHO 2017; EMA 2016; Hedman 2016). This study uses the supply-oriented WHO definition stating that a shortage occurs when medicines are "considered to be insufficient to meet public health and patient needs" (WHO 2017).

Medicines shortages are a global problem (WHO 2015), with Europe being no exception. Over the last decade, the European Union has seen a considerable increase in frequency of medicines shortages (Economist Intelligence Unit 2017;

Cherny et al. 2016, Pauwels et al. 2015; EAHP 2014). Multiple surveys have determined that medicines shortages have a significant impact on patient care (Pauwels et al. 2015; EAHP 2014; EMA 2012). Reported outcomes from medicines shortages include failure to treat conditions, delayed treatment, or the prescription of less effective treatment (Pauwels et al. 2015; Goldsack et al. 2014; Becker et al. 2013; Gogineni et al. 2013).

The underlying causes behind medicines shortages are complex and multifaceted (Economist Intelligence Unit 2017; Birgli 2013; Kweder and Dill 2013). In the absence of reliable, standardized data regarding shortages (Bogaert et al. 2015; Pauwels et al. 2014), this study relies on the existing qualitative studies and surveys described in the remainder of this section.

Birgli (2013), commissioned by the European Association of Euro-Pharmaceutical Companies (EAEPC), was one of the first European studies of medicines shortages and has often been cited in later literature. Birgli (2013) proposes a non-exhaustive list of reasons for medications shortages, grouped into three key categories: economic, business, and manufacturing. Economic reasons refer to reduction in governmental healthcare spending including any form of price reduction, reference pricing, or payment delays (Birgli 2013). Business reasons for medications shortages include any reactions by manufacturers to the change in governmental healthcare spending throughout the supply chain. Business reasons include the tightening of payment terms, reduced product introductions, as well as market withdrawals, parallel distribution, as well as quotas or other supply chain filters (Birgli 2013).

Further work by Bogaert et al. (2015) investigated problems underlying medicines shortages. Their work consisted of semi-structured interviews with representatives of national health authorities, pharmaceutical companies, and wholesalers, as well as hospitals and community pharmacists in Belgium and France. Our study uses the results obtained from the group of countries studied to reflect on the European Union's perspective on shortages. Bogaert et al. (2015) classified determinants in three broad categories: manufacturing problems, distribution and supply problems, as well as economic problems. Bogaert et al. (2015) also had their interviewees rank the relative prevalence of various causes of shortages, which had been previously identified by Birgli (2013). The perceived prevalence of the most common causes is listed in descending order of frequency below:

1. Manufacturing problems.
2. Limited manufacturing capacity.
3. Parallel distribution.
4. Raw material shortages and rationing/quotas.
5. Non-compliance with regulatory standards.

Another study by Pauwels et al. (2014) reports on oncological medicines shortages in the EU. They investigated medicines shortages from seven member state countries to present a quantitative analysis of shortages in Europe: Belgium, the Netherlands, the United Kingdom, Italy, Germany, Spain, and France. Our study has limited reliability in terms of the sample size of 71 oncological medicines and therefore should only be an initial indication of the scope of the problem.

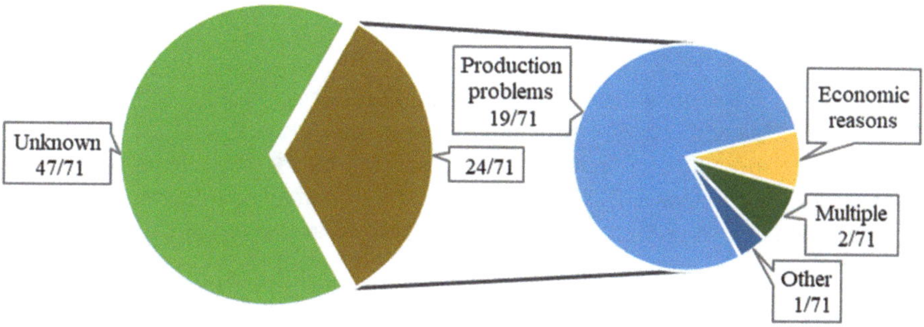

Fig. 9.1 Causes of oncological medicines shortages. (Pauwels et al. 2014)

The majority (74%) of oncological medicines facing shortages were administered through injection, while more than half (54%) were generic products. Figure 9.1 shows the causes of oncological medicines shortages[1]. The study's reliability is further diminished by the fact that the cause of 66% of the reported medicines shortages was unknown. Of the known causes, production problems were the most prevalent by a factor of nine (Pauwels et al. 2014).

Due to the fragmented nature of the information systems available to Pauwels et al. (2014), no further information is available on the specific types of production and economic complications. It is therefore assumed to correspond to the most common production and economic reasons for medicines shortages listed above. According to EAHP (2014), oncological medicines most commonly affected by shortages in Europe are 5-fluorouracil, carboplatin, cisplatin, doxorubicin (liposomal), etoposide, melphalan, methotrexate, oxaliplatin, and vincristine.

GIRP (2013) identified common underlying causes of medicines shortages from the pharmaceutical wholesalers' perspective. GIRP, the European Healthcare Distribution Association, is the umbrella organization for pharmaceutical full-line wholesalers and distributors of healthcare products and services in Europe. According to GIRP (2013), the occurrence of medicines shortages in the EU has increased over the past few years even though supply chain partners are obliged to comply with the EU legal framework ensuring the availability of medicines for patients. The primary underlying manufacturing cause of shortages, as perceived by the GIRP, is the difficulty in sourcing active pharmaceutical ingredients (APIs) caused by the introduction and subsequent changes in European Good Manufacturing Practice (GMP) guidelines that often result in onerous hunts for compliant API sources. Other reasons include parallel trade, economic incentive, or unforeseen demand.

In conclusion, this section defines the term medicines shortage and explains the negative effects medicines shortages have on patient welfare as well as treatment cost – translating to having negative impacts on social and economic sustainability. Our review concludes that supply shortages of generic, injectable oncological medicines are most commonly attributed to manufacturing and distribution hitches.

[1] Birgli (2013)'s headings "economic" and "business" correspond to "economic" causes of medicines shortages in Pauwels et al. (2014), while Birgli (2013)'s heading "manufacturing" corresponds to "production" causes of medicines shortages in Pauwels et al. (2014).

9.3 Model Description

This section describes the Oncological Medicine Supply Chain Model (OncoMedSCM) and starts by stipulating a clear model boundary, modeled causes of medicines shortages, as well as performance indicators. Next, the key assumptions underlying OncoMedSCM are made explicit. Finally, a description of the three echelons of OncoMedSCM is presented using an aggregate stock and flow diagram.

9.3.1 Model Boundary

Setting an appropriate model boundary is one of the principles for effective model development (Sterman 2000). The model boundary is therefore formalized as a prelude to the technical model description. The OncoMedSCM depicts a three-echelon supply chain. The echelons modeled are the active pharmaceutical ingredient (API) manufacturer, final product (FP) manufacturer, and hospital pharmacy (HP). The model is calibrated to Switzerland due to the availability of qualitative and limited quantitative data provided by a Swiss hospital pharmacist (Martinelli 2017). Oncological medicines that frequently experience shortages in Switzerland are carboplatin, cisplatin, and fluorouracil. The model description, including the model boundary, is purposefully kept broad so that OncoMedSCM can be applied to any other case given different input values, i.e., it could be used to model oncological medications production for the entire European Union instead of only Switzerland.

A theoretical representation of the model scope is shown in Fig. 9.2. A column is designated to each respective supply chain echelon: API manufacturers, FP manufacturers, and HP. Each entity is represented by a circle. An entity represents either a manufacturer or a hospital pharmacy, depending on the echelon in which the circle is placed. The model is calibrated to a specified market segment meaning the outer border of Fig. 9.2 contains all hospital pharmacies dispensing the specified oncological medicines within the market segment as well as all FP and API manufacturers assumed to be granted market access by the regulatory authority.

The OncoMedSCM has an organization-centric focus in which each modeled echelon represents a single entity within each link of the pharmaceutical supply chain. This means that OncoMedSCM models one API manufacturer, one FP manufacturer, as well as one hospital pharmacy – identified as solid circles in Fig. 9.2. The boundary of OncoMedSCM is depicted by the irregular shape overlaying a single entity per echelon.

Parallel trade is defined as the legal trade of legitimately produced goods between countries (Ganslandt and Maskus 2004). The model can simulate the total production size of both the API and FP manufacturers intended for the customers within the given market segment as well as customers of parallel trade even though the

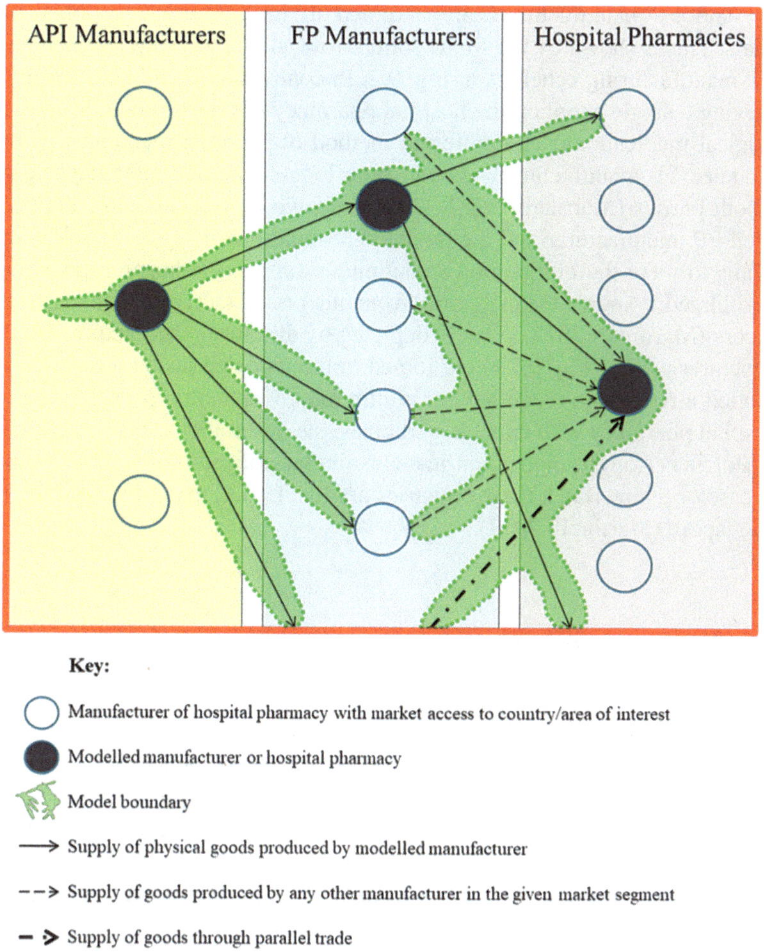

Fig. 9.2 Model scope showing three echelons of the oncology medicine supply chain

model depicts only a single downstream customer for each echelon[2]. This is shown in Fig. 9.2 by the irregular shape surrounding all arrows (representing the supply of physical goods) from each solid manufacturing echelon while excluding the corresponding receiving customers. The receiving customers, shown by a circular outline, are not depicted by the model. Outgoing parallel trade is depicted by the supply arrows pointing from each solid, modeled manufacturing echelon to the scope boundary. This indicates goods being shipped to customers outside the market segment studied.

[2]Due to the lack of data available, the total production sizes are estimated in the examples shown in this study.

API and FP manufacturers are assumed to have a single supplier of raw materials. This is shown by the single solid arrow pointing toward the modeled API and FP manufacturing echelons in Fig. 9.2. In contrast to the manufacturing echelons having a single supplier, the hospital pharmacy has three methods of sourcing oncological medications. The preferred method of sourcing medicines is through the modeled FP manufacturer – bound by a long-standing contractual obligation from both parties (Martinelli 2017). This is shown by a solid arrow between the modeled FP manufacturer and the modeled hospital pharmacy. If the FP manufacturer fails to meet the obligation of continuous supply, the hospital pharmacist is legally obliged to source the medicines from other manufacturers with existing market access (Martinelli 2017). This is depicted by all other manufacturers in the FP manufacturer area of Fig. 9.2 being joined to the modeled hospital pharmacy by a thin dotted arrow. If no manufacturer within the given market segment can supply the hospital pharmacy with enough medicines, the hospital pharmacy may then procure medicines from foreign countries with similar drug administration authorities. In the case of Switzerland, these countries are EC, EFTA, the USA, Canada, Japan, and Singapore (Martinelli 2017).

9.3.2 Key Assumptions

The assumptions underlying the model structure are listed below.

- **Order Amplification of Customers Not Explicitly Modeled**
  It is reasonable to assume that customers who are not explicitly included in the model react in the same way to supply shortages as those customers within the bounds of the model. This means that rather than having a static order rate for customers not modeled in the system, *other customer demands* exhibit the same amplifications of the modeled downstream customer. The net order rate is therefore amplified downstream of any disturbance in the supply chain but remains static with upstream disturbances. Upstream amplification could also be initialized, depending on the situation that the model user is depicting.
- **Permanent Change in Pharmaceutical Suppliers**
  Amplification is not the same as change in demand. Demand decrease as customers seek alternative permanent suppliers is beyond the scope of the model. It is assumed that all echelons modeled have the intention to keep their primary supplier – even during a period of temporary supply shortage.
- **Product Perishability**
  Most generic oncological medications are perishable (Martinelli 2017). Perishability of oncological medications is not considered due to their long shelf lives relative to the average inventory coverage at each echelon of the supply chain. For example, the shelf lives of carboplatin, cisplatin, and fluorouracil are 18 months, 24 months, and 24 months, respectively, while the average inventory coverage in a hospital pharmacy is around 2 weeks (Martinelli 2017).

- **Production Capacity Change**
  The OncoMedSCM models a single product even though our study assumes other medications are also produced in both manufacturing echelons. This is achieved by constraining production capacity to the production capacity of this specific product. When production capacity changes, it does not necessarily imply that the production capacity of the plant changes. Instead, it indicates a change in production capacity allocated to the medication modeled.

- **Plant Management**
  The opening and closing of production plants is not modeled to be endogenous to the system. This means that opening and closing production plants are not a result of variable interaction within the model; changes in production plants are initialized during model setup. The reason for excluding this element from endogeneity but not excluding it entirely is that the number of plants is dependent on changing legal regulations as well as executive-level business decisions rather than any causal relationships being addressed in the model. Even though the reasons for closure are beyond the scope of the model, naturally the effects of such closures have a palpable effect on oncological medication supply. It is therefore worth simulating the systems' reaction to changes in plant numbers.

Causes of Generic, Injectable Oncological Medicines Shortages

As mentioned previously, given the dearth of reliable data, this study has been forced to make assumptions about the root causes of oncological medicines shortages using literature study (Economist Intelligence Unit 2017; Cherny et al. 2016; Bogaert et al. 2015; Pauwels et al. 2014; EAHP 2014; Birgli 2013; Kweder and Dill 2013; GIRP 2013). Causes are classified as primary, secondary, or tertiary. Primary causes for medicines shortages are exogenous, often unpredictable, disturbances to the modeled causal structure. Primary causes of medicines shortages are shown in Table 9.1. The user has the freedom to initialize OncoMedSCM disturbances as a

Table 9.1 Primary causes of medicines shortages

Shortage causes	Echelon of origin	Disturbance name
Stoppages due to API shortage – extraction issues	API	*DISTURBANCE IN RAW MATERIAL AVAILABILITY*
Plant closures due to Good Manufacturing Practice (GMP) non-compliance	API and FP	*DISTURBANCE CAUSING PLANT CLOSURES DUE TO LOSS OF COMPLIANCE*
Closure or consolidation of plants by manufacturers due to financial incentive	API and FP	*DISTURBANCE IN PLANT NUMBER BY MANAGEMENT DECISION*
Shipping delays	API and FP	*DISTURBANCE IN TRANSIT TIME*
Change in overall demand as market access is gained in developing countries	API and FP	*DISTURBANCE IN OTHER CUSTOMER DEMAND*
Unavailability of inactive ingredient or packaging material	API and FP	*DISTURBANCE IN OTHER INGREDIENT AVAILABILITY*

pulse, sine wave, step change, ramp, and in some cases even random changes. Additionally, the user can initialize multiple disturbances at separate times throughout the simulation. In accordance to established naming convention (Rahmandad and Sterman 2012; Lai and Wahba 2001), the exogenous model variables are written in capital letters and *italicized.*

Secondary causes of medicines shortages are stakeholder decisions within the system that affect the resilience of the system structure to medicines shortages in the event of system disturbances. The secondary causes of medicines shortages are listed in Table 9.2 along with the corresponding OncoMedSCM policy variable.

Tertiary causes of shortages are causes of medicines shortages that do not have a tangible effect on the modeled system structure. Known causes of medicines shortages that do not have a direct effect on the oncological medication supply chain are defined as tertiary causes of medicines shortages. Tertiary causes of medicines shortages are beyond the scope of our study. Tertiary causes include pre-financing due to the financial crisis, the tendering process between European countries, and parallel trade.

9.3.3 Oncology Medicine Supply Chain Model Structure

This section contains an overview of the structure of both the manufacturing and hospital pharmacy echelons of OncoMedSCM. Mathematical equations can be found in a separate publication. An aggregated stock and flow diagram is presented due to the size and complexity of the three-echelon supply chain modeled. Additionally, the underlying structure driving behavior is explained with reference to the OncoMedSCM's feedback loops.

Ventana established a naming convention in Lai and Wahba (2001). OncoMedSCM documentation follows the convention of exogenous variables being in capital letters, converters (experiencing instantaneous causal effect) in lower case, and stocks capitalized in the first letter of each word. Additionally, causal connectors are curved if a link forms part of a feedback loop. Model parameters are identified by *italic* font in this document as prescribed in Rahmandad and Sterman (2012).

9.3.3.1 Manufacturing Echelons

The first two echelons of OncoMedSCM are manufacturing echelons and share the same structure but with a different calibration. The manufacturing echelons are based on the manufacturing supply chain proposed by Sterman (2000), with the inclusion of a basic production capacity management structure as well as additional order fulfillment structure. The intrinsic structure driving the behavior of the e production supply chain echelons in this model matches those of a typical industrial company involved in production and distribution. In terms of supply chain structure, OncoMedSCM is similar to edible, processed fast moving consumer goods such as milk powder.

Table 9.2 Secondary causes of medicines shortages

Shortage causes	Echelon of origin	Policy variable	Explanation
Supply less than demand due to capacity constraints	API and FP	*DESIRED CAPACITY UTILIZATION*	The desired capacity utilization is the percentage of total capacity the manufacturer wants to use for production. Spare capacity is used for maintenance or serves as a buffer to any change in either capacity or demand and therefore increases resilience toward disruptions
	API and FP	*DELAY IN PRODUCTION AUGMENTATION*	The lead time involved to any change in production capacity in response to a change in demand. This is usually a change in production scheduling rather than acquisition or decommissioning of production capacity. An agile reaction to a changing production rate will decrease the likelihood of shortages from lack of capacity at the expense of higher production cost
Stock out due to a policy of having low inventory levels of finished goods	API and FP	*SAFETY STOCK COVERAGE*	The desired amount of time existing inventory should last under the current demand. A larger safety stock implies greater resilience to shortages. This comes at the expense of higher storage costs as well as the accumulation of current assets that decreases liquidity
JIT manufacturing	API and FP	*SAFETY STOCK COVERAGE OF MATERIAL*	The desired amount of time an existing inventory of unprocessed material should last at the current production rate. A larger safety stock implies greater resilience to production disruptions at the expense of higher storage costs as well as the accumulation of current assets decreasing liquidity

Material Management

An aggregated overview of the material management system structure is shown in Fig. 9.3. The material management section of OncoMedSCM simulates the inventory levels of finished goods, unprocessed goods, and the acquisition of additional unprocessed goods. The primary premise of this section is that production can only begin if there is enough material and material are only replenished if acquired. This section consists of five stocks which are governed by nine feedback loops. Stocks involved in major system dynamics are stocks that form part of higher-order feedback loops and may therefore oscillate under certain conditions. The three stocks involved in major system dynamics are *Inventory*, *Material Inventory*, and *Material on Order*.

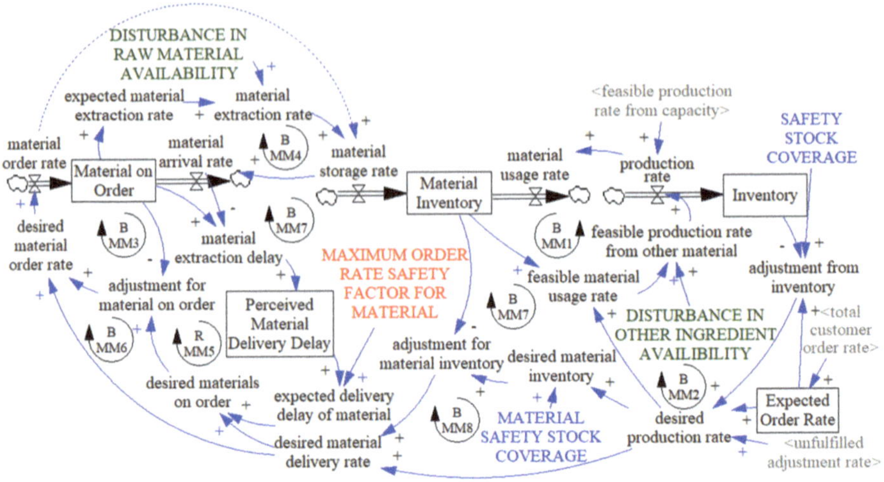

Fig. 9.3 Aggregate structure of the material management section of the manufacturing echelons

Inventory is defined as the total amount of goods in storage ready for shipment. *Material Inventory* depicts the amount of raw material ready to be processed. This stock only represents the primary ingredient within the pharmaceutical formulation. In the case of API manufacturers, this is often a biological ingredient occurring in nature, while the FP manufacturer considers API to be the raw material.

All other ingredients as well as miscellaneous goods such as pamphlets, packaging, etc. are not explicitly modeled. However, OncoMedSCM can simulate the effect of a shortage in any other ingredient or miscellaneous goods through the exogenous policy variable *DISTURBANCE IN OTHER INGREDIENT AVAILABILITY*.

Material on Order depicts the amount of primary ingredient the manufacturer anticipates receiving and represents an accounting value. In the case of the API manufacturer, *Material on Order* represents the planned extraction of raw ingredients. For the FP manufacturer, *Material on Order* depicts the amount of primary ingredient the manufacturer has ordered from the API producer and not yet received.

The final two stocks, *Perceived Material Delivery Delay* and *Expected Order Rate*, depict adaptive expectations of *material extraction delay* and *total customer order rate* over time. The adaptive expectations are modeled as first-order exponential smoothing of the respective goal functions.

First-order balancing feedback loops exhibit goal-seeking behavior. Since a first-order feedback loop contains a single stock or delay, no oscillations are possible, and these feedback loops have a stabilizing effect on the model's dynamics. First-order feedback loops in the material management section of the OncoMedSCM are loops labeled 1 to 4. In contrast, higher-order feedback loops with causal structures containing more than one stock or delay may well oscillate. Higher-order feedback loops are the source of complex model behavior; they are labeled 5 to 9.

The causal link shown by a dotted arrow in B_MM4 is only applicable to the API manufacturer. In the case of the FP manufacturer, *material order rate* and *material storage rate* interact with their upstream supplier of raw materials, the API manufacturer. Insights gained from the simpler API manufacturer structure remain valid for the FP manufacturing echelon since all causal paths of the downstream echelon between *material order rate* and *material storage rate* have the same net-causal-polarity as B_MM4.

Inventory and *Material Inventory* are related by the causal links shared by the two first-order balancing feedback loops: B_MM1 and B_MM2. The two feedback loops compete for dominance to control the behavior of the *production rate*, which is analogous to *material usage rate* and therefore drives both *Inventory* and *Material Inventory*.

The *Material on Order* inflow, *material order rate*, is regulated by first-order balancing feedback loop B_MM3 as well as the second-order reinforcing feedback loop R_MM5. R_MM5 reflects the confidence that purchasing managers have in their suppliers' ability to fulfill their agreed upon obligations.

Two major second-order balancing feedback loops regulate *Material Inventory* relative to *Material on Order*, that is, B_MM6 and B_MM7. The greater the difference between *Material on Order* and its' desired level, the more dominant B_MM6 becomes.

Two third-order feedback loops, B_MM8 and B_MM9, have a balancing force on all three major stocks. Loop dominance depends largely on the imbalance of the *Material Inventory* stock relative to *Inventory*. The greater the imbalance, the greater the dominance of B_MM9.

Order Processing

The order processing section of OncoMedSCM shows the structural interplay between orders from customers and subsequent fulfillment of orders. An aggregated overview of the order processing structure is shown in Fig. 9.4. The OncoMedSCM order processing section simulates the maintained backlog, product shipment, and subsequent delivery, corresponding to order fulfillment. The main premise of the order processing section is that only goods in storage can be shipped if ordered. This section consists of six stocks which are governed by six feedback loops. The four stocks involved in major system dynamics are *Inventory*, *Inventory in Transit*, *Unprocessed Backlog*, and *Shipped Backlog*.

Inventory in Transit forms part of the *Inventory* stock and flow structure introduced in Section 4.1.1. The stock *Inventory in Transit* is the recipient of *shipping rate*. The entire structure represents the flow of physical, finished goods from conception, represented by *production rate*, until reaching their delivery destination – ownership is transferred to the downstream customer by the *delivery rate*.

The backlog structure directly above the inventory structure keeps track of the unfulfilled customer orders. The backlog structure depicts *Unprocessed Backlog* and *Shipped Backlog*. The backlog structure represents an administrative value rather than physical goods. *Unprocessed Backlog* is the total number of customer

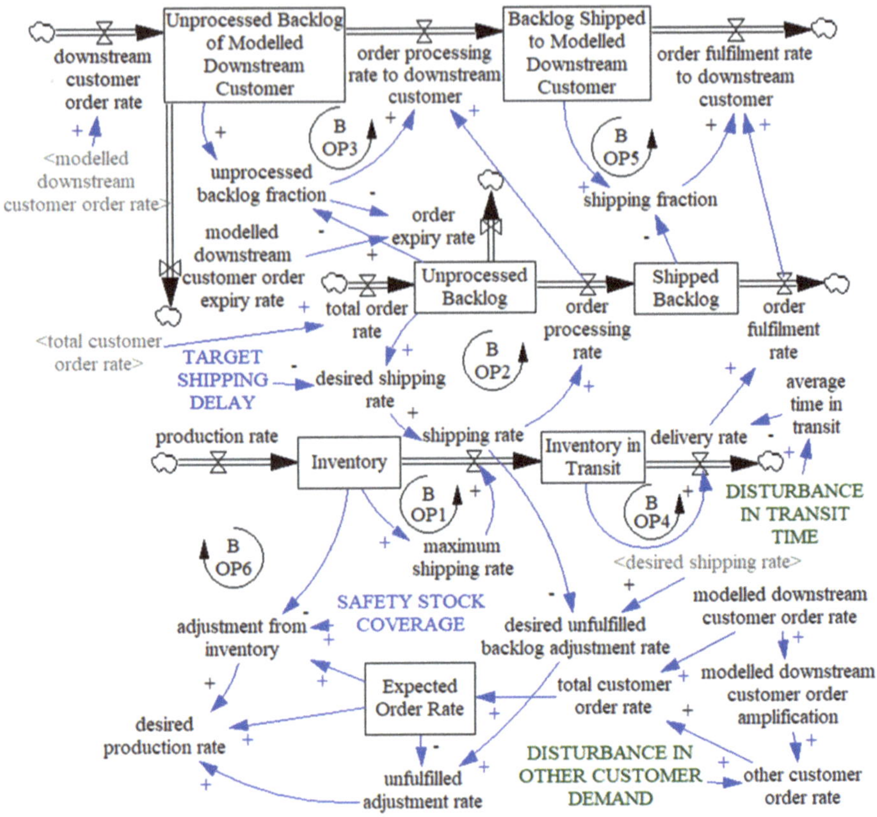

Fig. 9.4 Aggregate structure of the order processing section of the manufacturing echelon

orders received that are yet to be shipped, while the *Shipped Backlog* (always numerically equal to *Inventory in Transit*) is the recorded number of products in transit according to the inventory management system.

The final stock and flow structure introduced in the order processing sector of the manufacturing echelon is a co-flow of the backlog structure. A co-flow structures embodies a particular attributes of an item as it moves through a stock and flow structure (Sterman 2000). The co-flow is used to keep track of modeled downstream customer orders. Orders are assumed to be fulfilled using a relaxed First-In-First-Out (FIFO) policy. Older orders are generally processed first, without strict precision synonymous with discrete event-based modeling. Instead of strict FIFO policy, where orders would be fulfilled in the exact chronology of being received, the model shipment rates are proportional to the size of the total backlog relative to the backlog of the modeled downstream customer supplier. Such a formulation is compatible with the FIFO assumption except in short reactionary periods after large, singular changes in either the *modeled downstream customer order rate* or the *other customer order rate*. FIFO holds when both *modeled downstream customer order rate* and *other customer order rate* are amplified proportionally. This assumption is made since system dynamics modeling has no way to implement FIFO without extensive structural additions that would add trivial improvement in model accuracy. All downstream customers are assumed to have a standard agreement, and the OncoMedSCM therefore allocates supply shortages evenly among all customers. While the total backlog is used for the current echelon's decision-making, the modeled downstream customer backlog is required when assigning the amount of total orders processed to the modeled downstream customer.

The order processing section of OncoMedSCM is inherently stable. The dynamics are governed exclusively by first-order balancing feedback loops. Since there are no higher-order feedback loops in this section, the order processing section cannot be a source of continuous system oscillation.

B_OP1 and B_OP2 compete for dominance to determine the *shipping rate*. In the case where *Inventory* allows the manufacturer to supply the *desired shipping rate* in relation to the *Unprocessed Backlog*, B_OP2 maintains dominance. If *Inventory* is too depleted to ship the *desired shipping rate*, B_OP1 gains dominance, and *Inventory* is reduced at the *maximum shipping rate*. Complete dominance of B_OP2 is defined as an equilibrium condition.

B_OP3 fulfills the same purpose as B_OP2 in the co-flow structure, namely, to deplete the *Unprocessed Backlog of Modeled Downstream Customer* through shipping unprocessed orders. B_OP4 depletes *Inventory in Transit* as ownership is transferred to the downstream supply partner upon delivery. B_OP5 does the same job as B_OP4 in the co-flow structure. Finally, B_OP6 remains dormant in non-extreme conditions and prevents system paralysis from no new orders due to a major backlog in production (that would otherwise only start up in the event of new orders).

Plant Management

This section gives an aggregated overview of the plant management section of OncoMedSCM, shown in Fig. 9.5. The plant management section of the model depicts the number of production plants as well as the production capacity per plant. This model allows for long-term changes in production capacity of well-established production plants, depicted as changes in stock *Production Capacity per Plant*. The model also allows for near-instantaneous change in production level as demand or other resources fluctuate – constrained to the maximum value of the total production capacity available. The structural assumption underlying this section is that the production capacity can vary endogenously, while the number of plants fluctuates exogenously to the system. The plant management section is based on the premise that production can only happen when production capacity is available. This section is made up of five stocks, excluding the *Inventory* stock introduced in the material management section of OncoMedSCM that is also involved in plant management dynamics. The plant management system is governed by eight feedback loops. Two of the five stocks form part of major system dynamics: *Production Capacity on Order per Plant* and *Production Capacity per Plant*.

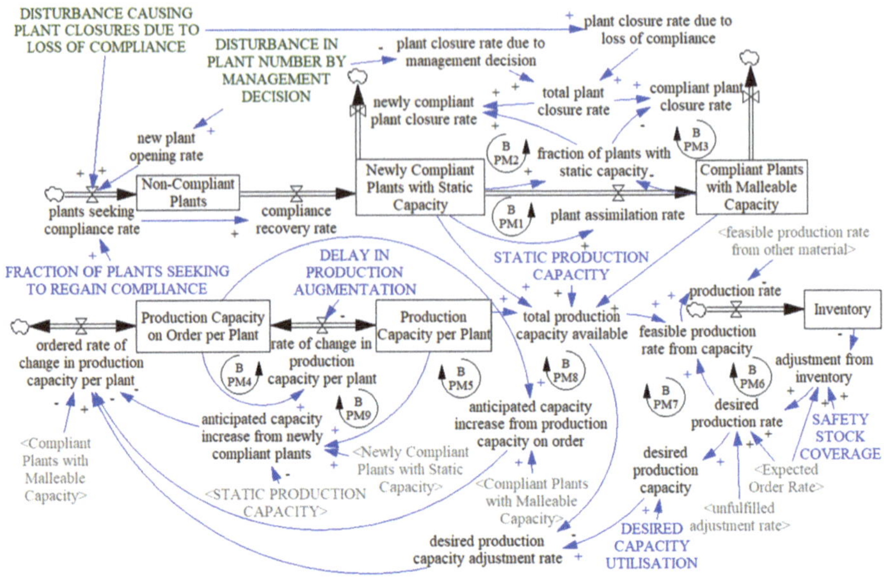

Fig. 9.5 Aggregate structure of the plant management section of the manufacturing echelon

The plant management section of OncoMedSCM is made up of two stock and flow structures. The first structure, shown in the top half of Fig. 9.5, represents the number of production plants. The production plant structure represents all production plants owned by the manufacturing echelon with interest in producing medicines for the identified market segment. Note that the manufacturer could own other plants that produce medicine sold in other market segments, not depicted in this model. The OncoMedSCM assumes that production plants have flexible production capacity only once they are certified and compliant to produce pharmaceuticals. Newly compliant plants have a static production capacity, while production processes are finalized. It is therefore necessary to distinguish between plants which could become compliant but not yet certified to produce for the given market segment, plants that are newly compliant with non-variable production capacity, as well as established plants with full compliance that can alter production levels. These categories are represented by the stocks *Non-Compliant Plants*, *Newly Compliant Plants with Static Capacity*, and *Compliant Plants with Malleable Capacity*, respectively.

The second stock and flow structure in the plant management section of the OncoMedSCM, shown at the bottom of Fig. 9.5, represent the production capacity per production plant of plants that are established enough in their production processes to allow variable production capacity. This structure is a depiction of a specific, dynamic attribute of the stock *Compliant Plants with Malleable Capacity*. The structure consists of two stocks: *Production Capacity per Plant* and *Production Capacity on Order per Plant*. *Newly Compliant Plants with Static Capacity* is assigned *ASSUMED STATIC PRODUCTION CAPACITY* while *Non-Compliant Plants* have no production capacity.

First-order feedback loops in this section of the OncoMedSCM are loops labeled 1 to 6 in Fig. 9.5. From all the feedback loops presented in this section, there are two second-order feedback loops and one third-order feedback loops, labeled from 7 to 9. Higher-order feedback loops have the potential of producing oscillations caused by multiple delays in the system. The structure depicting the number of plants in this model contains no higher-order feedback loops meaning that the production plant section of OncoMedSCM is inherently stable and cannot be a source of system oscillation.

B_PM1, B_PM2, and B_PM3 are first-order balancing feedback loops depleting stocks that represent production plants upon exogenous initialization. The first-order balancing feedback loop B_PM4 implements a planned change in production capacity.

Major dynamics in this section are led by the third-order feedback loop, B_PM7, which ensures mutual regulation between the inventory planning and plant management sections of the manufacturing echelon. B_PM7 is primarily dampened by the second-order balancing feedback loop B_PM8 that adjusts the capacity relative to the desired production rate as well as the first-order balancing feedback loop B_PM6 that constrains the production rate to the amount of production capacity available. Other adjustments to B_PM7's drive include B_PM9 and B_PM5, accounting for the *anticipated capacity increase from newly compliant plants* and the *anticipated capacity increase from production capacity on order*, respectively.

9.3.3.2 Hospital Pharmacy Echelon

The hospital pharmacy echelon shows the procurement and treatment process of a single generic oncology medicine within a hospital pharmacy. The model is based on an interview and written correspondence with the chief hospital pharmacist of a Swiss hospital pharmacy.

Medicines Sourcing

The medicine sourcing section of OncoMedSCM is described using an aggregated overview of the system structure shown in Fig. 9.6. Goods are acquired through three different channels of ordering, listed in order of preference: the modeled FP manufacturer, all other FP manufacturers with the given market access, and international suppliers with acceptable quality standards. This section consists of seven stocks, four of which are part of higher-order feedback loops that could produce oscillations: *HP Inventory*, *Primary Supplier Stock on Order*, *Second Supplier Stock on Order*, and *International Supplier Stock on Order*. The medication sourcing section behavior is governed by 12 feedback loops.

HP Inventory is defined as the total units of modeled oncology medicine stored in the hospital pharmacy ready for administration. It is assumed that *HP Inventory* is dispensed in a sterile area and patients are therefore not constrained to administration of an entire stock keeping unit (SKU): whatever medication is not given to the patient is stored for the next one.

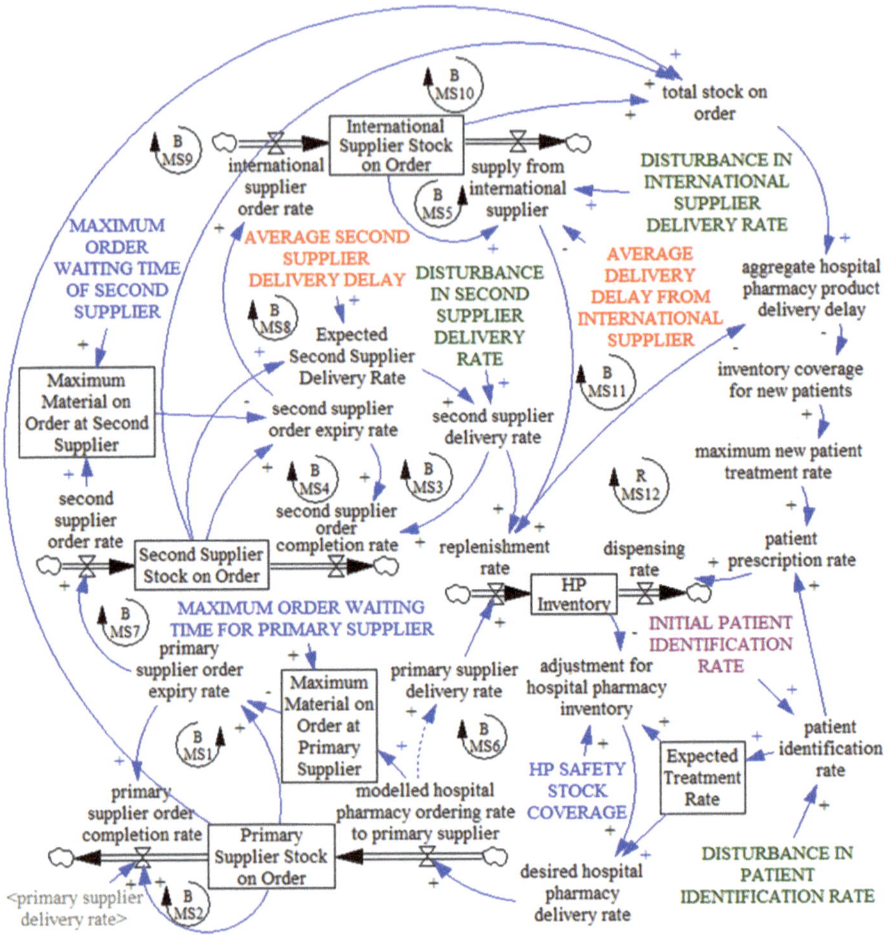

Fig. 9.6 Aggregate structure of the medicines sourcing section of the hospital pharmacy echelon

The stock on order at the modeled FP manufacturing echelon is described as the *Primary Supplier Stock on Order* in the medicines sourcing section of the OncoMedSCM. *Primary Supplier Stock on Order* numerically equals to *Unprocessed Backlog of Modeled Downstream Customer*. This stock represents an order number rather than physical goods. The hospital pharmacists have no guarantees that the stock orders will materialize before order expiry – implying a void in a contractual agreement after a specified waiting time.

In the case where orders placed at the primary supplier expire, the hospital pharmacists attempt to source medications from other producers. This is represented by the stock, *Second Supplier Stock on Order*, that represents the total stock ordered from all other producers with market access. As with the primary supplier stock, the

Second Supplier Stock on Order represents an order number with no guarantees that the stock will arrive before order expiry. If none of the producers with the required market access can supply the hospital pharmacy, hospital pharmacists source inventory from approved international sources with comparable quality standards; this is represented by *International Supplier Stock on Order*.

Two of the stocks not involved in any major system dynamics, defined using fixed delays, are *Maximum Material on Order at Primary Supplier* and *Maximum Material on Order at Second Supplier*. The purpose of both stocks is to keep track of the maximum amount of stock on order for a given order waiting period. This is used to determine the order expiry rate.

The final stock, *Expected Treatment Rate*, depicts adaptive expectations of the *patient identification rate* over time. This stock is modeled as first-order exponential smoothing of the goal function. In this case, the goal function represents the number of patients identified as candidates for the modeled oncology medication.

First-order feedback loops of the medicine sourcing structure of OncoMedSCM are labeled 1 to 5 in Fig. 9.6 while higher-order feedback loops labeled 6 to 12. First-order balancing feedback loops B_MS1 and B_MS4 remove expired orders from the *Primary Supplier Stock on Order* and *Second Supplier Stock on Order* stocks, respectively, while B_MS2, B_MS3, and B_MS 5 remove any orders received from suppliers from the three stocks containing medicines on order – these stocks are *Primary Supplier Stock on Order*, *Second Supplier Stock on Order*, and *International Supplier Stock on Order*.

B_MS6, which sources goods from the modeled upstream manufacturer of final products, is the preferred replenishment mechanism in this model. The dotted line in Fig. 9.6 represents the causal chain in the FP manufacturing echelon between when orders are received and when subsequent deliveries arrive. The mechanism replenishes *Hospital Pharmacy Inventory* relative to the *Expected Treatment Rate*.

B_MS7 sources goods from a second supplier once orders have expired at the primary supplier. B_MS7 remains dormant until it is activated by B_MS1. Similarly, B_MS8 sources goods from an international supplier once all the local options have been exhausted. B_MS8 remains dormant until activation by B_MS4.

When deciding whether to prescribe the first choice in treatment, medical practitioners consider the current stock levels available relative to the time it takes to source medicines (Martinelli 2017). The model therefore assumes medical practitioners would not prescribe the first choice in treatment if the *aggregate hospital pharmacy product delivery delay* exceeds the *hospital pharmacy inventory coverage*. The *aggregate hospital pharmacy product delivery delay* is determined through B_MS9, B_MS10, B_MS11, and R_MS12. B_MS9, B_MS10, and B_MS11 decrease *inventory coverage for new patients* as orders accumulate in *Primary Supplier Stock on Order*, *Second Supplier Stock on Order*, and *International Supplier Stock on Order*, respectively. In contrast, R_MS12 reinforces the process of prescribing the first choice in patient medication through a reliable replenishment of inventory.

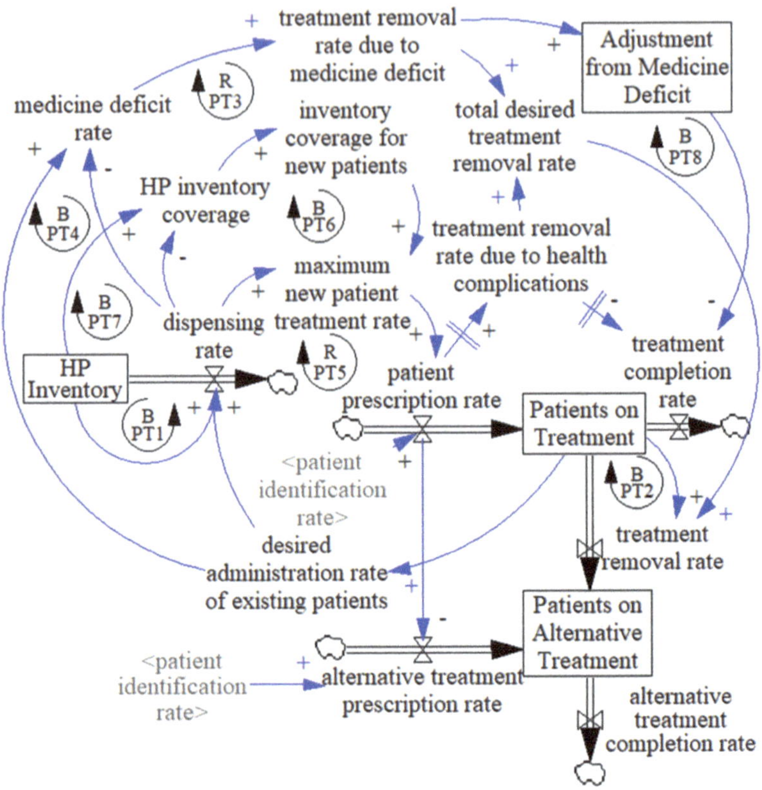

Fig. 9.7 Aggregate structure of the patient treatment section of the hospital pharmacy echelon

Patient Treatment

An aggregated overview of the patient treatment system structure of OncoMedSCM is shown in Fig. 9.7. The patient treatment section of OncoMedSCM simulates the patient management structure where patients are receiving continuous oncological medications for a prolonged time period. The primary premise of this section is that patients can only receive treatment if enough medication exists. If the inventory starts running low, new patients will not be prescribed the drug. In the event of a complete shortage, some patients will not be treated at all. This section consists of four stocks, three of which are involved in major system dynamics, *HP Inventory*, *Patients on Treatment*, and *Adjustment from Medicine Deficit*, and is governed by eight feedback loops.

HP Inventory has already been defined in the previous section as the total units of a specific medication in the hospital pharmacy inventory. *Patients on Treatment* represents all patients currently receiving the first choice in treatment. Finally, *Adjustment from Medicine Deficit* is the amount by which the *treatment completion rate* needs to be corrected when patients are removed from their treatment early. It is assumed that patients are removed based on priority and happens uniform to the time patients have spent in the stock.

Patients receiving treatment are routinely removed from treatment due to health complications related to the liver or kidney (Martinelli 2017). This removal occurs after a set number of treatments and is unrelated to medicines shortages but rather to avoid worsening the patient's health. Once a patient has been withdrawn from treatment, there is a fixed delay adjusting the treatment completion rate downward; this is represented by *treatment removal rate due to health complications*.

The only stock not involved in major system dynamics is *Patients on Alternative Treatment*. The stock depicts the number of patients whose first choice in treatment is the oncology medicine modeled but is receiving an alternative course of treatment due to health complications, missed opportunity of patient onboarding due to supply reliability concerns, or outright medicines shortages during the time receiving the first choice in treatment.

The first-order feedback loops in the patient treatment section of the OncoMedSCM are labeled 1 to 6 while the higher-order feedback loops are labeled 7 and 8. The two first-order balancing feedback loops, B_PT1 and B_PT2, deplete *HP Inventory* upon dispensing and divert patients to an alternative course of treatment.

The structure governing the decision to prescribe the first choice in medication, consisting of feedback loops B_PT5, B_PT6, and B_PT7, remains dominant in conditions where supply is so reliable that disruptions do not influence the number of patients on a specific treatment. The three feedback loops allow new patients identified for treatment to be prescribed treatment and ultimately receive treatment if there is enough *HP Inventory* for healthcare practitioners to assume that a supply shortage won't cause treatment disruptions in the foreseeable future, given the current *dispensing rate*.

B_PT4 is activated through the relaxation of R_PT3 in the extreme case where the hospital inventory must remove existing patients from their existing treatment program. Additionally, B_PT8 adjusts the regular *Patient on Treatment* depletion mechanism to avoid "double subtraction" of patients removed from treatment early.

9.4 Model Validation

Throughout the modeling process, it is crucial to apply model validation techniques. Exhaustive model validation is also a pre-requisite for making reliable interpretations of model behavior or performing robust policy analysis (Senge and Forrester 1980; Barlas 1996; Sterman 2000; Groesser and Schwaninger 2012; Pruyt 2013).

9.4.1 Model Testing

The model documentation tool "SDM-doc tool," developed by Argonne National Laboratory, was used for verification and documentation (System Dynamics Society 2017). Model documentation following the reporting guidelines of Rahmandad and Sterman (2012) ensures that the model is understandable, transparent, and fully replicable.

The model assessment results gives a quick overview of the basic elements and confidence-building tests of OncoMedSCM (Martinez-Moyano 2012). The model assessment results are organized into three categories in Fig. 9.8: model information, warnings, and potential omissions. A description of each table entry can be found in Martinez-Moyano (2012).

Model Information

The model information section shows the basic model information of OncoMedSCM being within the margin of reasonable model formulation for a supply chain. Information includes the size and complexity of the model, number of variables using metadata, model run specs, as well as general model completeness.

Model parameters should correspond conceptually and numerically with real life (Senge and Forrester 1980). Conceptual correspondence entails parameters matching elements of system structure and is therefore a combination of structure and parameter verification (Senge and Forrester 1980). The inclusion of parameters with no significance as independent structural components is often indicative of faulty model structure (Senge and Forrester 1980). The percentage of variables with dimensionless units is 15.5%, with the vast majority of dimensionless variables originating in the control panel view of OncoMedSCM. The control panel view produces model disturbances on demand for the purpose of model verification, validation, sensitivity analysis, or other testing procedures rather than generating any feedback.

Warnings

Figure 9.8 shows OncoMedSCM has two warnings, addressed below, as well as "equations with unit errors or warnings" given as indeterminate. Even though dimensional consistency cannot be used as an indicator of whether equations are correct, dimensional inconsistency proves equations are incorrect (Richardson and Pugh III 1981; Martinez-Moyano 2012). As a supplementary test, the Vensim® unit check determines all units to be correct.

The three final entries in warnings section of Fig. 9.8 identify functions commonly associated with discrete event-based or agent-based simulation models: "IF THEN ELSE," "MIN," "MAX," "STEP," "PULSE," etc. Such functions disrupt behavior otherwise produced endogenously from feedback. Since the purpose of system dynamics modeling is for feedback to drive behavior, serious verification and validation should justify opting to use such functions.

The "IF THEN ELSE" function has the power to alter the inherent behavior of the model given a prescribed qualifying condition. In reality, change is almost always gradual rather than abrupt, as is the nature of this function (Lai and Wahba

Model Information	Number
Total Number of Variables	583
Total Number of State Variables (Level+Smooth+Delay Variables)	66 (11.3%)
Total Number of Stocks (Stocks in Level+Smooth+Delay Variables) †	63 (10.8%)
Total Number of Macros	0
Variables with Source Information	0
Variables with Dimensionless Units	94 (16.1%)
Variables without Predefined Min or Max Values	583 (100%)
Function Sensitivity Parameters	0
Data Lookup Tables	0
Time Unit	Week
Initial Time	0
Final Time	150
Reported Time Interval	1
Time Step	0.015625
Model Is Fully Formulated	Yes
Modeler-Defined Groups	Yes
VPM File Available	- No -

Warnings	Number
Undocumented Equations	304 (52.1%)
Equations with Embedded Data (0 and 1 constants ignored)	31 (5.3%)
Equations With Unit Errors or Warnings	Unavailable
Variables Not in Any View	0
Incompletely Defined Subscripted Variables	0
Nonmonotonic Lookup Functions	0
Cascading (Chained) Lookup Functions	0
Non-Zero End Sloped Lookup Functions	0
Equations with "IF THEN ELSE" Functions	0
Equations with "MIN" or "MAX" Functions	50 (8.6%)
Equations with "STEP", "PULSE", or Related Functions	13 (2.2%)

Potential Omissions	Number
Unused Variables	17
Supplementary Variables	0
Supplementary Variables Being Used	0
Complex Variable Formulations (Richardson's Rule = 3)	30
Complex Stock Formulations	0

† Total stocks do not include fixed delay variables.

Fig. 9.8 OncoMedSCM assessment results using SDM-doc

2001). Therefore, we considered this function the most invasive to endogenous feedback and could not justify the implementation of any "IF THEN ELSE" functions in OncoMedSCM. Rather than altering behavior altogether, "MIN" and "MAX" functions limit the extent of effect of certain behaviors or function as boundary limits. We consider "MIN" and "MAX" functions to be less intrusive on dynamically created behavior than "IF THEN ELSE" functions. "MIN" and "MAX" functions were implemented using extensive verification and validation throughout the modeling process. "MIN" or "MAX" functions are present in 8.6% of model equations.

"STEP," "PULSE," and other related functions are time dependent rather than model behavior dependent. Such functions alter the model in a prescribed way at a set point during the simulation time horizon. "STEP," "PULSE," and related functions make up 2.2% of the total model equations. There are no "STEP," "PULSE," or related functions built into the model feedback loops to force certain model behavior through exogenous intervention. All "STEP," "PULSE," and related functions are in the control panel view of OncoMedSCM with the intention of potential activation of only one or two of the functions during model verification, validation, sensitivity analysis, or other testing procedures.

Potential Omissions

The last section in Fig. 9.8 reports on the potential omissions. Unused variables are any variables that do not provide any kind of input into any model structure. Unused variables in Vensim® are called "supplementary variables." These variables are used for reporting purposes in this project.

Complex variable formulations threshold is set to three input variables in accordance with Richardson (2000), who states that equations should have a maximum of three input variables to ensure model transparency (Martinez-Moyano 2012). No variables forming part of any model feedback loops or other endogenous behavior violate the Richardson (2000) guideline often referred to as the "Richardson rule." However, some variables in the control panel view or forming part of performance indicators contain more than three input variables.

Relying solely on the basic tests performed during automated documentation is not enough to promote deep understanding of complex systems; however, model documentation provides a broad overview of model structure which is the first step to enhanced model clarity and transparency (Martinez-Moyano 2012). As an additional means of validation, the System Dynamics Model Correctness Checklist was adhered throughout the model building process (Lai and Wahba 2001). OncoMedSCM adheres to all stipulations except point number 10, where "MIN" and "MAX" functions are used occasionally as explained previously.

9.4.2 Model Validation

Model validation is the process of determining whether a model meets the predefined objectives of the modeling study (Barlas 1996; Pruyt 2013). Unlike many other research areas where validity depends largely on the accuracy to which the model can reproduce historical data, system dynamics modeling is more concerned with the connection between system structure and behavior (Barlas 1996). Validation of system dynamics models is about building confidence in the usefulness of models for the purpose at hand (Sterman 2000). Valid models are therefore models deemed useful for their intended purpose (Pruyt 2013). Appropriate model validation tests as outlined by Barlas (1996) were applied at each iteration of the modeling process.

Pruyt (2013) identifies three categories of model validation tests: direct structure tests, structure-oriented behavior tests, and behavior reproduction tests. Direct structure tests assess model validity through comparing the model structure with what is known of the real system without any simulation (Barlas 1996). Structure-oriented behavior tests assess the validity of model structural composition by applying behavior tests to assess model-generated behavior patterns (Barlas 1996). Behavior reproduction tests focus on quantitative evaluation of model output relative to a historical reference mode (Pruyt 2013). The nature of this study required mostly direct structure tests as well as structure-oriented behavior tests. Only basic behavior reproduction tests were possible with the limited amount of numerical data – the project relies mostly on written and mental data as defined by Forrester (1980).

Data

System dynamics model structure and decision rules require three types of data during conception and development: numerical, written, and mental data (Forrester 1980). Most of the numerical data required for quantitative testing of OncoMedSCM was unavailable despite significant efforts made to collect such data. In the absence of quantitative data for statistical analysis, we made use of experiential data and thorough sensitivity analysis. We do not discount or minimize the necessity of statistical analysis and quantitative model validation expressed in Sterman (2000) but argue that data unavailability should not be a disqualifying factor for conducting a project that could benefit from an analysis involving feedback and dynamic behavior.

Validation Tests

Sterman (2000) summarizes the main model validation tests and assessments, their respective methods, as well as the required principle tools and methods required based on Senge and Forrester (1980), Barlas (1989, 1990, 1996). The method and extent to which we applied each method during the development of OncoMedSCM are summarized in Table 9.3.

Extreme condition tests were one of the validation tests conducted. According to Senge and Forrester (1980), crucial knowledge regarding the real system being modeled relates to consequences of extreme conditions. An example of an extreme condition test, depicting the extreme condition of a global halt in API extraction, is applied to the API manufacturing echelon to test the resulting system reaction. In reality, if the API raw material extraction process ceases the raw material, inventory will eventually be depleted resulting in a halt in production of API. With no new API being produced, inventory will eventually be depleted, resulting in zero shipping of API to downstream customers. This structural assumption holds true for OncoMedSCM, as shown in Figs. 9.9, 9.10, 9.11, 9.12, and 9.13, where the model calibrated to carboplatin's initial conditions. When the *API raw material extraction*

Table 9.3 Model validation tests performed on OncoMedSCM

Test	Method and degree of implementation
Boundary adequacy	Boundary chart, model scope sketch, aggregate causal loop diagrams, literature review, interview, e-mail correspondence, direct inspection of model equations
Structure assessment	Stock and flow maps, direct inspection of model equations, solicited expert opinion, partial model tests, development of disaggregated sub-models
Dimensional consistency	Inspection of suspect parameters, Vensim® dimensional analysis software
Parameter assessment	Partial model tests for subsystem calibration, judgment methods based on interviews, and e-mail correspondence
Extreme conditions	Equation inspection, testing response to extreme condition tests alone and in combination, subjection of model to large shocks, and extreme conditions to examine conformity
Integration error	DT-error test, comparison between different integration methods
Behavior reproduction	Qualitative comparison between model output and behavior, examination of responses to disturbances
Behavior anomaly	Loop knockout analysis, replacement of equilibrium assumptions with dis-equilibrating structures
Family member	Model calibration to the widest range or related systems
Surprise behavior	Qualitative analysis
Sensitivity analysis	Qualitative numerical sensitivity analysis (evaluating the relative significance of system change), behavioral sensitivity analysis, policy sensitivity analysis

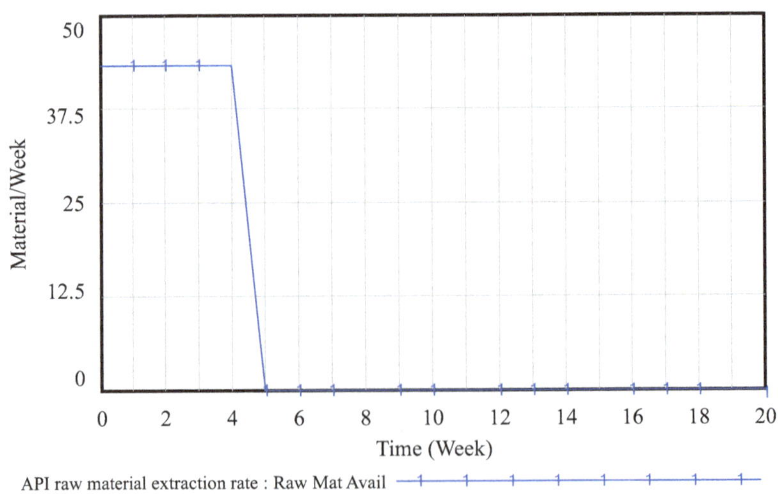

Fig. 9.9 API raw material extraction rate during extreme condition testing

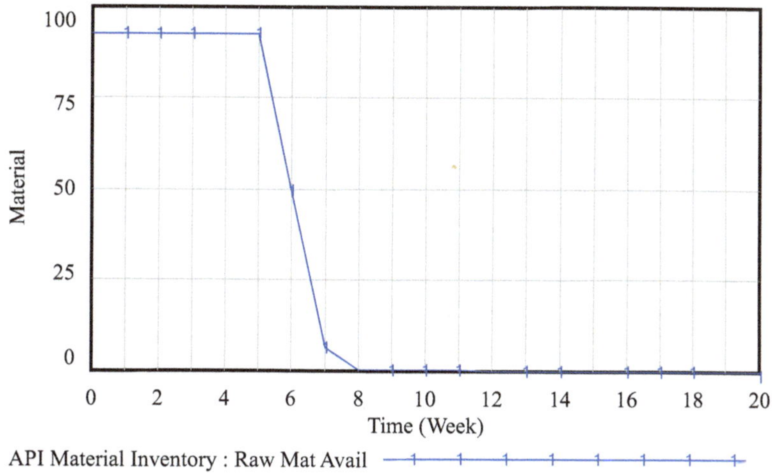

Fig. 9.10 API Material Inventory during extreme condition testing

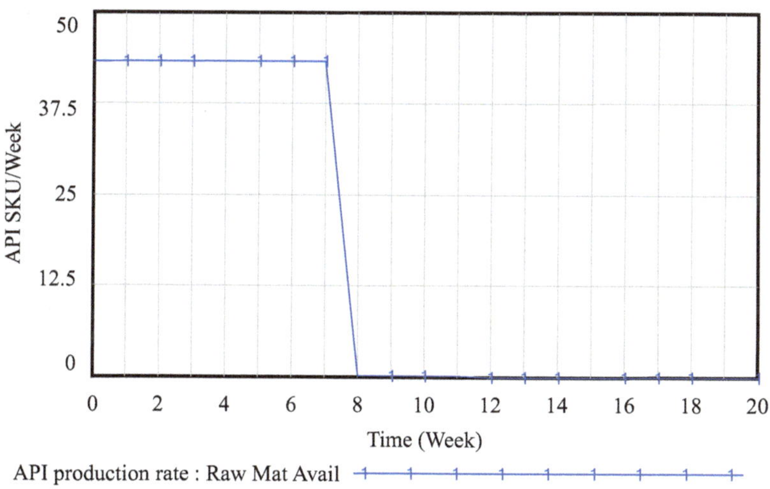

Fig. 9.11 API production rate during extreme condition testing

rate is reduced to zero in week 5, *API Material Inventory* will be depleted. If *API Material Inventory* reaches zero, the *API production rate* is halted. Furthermore, if *API Inventory* is depleted, the *API shipping rate* is reduced to zero.

From the validation tests described in Table 9.3 conducted throughout model development, we deemed OncoMedSCM to be valid and suitable for the qualitative exploration of future policies as well as studying the implications of different policies.

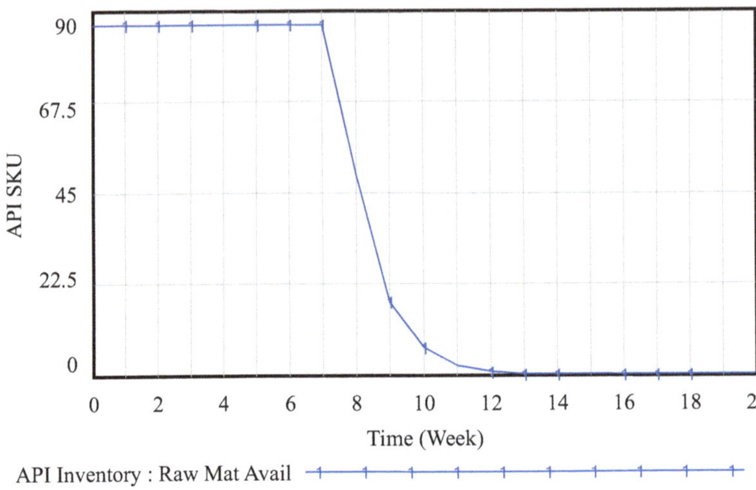

Fig. 9.12 API Inventory during extreme condition testing

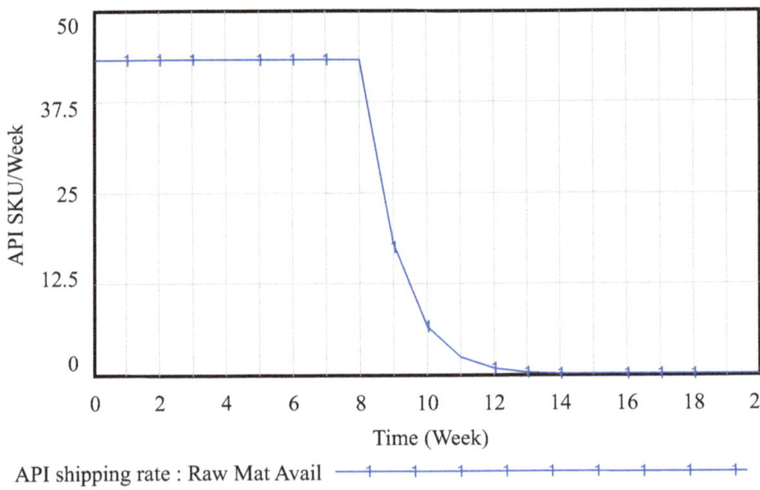

Fig. 9.13 API shipping rate during extreme condition testing

9.5 Simulation Results

The Oncological Medicine Supply Chain Model (OncoMedSCM) is the first quantitative model depicting the structure of the generic, injectable oncological medicines supply chain. Since there is no traditional reference mode to compare against the model output, this section presents two scenarios that are known causes of oncological medicines shortages. The scenarios are executed as controlled model tests accompanied by a causal explanation of how each shortage affects the oncological medicine supply chain.

When a model is initialized to replicate a reference mode, multiple, varying forces act upon the model throughout the simulation. It would not be scientifically accurate to attribute any model behavior to a single disturbance or to a single parameter, however major such the disturbance. Furthermore, it is impossible to deduce the true extent or long-term effect any parameter has on the system by examining how the system replicates the reference mode. Sterman (2000) prescribes model testing to be a process of controlled experimentation. The ideal condition for controlled experimentation is a model initialized in balanced equilibrium. Equilibrium refers to a state where the stocks in the model remain unchanged, implying that the net inflows should equal the net outflows. A balanced equilibrium has the added constraint that all stocks are equal to their desired values (Sterman 2000). OncoMedSCM was initialized in balanced equilibrated and calibrated for scenario testing. We refrained from using lookup functions in OncoMedSCM to ensure that a unique balanced equilibrium exists under any circumstance and can be defined using basic algebra.

The OncoMedSCM is initialized to represent the medicine carboplatin, as dispensed from a hospital pharmacy in Switzerland. The hospital pharmacy is assumed to treat approximately 1% of the total oncology patients in Switzerland. Additionally, the FP manufacturer is assumed to make up 1% of the total sales for the API manufacturer. The Vensim® model run specifications are listed in Table 9.4. The model can also be initialized to represent the medicine fluorouracil or cisplatin.

9.5.1 Model Output

The nature of the output produced by OncoMedSCM is given in this section. This section compares the relative reaction of equilibrated OncoMedSCM when confronted with two disturbances relative to the baseline of no disturbances. The simulation output shown was initialized to highlight various features of the model. The extent of disruption is not necessarily realistic. We therefore caution against drawing any conclusions from the example output shown in this section.

Table 9.4 Run specifications

Integration type	Euler
Simulation time	2 years
Unit of time	Week
Time step	$\frac{1}{(2)^7}$

Table 9.5 Description of disturbances

Run name	Implementation	Variable changed in model
1 Base	Model in equilibrium	No variable change
2 API	An instantaneous pulse outflow of *API Compliant Production Plants with Malleable Production Capacity*. Outflow set to 70% of total number of plants	API FRACTION OF PLANTS DEEMED NON COMPLIANT = −0.7
3 FP	A 10-week pulse reduction of the *FP feasible production rate* from other material. Reduction set to 70% of initial *FP production rate*	PULSE HEIGHT OTHER FP INGREDIENT AVAILABILITY = −0.7

OncoMedSCM was initialized in equilibrium as defined for carboplatin. The baseline scenario without disturbance is compared to two model runs. The disturbances are identified as primary causes of oncology medicines shortages that commonly affect oncology medicine supply chains. The three model runs are summarized in Table 9.5. Each disturbance takes place in a different echelon of the supply chain, and upstream and downstream effects of each disturbance are in Table 9.1.

The overall inter-echelon influence of each disturbance is presented using the general indicator, *inventory coverage*, before the social sustainability and economic sustainability indicators from the different scenarios are compared and discussed. *Inventory coverage* is considered the best indicator of overall echelon performance for each of the three echelons, respectively. *Inventory coverage* is defined as the amount of time; the existing inventory will last without further replenishment if the current usage continues. This value should be as close as possible to the equilibrium value: a higher *inventory coverage* could indicate loss in demand or an overproduction of inventory, while a low *inventory coverage* could indicate supply reliability issues. It reports on a combination of both the most important stock, *Inventory*, and the current supply expectations in each echelon. *Inventory coverage* is measured in weeks in all three echelons. Other performance indicators helpful during model interpretation, not shown in this paper, include *inventory*, *capacity utilization*, *material inventory coverage*, and *total average delivery delay*.

Run 1 Base: Equilibrated Model

As mentioned previously, the model was initialized in equilibrium and not subject to any disturbances. This scenario depicts the unlikely case in which there are no supply chain disturbances that could cause supply shortages. As shown in Fig. 9.14, the inventory coverage of all three echelons remains constant throughout the period of analysis since there are no disturbances causing any changes in the system.

Run 2 API: Manufacturing Plant Compliance Disturbance

Regular supply shortages are caused by API manufacturing plants being closed due to an update in the Good Manufacturing Practice (GMP) minimum standards or quality inspection failures (EMA 2012). In OncoMedSCM, this is simulated through

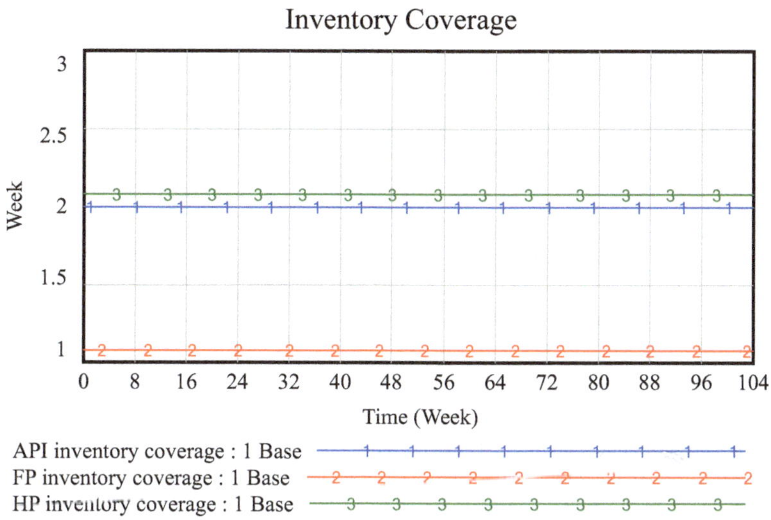

Fig. 9.14 Inventory coverage of three echelons during the 1 Base run

the activation of *DISTURBANCE CAUSING PLANT CLOSURES DUE TO LOSS OF COMPLIANCE* in the API echelon. A general overview of the effect of this disturbance for each of the three OncoMedSCM echelons is shown in Fig. 9.15.

The step reduction API production capacity causes a reduction in the *API inventory coverage* as shown by dataset 1 in Fig. 9.15. The *API inventory coverage* is constrained to the *API MINIMUM ORDER PROCESSING TIME*, defined as the minimum time it takes to process an order for shipment. Once *API inventory coverage* reaches the *API MINIMUM ORDER PROCESSING TIME*, the *API shipping rate* is reduced to below desired levels.

The delay between when the *API inventory coverage* reaches its minimum value and when *FP inventory coverage* starts decreasing is attributed to a shipping delay of 1 week as well as the unprocessed material safety stock. *FP inventory coverage* is shown by dataset 2 in Fig. 9.15.

As soon as supply to the FP manufacturer becomes unreliable, the ordering rates of FP manufacturers are amplified as trust erodes between supply chain partners. Order amplification is assumed to be the same for all downstream customers since all downstream customers are affected by the API manufacturer generated supply shortages. This amplification implies that customers order larger quantities than needed – ultimately meaning that the *API inventory coverage* stays at its extreme minimum value longer than necessary due to an erosion of trust between supply chain partners.

When *FP inventory coverage* reaches its minimum value, the hospital pharmacy orders start being affected by the supply shortage. Since there is a very small transit time (half-a-day) between the FP manufacturer and the HP, the *HP inventory coverage* (see dataset 3 in Fig. 9.15) seems to be affected almost immediately after *FP inventory coverage* reaches its minimum value. Given the resulting slight change in

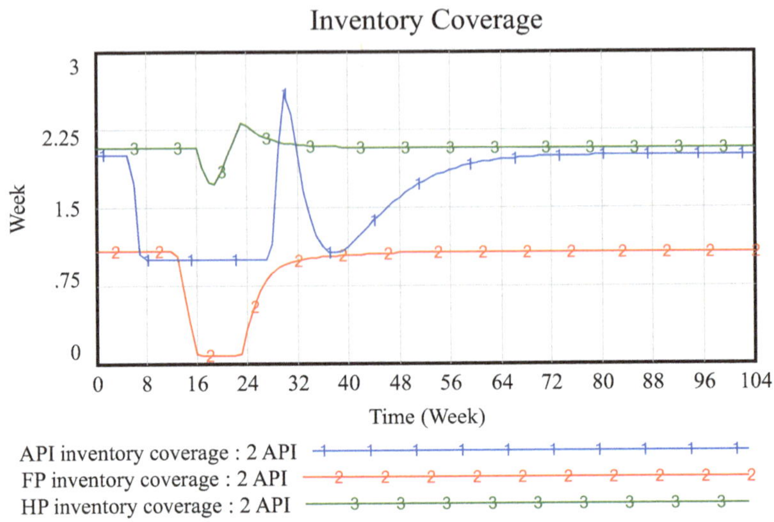

Fig. 9.15 Inventory coverage of three echelons during the 2 API run

the *HP inventory coverage*, the supply shortage appears to be minor for this scenario. After a brief period of order accumulation, the FP manufacturer starts catching up with orders around week 21. An indication of orders being fully caught up is the *FP inventory coverage* recovering from its minimum value. A fast recovery of *FP inventory coverage* is enabled by the fact that HP orders accumulate without amplifying like in the manufacturing echelons.

Run 3 FP: Other Ingredient Availability Disturbances

Another common source of supply shortages is the unavailability of inactive ingredients, packaging materials or even the dosage instructions which come in the packaging. In OncoMedSCM, this is indicated through *DISTURBANCE IN OTHER INGREDIENT AVAILABILITY*, which is manifested in the FP manufacturing echelon in this case. A general overview of the effect of this disturbance for each of the three OncoMedSCM echelons is shown in Fig. 9.16.

This disturbance acts on the FP manufacturer by reducing the production rate to a fraction of its initial value for a given period of time. During the time in which the *FP production rate* is less than desired, *FP inventory coverage* is reduced. *FP inventory coverage* is indicated by the dataset labeled 2 in Fig. 9.16.

A reduction in the *FP production rate* results in a reduction in FP order rate of raw materials from the API manufacturer. This increases the *API inventory coverage* as expectations try and keep up with reality. However, the influence of the disturbance on the API manufacturer is limited because it is a disturbance that only influences the ordering behavior of a single customer rather than the entire customer base, as in the previous example.

The severe nature of this supply shortage has a substantial effect on the hospital pharmacy. The *HP inventory coverage*, labeled as dataset 3 in Fig. 9.16, is supplemented by sourcing medication through alternative means, as well as by reducing

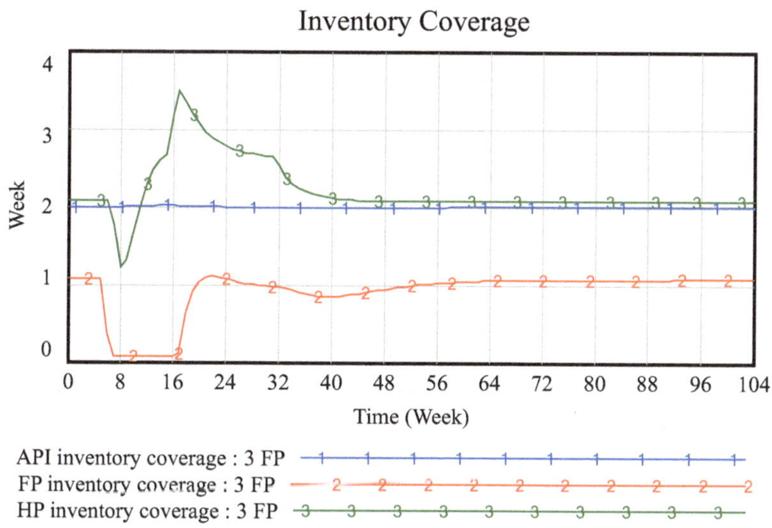

Fig. 9.16 Inventory coverage of three echelons during the 3 FP run

the number of patients being prescribed their first choice in treatment. The *HP inventory coverage* only returns to equilibrium once the dispensing rate realigns with the expected dispensing rate – assuming all patients receive their first choice in treatment. The *FP inventory coverage* shows a slight reduction after initial recovery. This is attributed to the misalignment of changing adaptive expectations from the reduction in patient orders, while some patients are not on their preferred treatment course.

9.5.2 Key Performance Indicators

A key part of a system dynamics modeling project is selecting the important metrics to measure performance in relation to the model objective. This is especially true for big models such as OncoMedSCM with 583 variables and 63 stocks. We selected the model key performance indicators with social and economic sustainability in mind. Another important requirement was that these indicators should be available in real life and in real time. The key performance indicators (KPIs) chosen for OncoMedSCM are listed below.

- *Fraction of patients receiving first choice in treatment.*
- *Total cost per patient per week.*
- *Cost of production per unit.*

Other performance indicators helpful during model interpretation, not shown in this paper, include *total potential patients on treatment*, *minimum patients on second treatment*, *hospital pharmacy unnecessary cost*, *total cost of patients identified for treatment*, and *treatment reassignment cost*.

This section introduces each KPI and compares the output for each of the three runs. The graphs in the remainder of the section are a comparison of the same variable in different runs, opposed to the previous section that compared variables from each model echelon during a single run. For each KPI graph, the baseline run is labeled "1 Base"; the disturbance originating in the API manufacturing echelon is labeled "2 API," while the disturbance originating in the FP manufacturing echelon is labeled "3 FP."

Fraction of Patients Receiving First Choice in Treatment

This is the primary social sustainability metric for the problem of oncological medicines shortages. The ideal fraction for this performance indicator is 1, implying that all patients are receiving the doctor's first choice in medication for their condition. This metric is only applicable to the hospital pharmacy echelon but is representative of the overall upstream supply chain performance. Medicines shortages lead to negative outcomes including failure to treat, delayed treatment, or prescription of less effective treatment (Pauwels et al. 2015; Goldsack et al. 2014; Becker et al. 2013; Gogineni et al. 2013). Considering the fraction of patients on the first choice of treatment rather than the absolute number of patients cancels out any noise from natural fluctuation in the number of patients identified for treatment. The fractional indicator has dimensionless units.

Of the example disturbances mentioned earlier, only the disturbance that originated in the FP manufacturing echelon was severe enough to compromise the social sustainability metric. The disturbance is manifested by a trough between weeks 7 and 35 in dataset "2 API" of Fig. 9.17. Even though the disturbance that originated in the API manufacturing echelon affected the *HP inventory coverage*, the safety stock was sufficient to prevent any changes in patient care.

A reduction in the *fraction of patients receiving first choice in treatment* implies that either potential patients were diverted to less optimal courses of treatment due to concerns about the availability of medication or in extreme cases existing patients being taken off treatment. Since medicinal treatment runs for a certain period, any recovery from a reduction in this fraction will happen gradually as the treatment courses of patients on less optimum treatments reach their conclusion.

Total Cost per Patient per Week

This is the primary economic sustainability metric for the hospital pharmacy echelon of OncoMedSCM. This value should be as low as possible. This is an estimate of the total weekly treatment cost incurred by the hospital pharmacy for all patients identified as candidates for a certain treatment, regardless of whether patients were successfully prescribed the preferred choice of treatment or receive an alternative treatment. A "per patient" indicator eliminates changes in the cost indicator caused by a change in number of patients identified for the treatment. This cost includes all sourcing costs, administration costs, alternative treatment costs (due to both health complications and shortages), and treatment reassignment costs. *Total cost per patient per week* is measured in CHF/(patient*week).

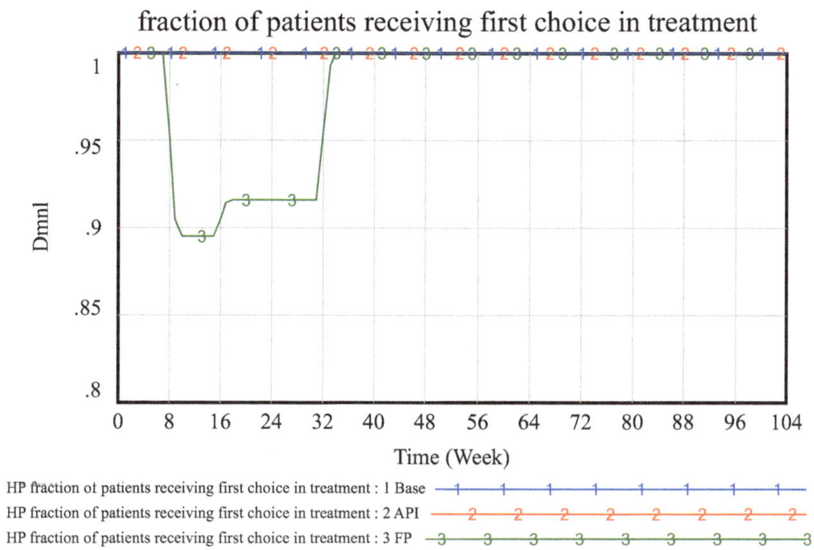

Fig. 9.17 Fraction of patients receiving first choice in treatment

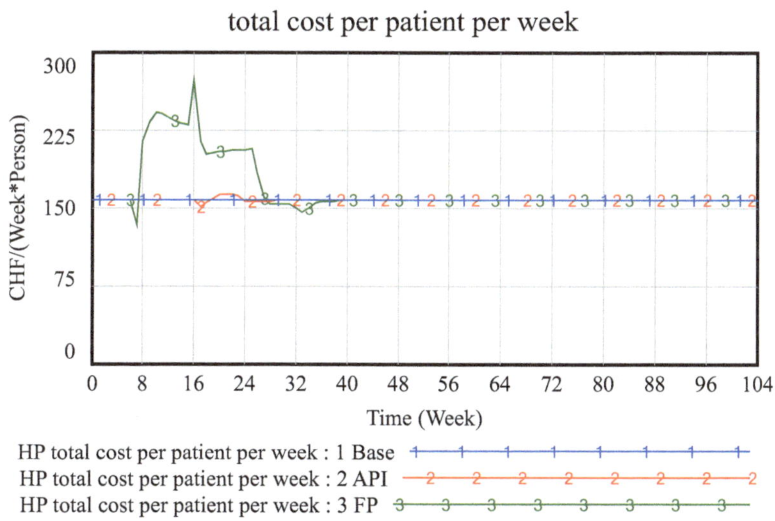

Fig. 9.18 Cost of patients identified for treatment per patient per week

Changes in the treatment cost are shown for both disturbances in Fig. 9.18. A change in treatment cost can occur even though the fraction of patients receiving first choice of treatment remains unchanged in "2 API." This indicates that the economic sustainability of a healthcare system is negatively affected by disruptions in supply even if the supply disruptions are successfully mitigated to preserve the system social sustainability.

The surges in treatment costs present in “3 FP” are attributed to administrative costs of sourcing medicines from alternative sources. Higher per unit medicines costs from non-regular suppliers drive up the cost in periods of supply instability. Reductions in treatment costs, exhibited in both “2 API” and “3 FP,” are caused by the absence of delivery (and therefore payment) of expected medication inventories.

The amplitude and wavelength of “3 FP” are greater than “2 API” indicating that the disturbance originating in the FP manufacturing echelon has a greater effect on the per patient cost than that of the API manufacturing echelon. Furthermore, a lag in the occurrence of the disturbance in “2API” is indicative of the FP manufacturing echelon serving as an inherent buffer to the API manufacturing echelon.

Cost of Production per Unit

This is the primary economic sustainability metric for the API and FP manufacturing echelons of OncoMedSCM. Similar to the previous KPI, this value should be as low as possible. Using a “per unit” performance indicator avoids any misinterpretation that could result from changes in production due to changes in demand. For the API and FP manufacturing echelons, the cost per unit produced encapsulates both fixed costs for manufacturing capacity and production variable costs per unit produced. This metric also exists for the hospital pharmacy echelon but is less insightful since it only captures costs associated with sourcing oncological medication. The *cost of production per unit* is measured in CHF/API SKU and CHF/SKU in the API manufacturing and AP manufacturing echelons, respectively.

The cost of production per unit for the API and FP manufacturing echelons is shown in Figs. 9.19 and 9.20, respectively. In both instances, the cost of production per unit is decreased as the production capacity utilization is maximized and increased if production capacity is left unutilized since it exceeds the desired or allowed production rate.

The run “2 API” causes a disturbance in the cost of all three echelons, while run “3 FP” has no effect on the per unit cost of its upstream supply chain partner. The initial reaction of the disturbance in “2 API,” originating in the API manufacturing echelon, manifests as a slight trough in the per unit cost of API production, shown in Fig. 9.19, indicating that production is at maximum capacity. The trough is followed by a peak overshoot as the available capacity exceeds production demand per product. In Fig. 9.20, run “2 API” shows a gradual increase in cost of production per unit as a shortage in raw materials inhibits production in the FP manufacturing echelon. The disturbance is ultimately dampened as the desired production volume and planned production capacity are equilibrated.

For both “2 API” and “3 FP,” the initial overshoot in per unit cost of production of the sector in which the disturbance takes place is comparable at slightly under 2. The run “3 FP” indicates an immediate disturbance in the per unit cost of production in the echelon in which the disturbance manifests: the FP manufacturing echelon. Unlike run “2 API” where a trough precedes the overshoot as production capacity is recovered, run “3 FP” immediately peaks as production capacity becomes underutilized due to a shortage of raw material.

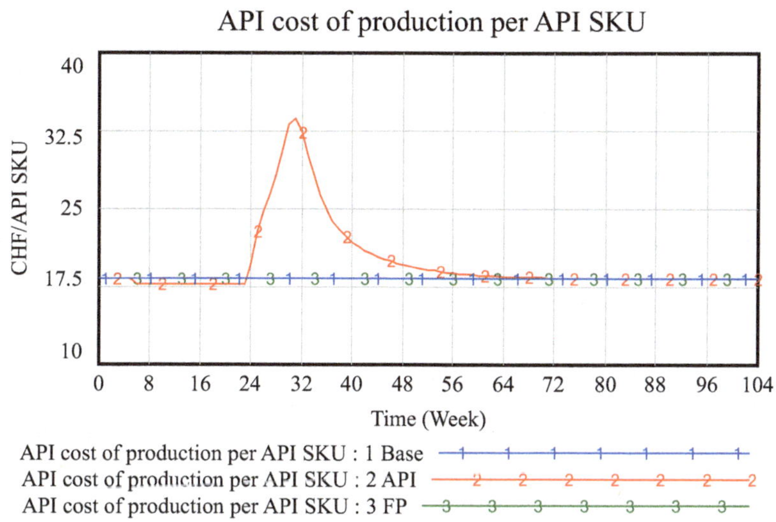

Fig. 9.19 Cost of API production per unit

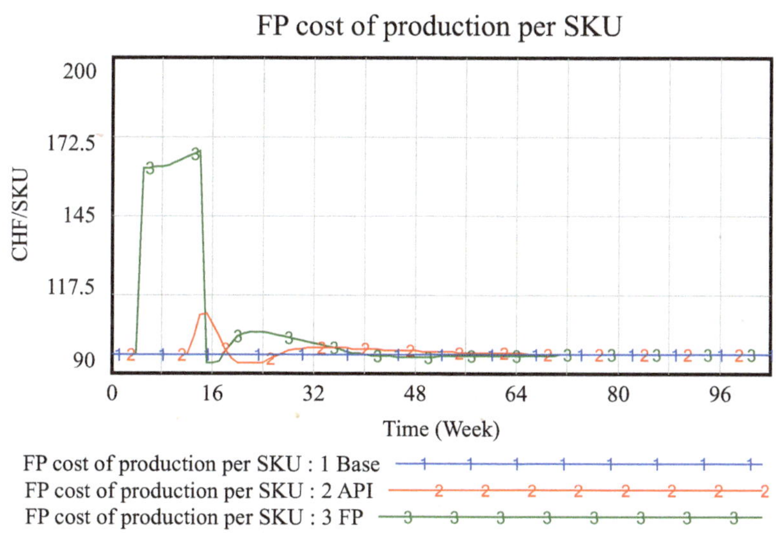

Fig. 9.20 Cost of FP production per unit

9.6 Conclusion

This study proposes the first causal model showing the underlying supply structure of the European essential consumer goods supply chain. Essential consumer goods are goods required to sustain health or life. Continuous, quality supply is therefore foremost in the interest of society. We took the case of inexpensive, generic, injectable oncological medicines shortages in Europe as they are proven to have a tangible negative impact on social and economic sustainability (Rider et al. 2013).

We identified the most common primary and secondary causes of oncological medicines shortages from literature. Then we presented a structural overview of the supply chain consisting of three echelons in the form of an aggregated stock and flow diagram. The three echelons described are an active pharmaceutical ingredient (API) manufacturer, a final product (FP) manufacturer, as well as a hospital pharmacy (HP). In the absence of a traditional reference mode due to data unavailability, structural model validation provided the necessary confidence in the validity of the Oncological Medicine Supply Chain Model (OncoMedSCM). As part of our study, an extensive model was developed to simulate the eleven primary causes and eight secondary causes of shortages as scenario variables and policy variables, respectively. Examples of the type of the output produced by the OncoMedSCM were generated by two scenarios. Finally, we proposed appropriate social and economic sustainability metrics as performance indicators.

We call for thorough quantitative investigation of the OncoMedSCM system output. Evaluation should include the comparison and sensitivity analysis of all primary and secondary causes of medicines shortages identified in this model. Focus should be on the determination of system leverage points or the identification of model segments most crucial to supply reliability. Thereafter, the existing policy recommendations relevant to the scope of OncoMedSCM presented by Economist Intelligence Unit (2017) should be evaluated using the model output and insights from the model results. Finally, new policy recommendations regarding management decisions that could be implemented as future updates to Good Manufacturing Practices should also be proposed.

References

Barlas, Y. (1989). Multiple tests for validation of system dynamics type of simulation models. *European Journal of Operational Research, 42*, 59–87.

Barlas, Y. (1990). An autocorrelation function test for out put validation. *Simulation, 55*, 7–16.

Barlas, Y. (1996). Formal aspects of model validity and validation in system dynamics. *System Dynamics Review, 12*, 183–210.

Becker, D. J., Talwar, S., Levy, B. P., Thorn, M., Roitman, J., Blum, R. H., Harrison, L. B., & Grossbard, M. L. (2013). Impact of oncology drug shortages on patient therapy: Unplanned treatment changes. *Journal of Oncology Practice, 9*, e122–e128.

Bhattacharjee, S., & Cruz, J. (2015). Economic sustainability of closed loop supply chains: A holistic model for decision and policy analysis. *Decision Support Systems, 77*, 67–86.

Birgli. (2013). An evaluation of medicines shortages in Europe with a more in-depth review of these in France, Greece, Poland, Spain, and the United Kingdom.

Bogaert, P., Bochenek, T., Prokop, A., & Pilc, A. (2015). A qualitative approach to a better understanding of the problems underlying drug shortages, as viewed from Belgian, French and the European Union's Perspectives. *PLoS One, 10*, e0125691.

Cherny, N., Sullivan, R., Torode, J., Saar, M., & Eniu, A. (2016). ESMO European Consortium Study on the availability, out-of-pocket costs and accessibility of antineoplastic medicines in Europe. *Annals of Oncology, 27*, 1423–1443.

De Weerdt, E., Simoens, S., Casteels, M., & Huys, I. (2015). Toward a European definition for a drug shortage: A qualitative study. *Frontiers in Pharmacology, 6*, 253.

Disney, S. M., & Towill, D. R. (2003). The effect of vendor managed inventory (VMI) dynamics on the Bullwhip Effect in supply chains. *International Journal of Production Economics, 85*, 199–215.

Dubey, R., Gunasekaran, A., Childe, S. J., Papadopoulos, T., & Fosso Wamba, S. (2017). World class sustainable supply chain management: critical review and further research directions. *International Journal of Logistics Management, 28*, 332–362.

EAHP. (2014). *Medicines Shortages in European Hospitals: The evidence and case for action*. Brussels: European Association of Hospital Pharmacists (EAHP).

Economist Intelligence Unit. (2017). *Cancer medicines shortages in Europe: Policy recommendations to prevent and manage shortages*. London: The Economist.

Elkington, J. (1998). *Cannibals with Forks: The triple bottom line of the 21st century*. Stoney Creek: New Society Publishers.

Elkington, J. (2004). Enter the triple bottom line. In A. Henriques & J. Richardson (Eds.), *The triple bottom line: Does it all add up?* London: Earthscan.

EMA. (2012). Reflection paper on medicinal product supply shortages caused by manufacturing/ Good Manufacturing Practice Compliance problems.

EMA. (2016). *Developing a proactive approach to the prevention of medicines shortages due to manufacturing and quality problems*. London. http://www.ema.europa.eu/docs/en_GB/document_library/Report/2016/01/WC500200281.pdf

Forrester, J. W. (1958). Industrial dynamics-a major breakthrough for decision makers. *Harvard Business Review, 36*, 37.

Forrester, J. W. (1961). *Industrial dynamics*. Cambridge, MA (now available from Pegasus Communications, Waltham, MA): MIT Press.

Forrester, J. W. (1980). Information sources for modeling the national economy. *Journal of the American Statistical Association, 75*, 555–566.

Fox, E. R., Sweet, B. V., & Jensen, V. (2014). *Drug shortages: A complex health care crisis*. Mayo Clinic Proceedings. Elsevier, 361-373.

Ganslandt, M., & Maskus, K. E. (2004). Parallel imports and the pricing of pharmaceutical products: Evidence from the European Union. *Journal of Health Economics, 23*, 1035–1057.

Georgiadis, P., & Athanasiou, E. (2013). Flexible long-term capacity planning in closed-loop supply chains with remanufacturing. *European Journal of Operational Research, 225*, 44–58.

Georgiadis, P., & Besiou, M. (2008). Sustainability in electrical and electronic equipment closed-loop supply chains: A system dynamics approach. *Journal of Cleaner Production, 16*, 1665–1678.

GIRP. (2013). Medicine shortages in Europe and their impact on patients: A reflection paper.

Gogineni, K., Shuman, K. L., & Emanuel, E. J. (2013). Survey of oncologists about shortages of cancer drugs. *New England Journal of Medicine, 369*, 2463–2464.

Goldsack, J. C., Reilly, C., Bush, C., Mcelligott, S., Bristol, M. N., Motanya, U. N., Field, R., Vozniak, J. M., Wong, Y.-N., & Schwartz, J. S. (2014). Impact of shortages of injectable oncology drugs on patient care. *American Journal of Health-System Pharmacy, 71*, 571.

Groesser, S. N., & Schwaninger, M. (2012). Contributions to model validation: Hierarchy, process, and cessation. *System Dynamics Review, 28*, 157–181.

Harland, D. (1987). The United Nations guidelines for consumer protection. *Journal of Consumer Policy, 10*, 245–266.

Havrilesky, L. J., Garfield, C. F., Barnett, J. C., & Cohn, D. E. (2012). Economic impact of paclitaxel shortage in patients with newly diagnosed ovarian cancer. *Gynecologic Oncology, 125*, 631–634.

Hayes, M. S., Ward, M. A., Slabaugh, S. L., & Xu, Y. (2014). Lessons from the leucovorin shortages between 2009 and 2012 in a medicare advantage population: Where do we go from here? *American health & drug benefits, 7*, 264.

Hedman, L. (2016). Global approaches to addressing shortages of essential medicines in health systems. *WHO Drug Information, 30*, 180.

Hutchins, M. J., & Sutherland, J. W. (2008). An exploration of measures of social sustainability and their application to supply chain decisions. *Journal of Cleaner Production, 16*, 1688–1698.

Kamath, N. B., & Roy, R. (2007). Capacity augmentation of a supply chain for a short lifecycle product: A system dynamics framework. *European Journal of Operational Research, 179*, 334–351.
Kweder, S. L., & Dill, S. (2013). Drug shortages: The cycle of quantity and quality. *Clinical Pharmacology & Therapeutics, 93*, 245–251.
Lai, D., & Wahba, R. (2001). System dynamics model correctness checklist. *MIT D-Memo, D-4851*(1), 15.
Lehr, C. B., Thun, J.-H., & Milling, P. M. (2013). From waste to value – a system dynamics model for strategic decision-making in closed-loop supply chains. *International Journal of Production Research, 51*, 4105–4116.
Lin, W.-J., Jiang, Z.-B., Liu, R., & Wang, L. (2014a). The bullwhip effect in hybrid supply chain. *International Journal of Production Research, 52*, 2062–2084.
Lin, W.-J., Jiang, Z.-B., & Wang, L. (2014b). Modelling and analysis of the bullwhip effect with customers' baulking behaviours and production capacity constraint. *International Journal of Production Research, 52*, 4835–4852.
Martinelli, E. (2017). Interview about medicine shortages (not published, personal communication).
Martinez-Moyano, I. J. (2012). Documentation for model transparency. *System Dynamics Review, 28*, 199–208.
Mehrjoo, M., & Pasek, Z. J. (2016). Risk assessment for the supply chain of fast fashion apparel industry: A system dynamics framework. *International Journal of Production Research, 54*, 28–48.
Pauwels, K., Huys, I., Casteels, M., & Simoens, S. (2014). Drug shortages in European countries: A trade-off between market attractiveness and cost containment? *BMC Health Services Research, 14*, 438.
Pauwels, K., Simoens, S., Casteels, M., & Huys, I. (2015). Insights into European drug shortages: A survey of hospital pharmacists. *PLoS One, 10*, e0119322.
Pruyt, E. (2013). *Small system dynamics models for big issues: Triple jump towards real-World complexity*. Delft: TU Delft Library.
Rahmandad, H., & Sterman, J. D. (2012). Reporting guidelines for simulation-based research in social sciences. *System Dynamics Review, 28*, 396–411.
Richardson, G. P. (2000). Richardson's rule. *Simulating dynamic systems (PAD 624 Course) lecture notes*. University at Albany, State University of New York.
Richardson, G. P., & Pugh, A. I., III. (1981). *Introduction to system dynamics modeling with DYNAMO*. Productivity Press: Cambridge, MA.
Rider, A. E., Templet, D. J., Daley, M. J., Shuman, C., & SMITH, L. V. (2013). Clinical dilemmas and a review of strategies to manage drug shortages. *Journal of Pharmacy Practice, 26*, 183–191.
Senge, P. M., & Forrester, J. W. (1980). Tests for building confidence in system dynamics models. In *System dynamics, TIMS studies in management sciences* (Vol. 14, pp. 209–228).
Sterman, J. (2000). *Business dynamics: Systems thinking and modeling for a complex world*. Boston: Irwin/McGraw-Hill.
System Dynamics Society. (2017). *Tools for system dynamics: SDM-Doc* [Online]. Available: http://tools.systemdynamics.org/sdm-doc/ Accessed 15 July 2017.
Van Boom, W., & Garde, A. (2016). *The European unfair commercial practices directive: Impact, enforcement strategies and national legal systems*. London: Routledge.
Vlachos, D., Georgiadis, P., & Iakovou, E. (2007). A system dynamics model for dynamic capacity planning of remanufacturing in closed-loop supply chains. *Computers & Operations Research, 34*, 367–394.
Wangphanich, P., Kara, S., & Kayis, B. (2010). Analysis of the bullwhip effect in multi-product, multi-stage supply chain systems-a simulation approach. *International Journal of Production Research, 48*, 4501–4517.
WHO. (2015). *Executive Board 138th session: Addressing the global shortage of medicines, and the safety and accessibility of children's mediciation*. Geneva: World Health Organization.
WHO. (2017). *Technical definitions of shortages and stockouts of medicines and vaccines*. Geneva: World Health Organization.

Chapter 10
Designing a Sustainable Tourism Supply Chain: A Case Study from Asia

Winai Wongsurawat and Vivek Shrestha

10.1 Introduction

While tourism is an increasingly prominent engine of economic growth for many economies, there is an ongoing debate about whether the benefits from the industry are fairly divided between local communities and powerful capitalists or large multinational corporations operating travel agencies, airlines, hotel chains, rental car services, and other accommodations (Mitchel and Ashley 2010). Local communities, with little access to capital and the rapidly changing technology infrastructure, critic of tourism argue, command little bargaining power, and are therefore pushed to the margins of this thriving tourism industry. It is possible that such lopsided distribution of benefits is contributing to worsening income and wealth inequality, a phenomenon that can erode the foundations of a well-functioning market economy.

Creating greater equity in the distribution of benefits from tourism requires a disruptive change to the current industry supply chain. This chapter aims to analyze the process in which a successful startup redesigned the tourism supply chain in multiple emerging Asian economies to redistribute greater benefits to disadvantaged members of local communities. *Peculiar Encounters*[1] is an online service providing authentic travel experience by matching tourists with locals who are willing to share native skills or traditional knowledge. Described as "the Airbnb of unusual travel experience," the startup was founded by a team of Singaporean undergraduate business students. With numerous offices offering services in cities across South and Southeast Asia (Cambodia, Laos, Myanmar, the Philippines, India, Nepal, etc.), the

[1] By request of the startup founders, the name of the startup as well as those of the founders have been changed to maintain anonymity.

W. Wongsurawat (✉) · V. Shrestha
College of Management, Mahidol University, Bangkok, Thailand
e-mail: winai.won@mahidol.ac.th

H. Qudrat-Ullah (ed.), *Innovative Solutions for Sustainable Supply Chains*, Understanding Complex Systems, https://doi.org/10.1007/978-3-319-94322-0_10

startup has so far proven to be commercially sustainable. With a social mission to uplift lives in developing countries, *Peculiar Encounters* is also growing in stature as a social enterprise. Using the experience of *Peculiar Encounters* thus far as a case study, this chapter will analyze the challenges involved in redesigning a new service supply chain that generates novel value for customers and at the same time shifts a bigger share of benefits toward previously marginalized players.

The central character in the case study is a social entrepreneur we will call Preston. Growing up in Singapore, Preston was bullied incessantly for being overweight. He had a poor command of English, stuttered, and performed miserably in school. Luckily, an uncle took the young Preston under his wings and began tutoring him in the subjects he had fallen behind at school. The coaching transformed Preston's life. He began to excel at school and took up sports seriously. By the time he graduated from high school, he had discovered self-esteem, won a place on the football team, and earned admissions to the business administration program at one of Singapore's top universities. During the summer before entering university, Preston went backpacking through Asia. In Vietnam, Laos, and Cambodia, he witnessed poverty for the first time. The experience planted in the young man a determination to do something to help the poor. The subsequent sections lay out this Preston's journey in building an innovative social enterprise within the tourism industry. The narrative will chronicle how Preston figured out a way to do for the poor and disadvantaged what his uncle had done for him as a boy – provide them with tools to gain self-esteem and better their lives. Insights and lessons from Preston's experience developing new tourism products and redesigning the industry's supply chain will be discussed and analyzed.

The structure of this chapter is arranged as follows. First, essential definitions and concepts including those of tourism supply chain, sustainable tourism, disruptive entrepreneurship and social enterprises will be introduced. A concise background of the histories and economies of the South and Southeast Asian countries featured in the case will then follow. The main material covering the *Peculiar Encounters* case study will then appear followed by an analysis and discussion of the major issues and lessons derived from the case. The chapter closes with a summary and some brief concluding remarks.

10.2 Definitions and Concepts

10.2.1 Tourism Supply Chain

A *tourism supply chain* is a network of organizations, both public and private, engaged in activities ranging from flights and accommodation to the distribution and marketing of the final tourism product at a specific destination (Zhang et al. 2009). A typical tourism supply chain involves suppliers, tour operators, travel agents, and customers. Suppliers can be further divided into two tiers. Tier 1

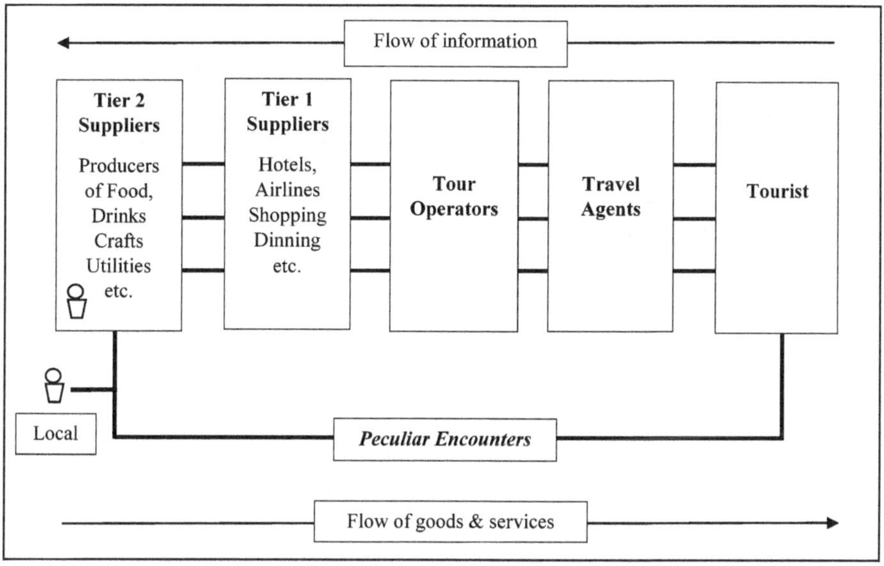

Fig. 10.1 Network structure of a typical tourism supply chain

suppliers include providers of accommodation, transportation, excursion, shopping, and dining. Manufacturers of food, equipment, and crafts along with basic utilities (power, water, garbage collection, etc.) are considered tier 2 suppliers. Similar to other types of supply chains, a tourism supply chain is characterized by a forward flow of services from suppliers to customers and a backward flow of information from customers to suppliers. See Fig. 10.1 for details.

Tourism is a coordination- and information-intensive industry (Zhang et al. 2009). It involves a complex, experience product whose quality is largely determined based on presentation and interpretation. Tourism services cannot be stored, forcing producers to depend on forecasted demand. Unfortunately, demand for tourism at a given location can be highly volatile, sometimes depending on difficult-to-control factors such as word of mouth. Zhang et al. (2009) identify the following seven critical issues in tourism supply chain management.

Demand management includes demand forecasting, marketing, and sales planning. Forecasting demand is critical for planning and executing important investments decisions, such as the construction of new infrastructure (airports, highways, rail links) by the government and expanding facilities or equipment (hotel rooms, aircrafts, cruise ships) by the private sector. Marketing efforts or government-led promotional campaigns, done well, can influence future demand for tourism at a particular destination. The number of travelers to a location may also shoot up due to serendipitous incidents such as a hit movie based in a city or country. See, for example, the impact of the hit comedy movie *Lost in Thailand* on the number of Chinese tourists to Thailand documented in Peel (2017).

Relationship management includes the management of vertical (between players in different parts of the supply chain, such as between agents and suppliers) and horizontal (between players occupying similar positions in the supply chain, such as between competing hotel chains) relationships. Both competitive and cooperative relationships have their benefits and drawbacks. While competition can create pressure for efficiency and quality improvements, it can also cause destructive price wars. Cooperative relationships are the bases for trust and fruitful collaboration but may also lead to inefficient cartels. Managing relationships with other key stakeholders such as governments, NGOs, and local communities must also be considered.

Supply management emphasizes supplier certification, selection, supplier-base reduction, and building long-term relations with suppliers. In the context of tourism, managing the supply of airline tickets and hotel rooms (both supplied by tier 1 suppliers) by travel agents and tour operators has been the main area of focus. There remain opportunities to understand and improve the supply of goods and services from smaller, tier 2 suppliers, which often involve local resources and communities.

Inventory management is critical for tourism supply chains, which are traditionally built on a *push* system where the production of tourist services is derived from demand forecasts (Zhang et al. 2009). With high fixed costs and perishable products, effective inventory management is vital. With the emergence of advance information technology, however, there is a possibility that a push system can be transformed into a scheme driven by customer pull. Supply chains based on a *pull* system, in which production is triggered by customer purchases, an arrangement that has become standard in many manufacturing and retailing environments (e.g., just-in-time production), are often more flexible and facilitate inventory cost minimization.

Product development is a key component of customer satisfaction, repeated business, and good word of mouth. In a world where technology is rapidly transforming ways, businesses can satisfy customer needs; the ability to quickly design and deliver new products or services that synchronize with emerging trends can become a source of competitive advantage. Designing new products in the information-intensive and coordination-intensive tourism industry requires both good understanding of the customer and joint effort among players along the value chain. Zhang et al. (2009) identifies product development as a significantly under-researched area of tourism supply chain management.

Supply chain coordination is often achieved by full or partial integration of the supply chain. Theuven (2004) points out that, to achieve seamless coordination, major European tour operators are highly integrated vertically with airlines, hotels, and other travel intermediaries. Yet, integration comes at a cost. Integrated supply chains require higher fixed costs and respond less flexibly to changing market environments (Sinclair and Stabler 1997). While the trade-off might have been necessary in the past, the rapid improvement of information technology, allowing players to conveniently monitor and communicate with each other, may soon make the attraction of full integration significantly less strong.

Information technology has transformed the tourism industry across many dimensions. As tourism is an experience good, information is the industry's "life blood" (Sheldon 1994). The growth of the internet and peer review websites has allowed the quality of tourism services to become more transparent. A tour operator that is slack in quality control will quickly damage its reputation and lose clients. Information technology also enhances coordination. Tour operators, for example, can access a hotel chain's room rates and availability and communicate an expected large group of visitors to all of its partners. Information technology may nevertheless turn into an emerging source of conflict. Travel agents, for example, are regularly threatened by hotel and airline's ability to offer their own booking services.

A number of critical issues identified by Zhang et al. (2009) above will be central to the analysis and discussion of the case study that will follow. Specifically, issues relating to stakeholder relationship management, product development, fixed costs, push vs. pull supply chains, and coordination through the use of emerging information technologies will emerge as recurrent themes in subsequent sections.

10.2.2 Sustainable Tourism

According to the World Tourism Organization (UNWTO n.d.), sustainable tourism encompasses environmental, economic, and sociocultural dimensions. The *environmental* dimension requires the prudent use of natural resources and the conservation of natural heritage and biodiversity. The central problem is to find a balance between the use and preservation of resources to maintain tourism development at the destination. Respecting traditional values of host communities and encouraging tolerance constitutes the *sociocultural* aspect of sustainable tourism. Finally, the *economic* dimension involves designing a tourism supply chain that enhances the livelihoods of all stakeholders, especially the host communities, and promoting a fair distribution of economic benefits to alleviate poverty.

The case study in this chapter is mainly concerned with the economic aspect of sustainable tourism in developing economies. How can a more equitable distribution of income generated from the tourism sector be achieved? Can indigenous people wrestle a larger share of the earnings from corporations? While using laws and regulations (e.g., redistributive taxation or minimum wage laws) is a widely discussed approach, such policies may entail unintended consequences (e.g., lower investment, unemployment) that can reduce economic growth in the long run, an outcome indisputably bad for the poor. An approach that is more decentralized and potentially more fruitful is harnessing the creative powers of innovation and entrepreneurship to help bring about sustainability. The following subsections review the influential concepts of *disruptive innovation* and the increasingly popular idea of a *social enterprise*, notions that are both important for realizing sustainability through decentralized, private action.

A key theme that will emerge from the case study is that, rather than giving locals more capital or resources, economic sustainability may be better achieved by providing communities with better means to utilize capabilities that they already possess to generate additional income. Current local capabilities may need to be tweaked and improved. The local people, however, are often eager to work with social entrepreneurs to adjust their knowledge and skills to match the evolving demands of the modern world.

10.2.3 Disruptive Innovation

Ever since the publication of *The Innovator's Dilemma* (Christensen 1997), "disruption" has gained a prominent place in the business management lexicon. In essence, by earnestly focusing on improving products or services delivered to its current customer base, a large and successful incumbent can fail to prepare itself for a major industry transformation that is soon to be brought about by a currently inferior but rapidly improving innovation. The incumbent's blind spot can allow a lean and hustling startup, built upon the emerging technology, a foothold in the down-market. By the time the incumbent realizes the impending industry transformation, the startup entrant has already gained an insurmountable lead in the new technology, allowing it to move up-market to completely overtake the once dominant firm.

The stunning rise of Airbnb, the online marketplace for lodging and hospitality services, is an illustrative example of the disruption process. Initially targeting budget travelers who were willing to sleep on an airbed, the entrepreneurs who founded Airbnb were pioneering a new service – based on a sharing economy[2] – that at the time would not have been appealing to the core customers of the giant hotel chains such as the Marriott or Hilton. Yet, within less than 10 years, Airbnb has grown to become one of the largest players in hospitality services, successfully moving up-market to offer bookings that include luxury accommodations. As of 2017, Airbnb has over 3 million listings, is valued at US$30 billion, and among hotel chains is second only to the Marriott (US$34 billion) in market capitalization. Other widely discussed examples of disruptive startups include Uber in taxi services, Netflix in movie rentals, and Amazon.com in retail.

Christensen et al. (2006) coined the term "catalytic innovation" to describe a subclass of disruptive innovation that can catalyze social change. By creating scalable, sustainable, system-changing solutions to social-sector problems, entrepreneurs can overcome societal challenges that established incumbents wedded to the status quo are unlikely to accomplish. MinuteClinic, an American, for-profit clinic chain operating out of CVS drug stores, is identified by Christensen et al. (2006) as an example of catalytic innovation. Created as a low-cost substitute for traditional healthcare providers such as a doctor's clinic or hospital, MinuteClinic counters are operated by nurses who use software-based protocols to serve walk-in patients with basic diagnosis and treatment. While the service offered is clearly inferior to what a

[2] According to the World Economic Forum (Rinne 2017), the sharing economy model focuses on the sharing of underutilized assets, monetized or not, in ways that improve efficiency, sustainability, and community. The underutilized asset in the case of Airbnb would be, for example, apartment rooms that are available for temporary rental. Uber is another example of a sharing economy applied to transport services (underutilized cars and drivers).

patient would receive from a hospital, MinuteClinic has made valuable contributions to one of America's biggest social challenges – healthcare affordability.

A second example of a catalytic innovation is online education. While American public schools struggle to offer regular foreign language and advanced placement courses to their students due to budget problems, for-profit and nonprofit startups have made great strides in providing thousands of students the opportunity to enroll in specialized courses through their online curricula. While an online course is an inferior alternative to the learning experience in a real classroom, the quality of the online material is rapidly improving, and the innovation so far has contributed to the expansion of educational opportunities to students without reliance on governmental sponsorship. A third example of a catalytic innovation put forth by Christensen et al. (2006) involves investing in economic development and furnishing the poor with microloans. The case of the Grameen Bank social enterprise is detailed in the following subsection.

10.2.4 Social Enterprises

A social enterprise is a financially self-sustaining organization with a clearly defined social mission. Unlike a traditional business, social enterprises do not aim to maximize profits for their shareholders but strive to tackle a societal problem, be it crime, poverty, unemployment, or environmental degradation. However, unlike a charity organization, a social enterprise does not rely exclusively on donations or government support. The enterprise can survive and grow on a steam of revenue generated through its initiatives and activities. See Haught (2005) for a discussion of the definition and scope of social entrepreneurship.

Founded in 1983, the Grameen Bank is a widely admired example of a social enterprise. Concerned that a lack of access to loans is a significant impediment for many disadvantaged but industrious individuals to escape poverty, Muhammad Yunus, an economics professor at the University of Chittagong in Bangladesh, initiated a research project aimed at figuring out a way to lend to the poor in a cheap but profitable way. The Grameen Bank was a business enterprise that grew out of the important insights from the project – that small, standardized loans made to groups of women without collateral but with strict repayment schedules came close to eliminating loan defaults by poor borrowers. Yunus's microcredit model has proven to be a great success story, extending access to credit to the poor, without any reliance on charitable donations or government handouts. In 2006, the Grameen Bank and Yunus were jointly awarded the Nobel Peace Prize "for their efforts to create economic and social development from below."

Microfinance enterprises have had their fair share of controversies. In 2010, news surfaced that scores of farmers who had taken microloans in the state of Andhra Pradesh in Northern India committed suicide (Mader 2015). The tragedy was a consequence of bad weather, failed crops, and high-pressure loan collection tactics by microfinance lenders. While the microfinance disruptive innovation is not short of critics, experts generally agree that it has promoted economic welfare for many of the world's poor who would otherwise have had little hope of accessing such financial services (Banerjee and Duflo 2011).

10.3 Country Backgrounds

The startup at the center of the case study analyzed in this chapter operates in numerous countries across South and Southeast Asia. These countries are of various sizes, from a small city state of 5 million citizens to the second most populous nation in the world with over a billion people. They are also at diverse levels of economic development. While most countries featured in the case are lower-middle-income economies with per capita GDPs of a couple of thousand US dollars, the case will also take us to one of the richest (Singapore) and poorest (Nepal) countries in the world today. See Table 10.1 for details.

This section supplies brief backgrounds of the recent history and political economy of six countries appearing in the subsequent case study. As will be seen, many

Table 10.1 Economic and social indicators for countries mentioned in the case study

South Asia	Population[a] (2016, mil.)	Foreign visitors[b] (2015, mil.)	International tourism receipts[c] (2015, mil. USD)	GDP per cap[d] (2015, USD)	HDI[e] (2014)	HDI level
India	1326	8.0	21,013	1617	0.609	Medium
Nepal	28	0.6	2567	751	0.548	Low
Southeast Asia						
Singapore	5	12.1	16,743	52,888	0.912	Very high
Vietnam	94	7.9	7301	2088	0.666	Medium
Laos	7	3.5	679	1779	0.575	Medium
Cambodia	15	4.8	3130	1168	0.555	Medium

Note: The Human Development Index (HDI), compiled and published by the United Nations, reflects the average life expectancy, education attainment, and GDP per capita of a country. (United Nations Development Program (2015))

[a]United Nations Department of Economic and Social Affairs (2015)

[b]UNWTO (2016)

[c]According to the World Economic Forum (Rinne 2017), the sharing economy model focuses on the sharing of underutilized assets, monetized or not, in ways that improve efficiency, sustainability, and community. The underutilized asset in the case of Airbnb would be, for example, apartment rooms that are available for temporary rental. Uber is another example of a sharing economy applied to transport services (underutilized cars and drivers).

[d]International Monetary Fund (2016)

[e]United Nations Development Program (2015)

countries have suffered from war, tragic social upheavals, and natural disasters, contributing to low levels of economic development and highly uneven distributions of income. The overview will provide some background on the state of each country's tourism industry, laying the groundwork for further discussion about the possible contribution of tourism to future development that may be both more equitable and sustainable.

10.3.1 South Asia: Nepal and India

Despite India's remarkable record of development, the world's largest democratic nation has often found itself overshadowed by the stunning growth achieved over the past three decades of China, its neighbor to the northeast. With comparable population sizes, China's nominal GDP today is more than four times as large as that of India's. While the government has played a key role in China's manufacturing and exporting success, India's service-based economy has thrived in spite of an ineffective, even obstructive state (Das 2012). Liberalization of international trade only began in India during the early 1990s. Today, the country's major exports include petroleum and petrochemicals, jewelry, textiles, auto parts, and pharmaceuticals. India is also well known as a destination for the outsourcing of software, information technology, and other business process services.

The lack of government initiative may as well have contributed to India's underperforming tourism industry. Each year, China welcomes approximately seven times more foreign visitors than India and generates five times greater annual tourism receipts. Regardless, tourism accounts for roughly 6% of India's GDP and is responsible for about 9% of total employment. Major sources of international visitors include the United States, Bangladesh, and the United Kingdom.

Sandwiched between two global giants, China and India, Nepal is a mountainous, landlocked country of 28 million people. Agriculture is the main economic activity in Nepal. Due to a lack of economic opportunities, a significant portion of the country's labor force works abroad, making remittance an important source of income for many Nepali households. Much of Nepal's economic difficulties stem from a decade-long political struggle involving communist insurgents, ending in the abolition of the monarchy and a new constitution establishing a federal republic in 2007. With its political challenges barely settled, the country was hit with a devastating earthquake in 2015, resulting in over eight thousand deaths and a shattered domestic economy. The tourism sector, an important foreign income generator for Nepal, was also severely hit by the 2015 disaster. While the number of foreign visitors to Nepal has rebounded in 2016, they have not yet fully recovered to the pre-earthquake levels (Carswell 2017).

10.3.2 Southeast Asia: Cambodia, Laos, Vietnam, and Singapore

The modern history of Cambodia is deeply colored by the atrocities of the Khmer Rouge genocide of the late 1970s, when a tragic social-engineering revolution resulted in widespread famine and paranoiac mass executions. Since the end of the Cold War, enormous amounts of foreign aid have poured into the country, with the aim of building a prosperous and democratic Cambodia. While the country has experienced strong economic growth since the turn of the millennium, the visions of its aid donors have not been fully realized. Most Cambodians depend on agriculture for their livelihood. The country's main industries generating most income for the government are textiles and tourism. In 2015, close to 5 million international tourists visited Cambodia, generating receipts of over 3 billion US dollars. Cambodia remains one of the poorest countries in Southeast Asia with high levels of income inequality.

Vietnam is a Southeast Asian nation profoundly affected by a recent history of war. After successfully repelling the French in the First Indochina War, ending in 1954, the Vietnamese sustained two more decades of fighting to unify the country, enduring heavy bombing from by the American forces in the 1960s. By the end of the Vietnam War in 1975, the human casualties and economic devastation were immense. Following a decade of experimenting with central planning, the government introduced reforms shifting the economy toward a market system in 1986. The reforms have sparked a sustained episode of growth lifting millions out of poverty. Today, Vietnam is a thriving emerging economy with growing exports in textiles, electronics, crude oil, rice, and coffee. Hanoi, the capital, is the largest city in the north, while Ho Chi Minh City is the commercial center in the South. In 2015, Vietnam welcomed close to 8 million foreign tourists who brought more than 3 billion dollars to the country. Large numbers of visitors arrive from China, Korea, Japan, the United States, and Taiwan.

Laos is a landlocked country of approximately 7 million people. Similar to its neighbors to the east (Vietnam) and south (Cambodia), it fell under French rule in the late nineteenth century. After gaining independence in 1949, Laos was briefly a constitutional monarchy before the communist party seized power in the Laotian Civil War fought between 1953 and 1975. Today, Laos is one of the poorest countries in Southeast Asia with roughly one-fifth of its population living under the international poverty line. Although its economy is rapidly growing, the country remains highly dependent on natural-resource reliant industries such as mining and hydroelectric generation. Tourism is a fast-growing sector in the Laotian economy, with over 3 million foreign visitors in 2015, about half arriving from Thailand. Vietnam, China, and South Korea are the other sources of tourists arriving in Laos.

One of the four Asian Tigers, Singapore achieved rapid industrialization since 1965 under its visionary leader, Lee Kuan Yew. Today, the city state is well known for its high-tech industries and financial services and is ranked as one of the most competitive and corruption-free economies in the world. Singaporean students consistently rank at or near the top of international scholastic assessment tests (Davie 2016), and the country is home to some of the very top universities (Times Higher Education 2017) and business schools (Ortmans 2017) in Asia. Future challenges with which the Singaporean government must grapple include the country's very limited natural resources, a very low birth rate, and high dependence on foreign labor in its domestic economy. In 2015, over 12 million foreign visitors arrived in Singapore, generating close to US$17 billion in international tourism receipts.

10.4 Case: Peculiar Encounters

Information for the following case study is derived from a series of interviews with *Peculiar Encounters*' founding team conducted throughout 2016. Observations from firsthand participation in the startup's activities will also be included. Finally, secondary data from numerous interviews and features in international media will be used as background.

10.4.1 Origins

The founders of *Peculiar Encounters*, Preston and a pair of brothers from South Asia, were business school classmates in Singapore. In 2009, as part of their bachelor of business studies, the trio jointly founded an overseas community service project in Nepal. The project aimed to provide training and funding – about US$ 300–500 with no strings attached – to Nepali entrepreneurs looking to start a business with an aim to make a social impact. The team later decided to initiate a similar project in India. While the experience was informative and gratifying, obtaining adequate financing was a constant challenge.

On their way to work 1 day through a busy part of Kathmandu, Preston and his partner had a chance encounter with a craftsman selling handmade, traditional Nepali wooden masks. Impressed by the craftsmanship, Preston asked if the man would be willing to show him and his friend how the masks were carved. For a small fee, the man agreed, and both parties thoroughly enjoyed the ensuing activities. There was something especially authentic about the experience, unlike the usual bland and sanitized tourist offerings. More importantly, Preston could sense the pride the man felt in sharing his knowledge with a curious and appreciative stranger. The extra income earned, Preston knew, would also go a long way for a Nepali street hawker. When the impromptu lesson was over, Preston inquired whether the craftsman would be willing to offer a similar service to other visitors. Given an affirmative answer, a new business idea was hatched.

The plan to replicate Preston's unique experience for other tourists was extensively discussed among the team. The objective of the new venture they called the "tourism model" subsequently emerged. First and foremost, the venture would create opportunities to generate additional income for the local people, many of whom were marginalized by the big players in the tourism sector. The beneficiaries would not be limited to craftspeople, as other services such as coordination, transportation, and language translation could also be provided by locals. Additional fees charged by the new venture could be reinvested in local social initiatives that Preston and his team had come to admire and support.

The new venture would take the form of an internet portal linking backpackers seeking an authentic travel experience with local residents willing to share their skills and lifestyle. Preston and his team would scout out local talent willing to serve as "hosts." Additional accommodations such as transportation and an interpreter would also be provided. Each client would pay an online booking fee and, after arriving at the location, pay the driver, translator, and host in cash for the services rendered. The name of the startup would reflect the origins of the idea hatched in the back alleys of Kathmandu.

As the plans for the new project began to take shape, Preston and his team decided to pitch their idea for funding in business plan competitions. In January 2013, they won a business plan competition in Singapore. The sizable prize money, together with their own investment was sufficient for the launching of their preliminary operations. Friends were recruited to try out the bare-bones version of the website and services. With positive feedback coming in, the team knew they had a viable business. In 2014, the first version of the *Peculiar Encounters* website opened for booking by travelers from around the world.

10.4.2 Growth

The team's familiarity and local knowledge of Nepal led them to choose Kathmandu as their initial base. They established a relationship and began working closely with a local NGO to recruit local artisans who could act as hosts. The NGO also helped link the *Peculiar Encounters* team with university students eager to land part time jobs as English translators. The experience taught the team the tremendous value of working with local organizations, a practice they would repeat again and again as they expanded to other locations around Asia.

Being avid travelers themselves, Preston and his teammates were aware of popular backpacking routes through Southeast Asia. Riding the Southeast Asia backpacking circuit was the logical expansion strategy for their startup. The ability to scale quickly would therefore be a critical requirement for growing and sustaining *Peculiar Encounters*. After hiring a full-time local manager in Kathmandu, the team decided to split up to establish bases in various locations simultaneously. One of the brothers headed to Pokhara in Central Nepal, while the other traveled to Phnom Penh to setup their first operations in Cambodia. A similar strategy was followed in setting up operations in different locations. In Phnom Penh, for example, the project leader quickly got in touch with a local NGO to identify crafts people and students who could work as translators. He also discovered that mingling with backpackers in hostels and guesthouses offered a unique opportunity to learn about the needs of their target clients and new trends emerging in the backpacking community. A similar strategy was employed when *Peculiar Encounters* later expanded into Siem Reap, the Northwestern Cambodian town known as the gateway to the ruins of Angkor. See Fig. 10.2 ("Scouting for hosts in Phnom Penh") for a further detail.

Once an operation was setup, responsibilities would be handed over to a full-time local manager, and the founding team member would move on to establish a base in a new city. The ability to communicate decisions and execute financial transactions from any location facilitated the founding team's strategy to operate as *digital nomads*. *Peculiar Encounters* gained a foothold in Laos when Preston setup an office in Luang Prabang. The tourist cities of Hue and Hoi An were then chosen as points of entry into Vietnam. While the team found most cities in Southeast Asia to be business friendly, their experience in India was entirely different. One of the South Asian brothers was tasks with launching a new office in Udaipur, a city in the western Indian state of Rajasthan. The bureaucratic red tape he found himself up against was immense. Just to rent an office required a court permit. Such challenges significantly stalled their progress in India and came close to forcing *Peculiar Encounters* to abandon plans for operations in India entirely.

By 2016, the *Peculiar Encounters* founders were aiming to establish bases in cities across Southeast Asia. Preston is set to explore Chiang Mai, Thailand, while the two brothers have set their eyes on the big island nations further afield. One of the brothers has flown to Cebu in the Philippines, while the other will try to set up an office in Yogyakarta, an important tourist destination in Indonesia.

10.4.3 Impact

Peculiar Encounters has demonstrated an ability to financially sustain itself during its early years of operation. Up to the end of March 2016, it has served over seven thousand clients through its web portal, generating revenue of more than US$ 70,000. The startup has signed up over 400 hosts and has trained and hired about 500 students as part-time interpreters. The hosts, often disadvantaged members of the local communities, take half of the total fees charged to the clients. Student interpreters can earn a significant income supplement while putting their English skills to work. A Cambodian student claims she can earn about US$ 100 a month working as a part-time interpreter for *Peculiar Encounters* during the tourist high-season, a substantial pay check in a country where the average income per person is just over a US$ 1000 per year.

Perhaps equally as important as the economic impact is the positive influence the startup has had on the confidence and outlook on life of its hosts. The founders of *Peculiar Encounters* recounted an example of one of their hosts in Phnom Penh who teaches tourists how to carve coconut shells into keychains. The man suffers from

At 25, Preston, a graduate from a top Singaporean University, pays himself a meager salary of US$500 so he can pursue something he really values. With parents working overseas, Preston spends most of his time living away from home. Today he is visiting the Phnom Penh office of *Peculiar Encounters*, a social enterprise he help founded in 2010.The office is rented for $150 a month and is run by a full-time, local manager.

After a quick chat with the manager, Preston heads out to the local fresh market looking for hosts and activities to drive *Peculiar Encounters'* quick expansion. Wandering the streets of Phnom Penh, Preston photographs a street hawker cleaning a string ray and checks out the dried squid on display.

"The craft itself is very important", explains Preston. "It must be interesting to the tourist and intrinsic to the country or city's tradition and culture". Today, Preston thinks he has identified a perfect activity for backpackers visiting Phnom Penh – learning how to cook and eat tarantulas.

Eating insects became common in Cambodia during severe food shortages that occurred in the 1970s when the country was controlled by the Khmer Rouge. Before the activity can be offered, Preston must put himself in the shoes of tourists. A course structure is designed and the host is trained to ensure that the best bits are in and the boring bits are out.

Donning rubber gloves, Preston begins cleaning the hairy spiders with the help of his host. Parts of the arachnid must be removed before the spiders are seasoned and fried. Preston proudly holds up a tarantula he has helped prepare – "fried to perfection" he exclaims. Moments later, he bravely bites off one of the legs, squinting noticeably. "I don't know what to make of this" he murmurs.

This activity Preston dubbed "Fear Factor Challenge" will go on to become one of the most popular experiences in Phnom Penh. Lasting about two hours, the activity costs tourists $16. Half of the revenue goes to the host. *Peculiar Encounters* retains 10-15%, with the remaining going to the translator and tuk-tuk (auto rickshaw) driver whisking the tourist back to their hotels.

Activities whose structures have been vetted and hosts have been trained get a special designation on the company's website. There are detailed descriptions for each activity, photographs, user reviews and tips for preparing available. Preston and his team take very seriously transparency and the safety of their clients.

Most importantly, Preston realizes that tourists have very limited time to spend in each city they visit. His goal is to make each experience an unforgettable memory. So far, the team appears to be accomplishing its goal. *Peculiar Encounters'* courses have received rave reviews on leading travel websites. Its Phnom Penh offerings have been voted among the top 5 attractions for visitors to the city by users of a popular travel review and hotel booking website.

Back at the office, Preston emphasizes the benefits the hosts and translators enjoy by joining *Peculiar Encounters*. Not only can the additional income transform their lives, but the confidence and language skills they gain from interacting with foreigners often open up new hopes and opportunities for these local people as well.

Fig. 10.2 Scouting for hosts in Phnom Penh. (Source: On the Red Dot by CNA (2014))

Fishing on the Mekong

Dwelling on the banks of the Tonle Sap (Cambodia's Great Lake), "Ramorn" Less Sary heads out on his boat with his wife and children to fish on the Mekong river every day at the crack of dawn. The 29-year old claims his best catch ever was 30 kg, worth about US$ 150. That was during the high season running from December to June. During the off season, the fish lay eggs in small streams and hide among the roots of the trees. The strong wind and rain make fishing even more difficult. In these periods, Ramorn manages to earn less than US$ 5 a day, barely enough to feed his family.

Ramorn is a son of a fisherman. He was born on a fishing boat and began fishing at the age of 15. Now with kids of his own, he constantly worries about his youngest son falling off the boat and drowning in the river. Knowing well the hardships of a life as a fisherman, he hopes for a different and better life for his children. He says he has a longing to live on land, and would do so if he had the money.

To sustain his family during this off season, Ramorn is focusing on a different kind of catch. Today, as a *Peculiar Encounters* host, he is teaching Carl Polen, an American tourist, how to fish using a net. Carl used to fish as a child, but with a rod and reel. With the help of a translator, Ramorn starts Carl out on a small net weighting about 7 kg. The trick to a good throw is in the intricate layering of the net to ensure it opens fully. On Carl's first throw, the net gets tangled under water prompting Ramorn to dive into the river to free it. After five minutes, the net is free and the team carries on with activities that mirror a Cambodian fisherman's daily routines, but are doubtlessly uniquely foreign to an American.

A pleasant surprise occurs while Ramorn is demonstrating to Carl the use of a larger net that requires a team of two working together. From the net emerges a rare Mekong prawn that can fetch up to US$5 at the market. The master and apprentice celebrate with lunch prepared on board by Ramorn's wife. The offerings are meager, comprising small fish and vegetables gathered nearby. The experience gives Carl a deep appreciation for what it takes to make it in this part of the world.

For Ramorn, the extra US$15 earned during the 4-hour experience will go a long way in building a better life for his family. During these lean months, income from working with *Peculiar Encounters* can double the family's income. The earnings have allowed him to purchase a new battery to power his boat. The authentic interaction with tourists like Carl also broadens the Cambodian fisherman's worldview and gives him a newly found sense of confidence.

Fig. 10.3 Peculiar Encounters' *Experiences*. (Source: On the Red Dot by CNA (2014))

dwarfism and prior to becoming a host was known to be extremely shy and often depressed. Through his regular interactions with foreigners, the man has gained confidence and a sense of optimism. He now looks people in the eye when talking to them and strives to have his own shop 1 day (see Fig. 10.3).

The Coconut Carver

Sok Chea suffered multiple accidents early in life. Two falls during his childhood resulted in his stunted growth. When he was six years old, a cow trampled over him leaving him deformed and unable to do hard work. He was jobless in the first twenty odd years of his life.

Driven by a desire to be independent, he join a job training program with a local NGO, Khmer Life, and learned how carve coconut shells into various handicrafts such as keychains and ornaments. After completing his training, he got a job at the Khmer Life handicraft shop. Working nine hours a day, he earns US$50 a month plus meals and lodging. It is hard work, but Sok Chea values the autonomy he has gained.

Becoming a host for *Peculiar Encounters*, teaching tourist to carve coconut shells, has made an appreciable difference in the craftsman's life. Although the host's earnings of US$6 goes to Khmer Life – Sok Chea does his teaching during regular working hours at the NGO's craft shop – *Peculiar Encounters* pays Sok Chea a few extra dollars per session. The supplementary income will contribute to the man's dream of one day owning his own crafts shop. A local student serving as a translator and the tuk-tuk driver who picks up and returns the tourists to their Phnom Penh hotels also appreciate the chance to earn some extra cash.

The impact generated by *Peculiar Encounters'* initiative does not end with the opportunity to make money. Interacting with foreigners gives the hosts and local student the chance to improve their English and develop self-esteem. While people like Sok Chea carve out a better living for themselves, tourist also get a glimpse into the real Cambodian life – a win-win arrangement for all parties involved.

Fig. 10.3 (continued)

10.5 Discussion and Analysis

There is clearly a demand from people around the world who want to do good. Unfortunately, well-meaning efforts as well as resources are often wasted on ineffective charities or lured into counterproductive scams (see Davidson (2014) on the notorious "orphanage tourism" in Cambodia). One of the motivations behind *Peculiar Encounters* was to provide people with an alternative in helping communities they visit through "impact travel." In today's globalized world, there is an ever-growing opportunity for travel. By arranging genuine exchanges between tourists and locals, travelers can contribute to a better livelihood for less privileged people while learning to appreciate a different culture, a rare skill, or a fleeting lifestyle.

Figure 10.4 illustrates the basic idea behind *Peculiar Encounters*. The portal allows tourists to bypass agents and tour operators and interact more directly with local people, who previously were either small, lower-tier suppliers in the tourism supply chain or not part of the chain at all. As services from local people flow to travelers with less intermediation, they have the opportunity to learn more about tourists' demand and preferences. They also enjoy a larger share of the value that is created from the trade – between 70% and 80% of revenue generated from each booking remains in the local community. The share of benefits that go to local

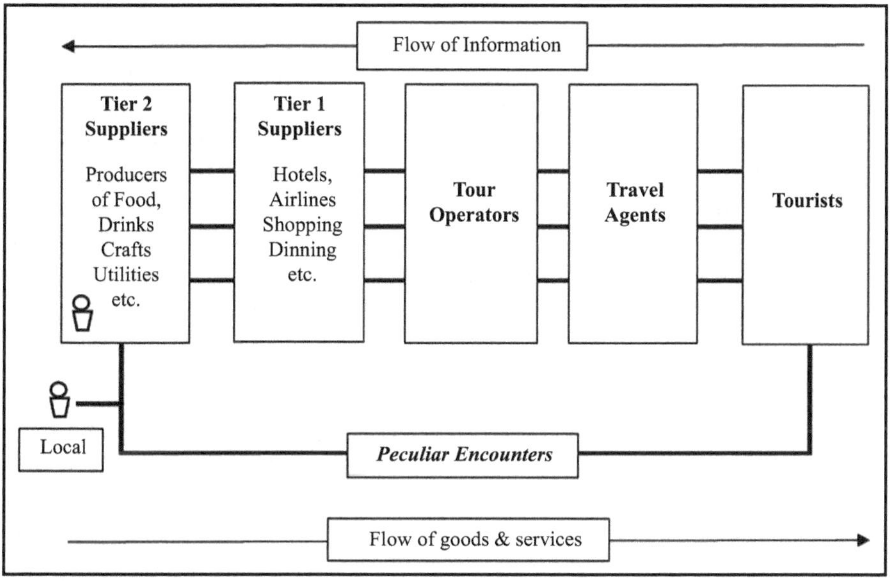

Fig. 10.4 *Peculiar Encounters* and the tourism supply chain

communities when tourism is organized through large, powerful players in typical tourism supply chains (Fig. 10.1) is likely much smaller. As for the tourists, it is easy to see how arranging part of their trip through *Peculiar Encounters* will afford them a more authentic experience with the local people and culture. The outcome is therefore a win-win arrangement.

Making the inspirational vision a reality required Preston and his team to overcome some difficult structural impediments. Most challenging were the entry barriers to the typically high-fixed-cost industry. As a small player with severely restricted capital, Preston needed to design a clearly differentiated product, communicate its value, gain the trust of new potential customers, and finally deliver the service seamlessly – all while maintaining a small team and ensuring the local people were offered a good deal. The following subsections discuss in further detail a few key issues that were critical in *Peculiar Encounters*' early survival and later success – dealing with fixed costs, designing the innovative tourism products, building unconventional networks, and exploiting information technology and globalization.

10.5.1 Push Versus Pull Supply Chains and Fixed Costs

Tourism supply chains often resemble a *push* system, where demand forecasts based on past statistics drive new investment and expansion plans (Zhang et al. 2009). Such a configuration is necessary when the industry exhibits high fixed costs. In the case of tourism, such fixed costs might include prime real estate rental, investment in construction and vehicles, and training and retaining a large number of staff. These costs must be sunk before tourism services can be rendered, thus heightening the importance of accurate demand forecasts. Large fixed costs carry a few important consequences. First, they reduce the ability of players to flexibly respond to changes in market conditions. Once the investment is sunk, the firm has little room to respond in the short run to demand that is stronger or weaker than expected. Second, fixed costs serve as a strategic entry barrier, making it difficult and risky for new players to challenge well-established incumbents. Finally, if customers are very sensitive to price changes, then the producer faces increasing pressure to find ways to lower other variable costs and ensure profitability. In the tourism industry, these variable costs may include contracted services and wages paid to non-permanent, unskilled labor.

The consequences of high industry fixed costs discussed above generally spell bad news for local communities. In developing countries, a surplus of unskilled labor together with the absence of labor unions means hotel chains and tour operators can pay very low wages. The necessity to keep wages low is heightened when travelers enjoy a wealth of destinations to choose from and therefore make their choices based on price, putting pressure on sellers to keep rates low. The absence of local competition, due to the high (local) entry barrier, also eliminates any chances that new firms entering the market will bid up labor compensation. It is thus easy to understand Preston and his team's pessimism about the prospects of conventional tourism contributing to the improvement of the livelihoods of the members of local communities.

Peculiar Encounters' key innovation can be viewed as redesigning the supply chain to drastically reduce fixed cost, thus allowing a *push* system to be replaced with a *pull* system. The startup has no need for access to prime tourist attractions sites – its activities are mainly held in the back alleys of a city anyway. Offices it rents in each city, always in non-prime locations, can be had at very low rates, usually a few hundred dollars a month. Apart from a few laptop computers, *Peculiar Encounters* owns very little capital equipment. The tools used for hosting the signature experiences – fishing boats and nets, carving tools, punching bags, etc. – represent instruments the hosts actually use to make a living. Finally, the hosts and translators who work together to provide the company's unique experiences work part-time and on-demand basis. During low-demand periods, the hosts fall back on their professions as fishermen, crafts persons, shopkeepers, or family makers, while student translators focus on their studies – thus generating no extra cost for *Peculiar Encounters*. When there is tourist demand, these spring into action, each coming together for mutually beneficial cooperation.

The key advantage of *Peculiar Encounters*' low fixed costs, *pull* supply chain is that it facilitates entry into the market for tourist services. Locals who previously had no chance of putting their skills and knowledge on the tourism market now have a channel to reach out to potential clients to offer services that clients genuinely value and appreciate. Most importantly, because *Peculiar Encounters* incurs very low fixed costs, it can afford to cut its local suppliers a bigger share of the revenues – up to half of the total fees charged to clients. These advantages, however, do come with some drawbacks. Services whose supply mainly generates variable costs do not enjoy benefits from economies of scale. Delivering each experience requires a fixed 2–4 h of time of the host. There is thus a natural limit to how many clients a popular host in each city can accept each day. Identifying a talented host with suitable skills and personality as well as providing the training required is not straightforward either. The complication lies in the fact that each experience is a unique and complex product. A capable trainer, usually one of the *Peculiar Encounters* founders, needs to sit down and design a course structure and run through the activities, making sure everything is safe and that things don't drag. Add to that the heterogeneity of language, culture, rules, and regulations across the developing Asian countries and it becomes clear that there can be no standardized process to achieve the scaling in the same speed and magnitude as disruptive start-ups like Facebook or Airbnb.

Preston and his team are prescient about the fact that their social enterprise is unlikely to make them millionaires in the near term – their service is in fact still focused on a niche market of backpackers and socially conscious tourists. *Peculiar Encounters*, however, has proven to be financially self-sustaining and is growing day by day. The impact the startup can make on the lives of the disadvantaged members in the communities it works in is potentially immense. Early in his journey as a social entrepreneur, Preston often refers to the "beacon effect," the hope that is spread throughout a community when one of its members successfully lifts herself out of poverty. The team's preferred measure of success for their innovative tourism supply chain mode is based on the number of lives touched and families who, as a consequence of their efforts, find a way to escape poverty through their own actions.

10.5.2 Product Development

The original insight for *Peculiar Encounters*' product harks back to the encounter Preston had with the craftsman selling traditional Nepali masks in the backstreets of Kathmandu. Recalling the experience, Preston describes his "five hours working at the feet of a master" as a "phenomenal experience" that he wished he could have at every city he visited. Notably, the craftsman, in addition to appreciating the extra income, also enjoyed the exchange so much he asked if Preston could bring more people for similar future interactions. Preston immediately realized that a great deal of value is frequently wasted in superficial transactions between tourists and locals. The Nepali craftsman he befriended had such wealth of knowledge, skill, and hospitality, but his interactions with tourists each day was sadly reduced to responding to: How much?

There are a few notable features in the emergence of *Peculiar Encounters*' product prototype. First, the idea surfaced out of the founder's long-time passions of traveling and trying to improve the lives of the poor. Without his previous experience of backpacking through Asia and initiating an overseas community service project, Preston would not have had the audacity to engage the street craftsman in such an unusual way. Second, as a student of business, Preston was well-equipped to recognize unexploited value he accidently came across. Having enjoyed his exchange with the craftsman so much, he saw an opportunity to design a tourism product that he himself would love to buy. Moreover, he deeply cared about the struggles of the poor in developing Asia and was ready to contribute his energy and creativity to build a business that could help people like the craftsman.

Passion and a hunch, however, are rarely adequate ingredients for developing a successful new product. After brainstorming the new business idea with his team, Preston and his partners quickly turned their attention to building a bare-bones service whose viability they could test with real users. Artisans were identified and course structures were designed. All activities needed to be vetted by the founders to ensure they were both interesting and safe. When a course was ready for offering, descriptions were drafted and provided on the website with accompanying photographs. Note that the funding from their win in a business plan competition came a few years after the idea was hatched. The first couple of years during which they were working to build *Peculiar Encounters* were therefore characterized by lean hustling. The team only became confident they had a viable new product to offer in the market when positive feedback from their friends, the very first users of the prototype service, began to trickle in.

Peculiar Encounters' product portfolio continues to evolve today, as the company enters new cities and learns from its customers and hosts the kinds of activities that generate the best experiences. In addition to its trademark activities priced around US$20–30, longer and more expensive experiences priced over one hundred dollars have been added. Perhaps this is the beginning of the startup's attempt to go up-market, a natural evolution in the disruptive process (Christensen 1997). The company has also begun partnering with tour operators in some cities to diversify its offerings and appeal to a broader range of clients. The advantage of its low-fixed-cost operation model is that new experiences that fail to become popular do not cost the company a great deal and can be easily retired from the listings. *Peculiar Encounters* must however continue to carefully balance the needs of diversification and maintaining its unique identity as a provider of authentic, non-touristy experiences. Doing so will be critical in staying true to the founders' original mission of the social enterprise.

10.5.3 Building a Network of Complementary Partnerships

The concept behind *Peculiar Encounters* was hatched in 2009 when Preston, an undergraduate from a Singaporean university, was on his way to work at a social venture in Kathmandu, Nepal. Preston approached an elderly craftsman he regularly passed on his daily commute and asked if the man would be willing to teach him how to make the traditional masks that were on sale. For a small fee, Preston won a memorable learning experience. He felt confident many other tourists would value a similar encounter and immediately began work with his team on launching a startup based on the idea.

Success would crucially depend on the startup's ability to recruit many diverse local artisans to join the initiative. The challenge became even more daunting given Preston's ambition to rapidly expand the service to numerous locations across South and Southeast Asia. How could such an unconventional recruitment be carried out given the startup's limited resources? The key to overcoming the challenge would lie in the team's ability to cultivate relationships with local NGOs. In addition to their knowledge about conditions on the ground, NGOs fundamentally shared the same goal as Preston's team – to better the lives of disadvantaged members of their communities. Through the partnership, *Peculiar Encounters* would gain invaluable assistance from recruiting local staff to hiring university students to act as interlocutors. The partnership would turn out to be a critical success factor in redesigning the tourism value chain to redistribute more benefits to local communities.

In addition to helping *Peculiar Encounters* achieve rapid expansion, NGOs have also given the social enterprise a clearly differentiated identity. Five years after *Peculiar Encounters*' launching, many similar websites offering tourists "local experiences" have emerged as competitors (Caruana 2016). Browsing the offerings on these competitor websites, one quickly notices that they have a very different "feel" than that of *Peculiar Encounters*. First, the majority of their services are based in popular tourist destinations in Thailand, Japan, Korea, and Europe. Second, the offerings advertised often involve typical tourist activities such as visits to temples, city tours, nature adventures, or spas. Finally, the services are provided by professional tour guides or middle-class locals who speak fluent English. While the services marketed by these competitors will obviously appeal to some segments of the traveler market, they probably offer little direct competition to *Peculiar Encounters*' core customers – backpackers and people looking for impact travel opportunities. The main reason *Peculiar Encounters* is able to offer such a differentiated product – direct contact with the regular, unprivileged locals – is its unique relationship with NGOs. Although not seasoned players in the tourism industry, NGOs have deep information about the underprivileged and have often gained the trust and respect of the communities in which they operate. Seeing that *Peculiar Encounters*' mission match with their own goals, these NGOs have whole-heartedly offered Preston and his team their support. Collaboration with local NGOs is evident in many *Peculiar Encounters*' offerings – from learning traditional crafts to experiencing the livelihood of a poor Cambodian fisherman. On the other hand, the offerings of *Peculiar Encounters*' competitors suggest their main partners are hotels, spa and tour operators, and collaborators who are established players in the tourism sector and bring an entirely different set of resources and knowledge to the table.

Partnership with traditional educational institutions is another type of collaboration showing good promise. *Peculiar Encounters* has collaborated with schools and universities to provide their students with authentic learning opportunities in the developing world. Students from the Nanyang Technological University in Singapore were treated with a traditional Apsara dancing class during their visit to Cambodia. Master degree students in the international development program at

Concordia University in Irvine, California, enjoyed a "customized, full village experience" during their visit to Siem Reap. They also were given an inside look into the operations of social enterprises in Cambodia, including how *Peculiar Encounters* Cambodia works. Secondary school students from the Hong Kong International School had their visit to Phnom Penh enriched by classes on pottery, leather carving, and bark art organized by *Peculiar Encounters*.

Unlike many NGOs and charity organizations, *Peculiar Encounters* has so far shunned the prospect of partnering with governments or international aid organizations. The government's absence is perhaps surprising, given it is one of the key stakeholders in the tourism industry and poverty alleviation. *Peculiar Encounters*' experience, however, has shown that in many developing countries, regulatory red tape is a major hindrance to the work of the most audacious and well-intentioned social entrepreneurs. Preston's determination to create social change through business – not through government or charity – may have taken root much earlier. In a 2011 public talk, a young Preston, who at the time was still working at the social venture he helped founded, described his experience as a backpacker through Asia seeing poverty with his own eyes for the first time. He described how he was influenced by the thesis of C.K. Prahalad's famous book, *The Fortune at the Bottom of the Pyramid* (Prahalad 2004), that there existed a tremendous amount of untapped opportunities in the world's poorest communities. Tapping such opportunities would require creativity, flexibility, and the business savvy of private entrepreneurs unrestrained by rigid bureaucracy. Today, *Peculiar Encounters*' website clearly proclaims this same viewpoint – that profit making and social impact go hand in hand.

10.5.4 Capitalizing on the Transformative Powers of Technology and Globalization

Working as "digital nomads" was the *Peculiar Encounters*' founding team's modus operandi. Preston and his co-founders would post themselves in target cities for a few months to establish partnerships with NGOs, setup an office, hire a local staff, and recruit artisans. Once the basic infrastructures were in place, each would move on to another location. Such rapid expansion under exceptionally lean operations was made possible by information technology. Team meetings and follow-ups with local staff were all conducted through Skype. Salaries and most expenses are transferred through PayPal. This low-cost communication and operations would not have been possible without the availability of cheap and reliable internet connection in urban areas throughout Asia.

Technology plays its most central role in helping the startup reach its customers. *Peculiar Encounters*' marketing is exclusively through its website, where activities (called "experiences") are listed by location together with key details such as group size, duration, and costs. In addition to short, high-quality video clips of the experiences, customer testimonials are prominently displayed on the website. The startup's marketing strategy has proven to be effective in establishing its beachhead market – backpackers on the Southeast Asia circuit. By 2016, *Peculiar Encounters* had served over 7000 thousand customers, trained upward of 500 university interlocutors, and made a positive economic impact on over 400 micro-entrepreneurs.

If information is the "life blood" of the travel industry (Sheldon 1994), then credibility is the currency industry participants deal in. One of the biggest challenges in winning new business for a startup like *Peculiar Encounters* is establishing trust with potential clients. In a talk given at his alma mater in 2016, Preston deplores the negative stereotypes imposed upon people at the bottom of the income pyramid. Foreigners often expect the local poor to be unprofessional, dishonest, and even dangerous. A basic instinct of a visitor to a new city is to avoid interaction with strangers on the streets. While Preston tries his best to convince his audience that the fulfillment and reward from having a little faith and stepping out of one's comfort zone is more than worth the initial discomfort, the *Peculiar Encounters* website may be the more effective tool for generating credibility and trust. Because consistency builds trust, the company invested in making the *Peculiar Encounters* website a model of consistency. Each experience is accompanied by a description, photographs, and client reviews. The information a viewer derives from each source impressively matches up. The information page for the Mekong Fishing experience (see Fig. 10.3) describes the 4-h activity as including fishing lessons on a traditional "Cham fisherman's boat" using a net. The lesson is followed by a homemade dinner prepared and consumed on the boat with the fisherman's family while enjoying the view of the sunset on Tonle Sap (Cambodia's Great Lake). The photographs accompanying the description show tourists – men and women, children and adults – throwing fishing nets on board a narrow wooden boat. There are also photographs of the fisherman, Ramorn, and his family preparing and serving dinner to their guests and a beautiful photograph of the sun going down on the lake. Finally, the thirty client reviews describe the same experience with lavish praise for the friendliness of the hosts and the humble-looking but highly tasty meal. All comments are dated, giving assurance to the browser that the reviews are current. A live chat option automatically pops up offering to answer any questions or clarify any confusion. Finally, the links to press coverage of *Peculiar Encounters* and its founders on the BBC, CNN, Channel News Asia, and TripAdvisor reinforce the potential client's confidence that the web portal is trustworthy and the activities offered are safe.

Information technology contributes to a seamless and reassuring experience after a browser becomes a paying customer. Once an activity is booked, the client receives an email confirmation containing further details and recommendations. She is informed about the pickup location and what to expect on the day of the activity. The host and translators are expected to dress in a specific uniform easily recognized by the client. There are different designated uniforms for different cities, information the client is given in the confirmation email, adding to the feeling of security and the uniqueness of the experience on the day of the activity.

The rise of *Peculiar Encounters* is undeniably aided by globalization. As a citizen of Singapore, a member of the ASEAN Economic Community, Preston is privileged to travel conveniently through Southeast Asia, not requiring an entry visa for many countries. The startup enjoys ever-growing opportunities as students and backpackers are increasingly drawn to Asia. As travelers become more sophisticated, they are less likely to be satisfied with sanitized, regimented tourist activities. Like Preston and his co-founders, the sophisticated travelers will demand a more authentic experience that includes genuine interaction with the local people and exposure to the local culture. Many undergraduate and master degree programs in Europe and America are beginning to incorporate overseas exposure as part of their curricula. All these developing trends suggest that, despite the possibility of new competitors entering the market, *Peculiar Encounters*' services will continue to enjoy increasing demand in the coming years.

10.6 Summary and Conclusion

The master craftsman with whom Preston and his partner had their first carving class in the back alleys of Kathmandu today hosts two successful experiences for *Peculiar Encounters* – a 4-h wooden mask carving experience, which costs US$25, and a day-long homestay experience costing US$65. The earnings from hosting a few tourists groups each week are enough to more than double his monthly income. Equally important, tourists now see and treat the wood carver as a master of an ancient and beautiful craft, not merely a street side peddler. Stories like this one illustrate the power of innovation and entrepreneurship in helping a major service industry like tourism contribute more to bettering the lives of the world's poor.

Peculiar Encounters today operates in numerous countries across South and Southeast Asia. These countries are of various sizes and are also at diverse levels of economic development. The startup has demonstrated an ability to financially sustain itself. It has served over seven thousand clients through its web portal. The startup has signed up over 400 hosts and has trained and hired about 500 students as part-time interpreters. Student interpreters can earn a significant income supplement while putting their English skills to work.

The online portal has allowed tourists to bypass traditional travel agents and tour operators and interact more directly with local people, who previously were either small, lower-tier suppliers in the tourism supply chain or not part of the chain at all. As services from local people flow to travelers with less intermediation, they have the opportunity to learn more about tourists demand and preferences. Locals also enjoy a larger share of the value that is created from the trade – between 70% and 80% of revenue generated from each booking remains in the local community.

Bringing their idea from the drawing board to reality required Preston and his team to overcome some difficult structural impediments. Most daunting were the entry barriers to the typically high-fixed-cost industry. As a small player with severely restricted capital, the *Peculiar Encounters* team designed a clearly differentiated product, communicate its value, gain the trust of potential customers, and finally deliver the service seamlessly – all while maintaining a small team and ensuring the local people were offered a good deal.

Earlier in the chapter, we reviewed seven critical issues in tourism supply chain management identified in Zhang et al. (2009): *(1) demand management, (2) relationship management, (3) supply management, (4) inventory management, (5) product development, (6) supply chain coordination,* and *(7) information technology*. Below, we conclude the chapter by relating the startup's success to the founding team's ability to address, in very creative ways, many of these critical issues.

Keys to the startup's success were its ability to:

1. *Redesign the tourism supply chain* to drastically reduce fixed cost, thus allowing a *push* system (based on demand forecasts) to be replaced with a *pull* system (based on emergent customer demand). The newly designed supply chain is more flexible, less dependent on demand forecasting (*issue 1*), and requires low inventory costs (*issue 4*). As a consequence, *Peculiar Encounters* is able to offer its hosts an unusually large share of the revenues – up to half of the total fees charged to clients – consistent with its social mission.
2. *Develop tourism products* (*issue 5*) based on the insight that a great deal of opportunity for authentic enjoyment for both tourists and locals was being wasted due to their superficial transactions. In order to develop their new offerings, scouting, training, and developing local hosts (*issue 3*) were of great importance. The products were then extensively tested before launching and continually improved from customer feedback.
3. *Build networks with local NGOs* (*issue 2*) that resulted in mutually beneficial cooperation. NGOs have help *Peculiar Encounters* locate hosts and identify potential high-quality translators. In addition to helping *Peculiar Encounters* achieve rapid expansion, NGOs have also given the social enterprise a clearly differentiated identity as a portal for authentic, non-touristy experiences.
4. *Exploit advances in information technology and globalization* in all aspects of its operations and marketing. Preston and his co-founders operate as digital nomads, moving from one location to another in rapid succession to drive quick expansion. They take advantage of new technology to communicate and coordinate with their geographically dispersed team and execute financial transactions (*issues 6 and 7*). Their marketing is 100% digital.

There is clearly a demand from people around the world who want to reach out and help individuals in economically disadvantaged communities. Sadly, well-meaning efforts as well as resources are often wasted on poorly run charities or lured into damaging scams. By arranging genuine exchanges between tourists and locals, travelers can contribute to a better livelihood for less privileged people while learning to appreciate a different culture, a rare skill, or a fleeting lifestyle.

Tourism is a coordination- and information-intensive industry. It involves the arranging of complex experiences whose quality is largely determined based on presentation and interpretation. Through a case study analysis, this chapter has demonstrated how innovation in supply chain management can enhance the economic aspect of supply chain sustainability to help improve the livelihoods of all stakeholders, especially the host communities, and promote a more balanced distribution of economic benefits to contribute to poverty alleviation.

Bibliography

[Documentary] (2014, November 13). *On the Red Dot by CNA* [Video File].

Banerjee, A., & Duflo, E. (2011). *Poor economics: Rethinking poverty & the ways to end it.* London: Random House.

Carswell, H. (2017, April 21). How Nepal's Tourist Industry is bouncing back two years on from devastating earthquake. The Independent. Retrieved from http://independent.co.uk.

Caruana, B. (2016, June 11). 5 Websites You Need To Know For Finding Local Experiences When You Travel [Blog post]. Retrieved from http://www.huffingtonpost.com/entry/5-websites-you-need-to-know-for-finding-local-experiences_us_575bd241e4b0b6c49600c69f.

Christensen, C. M. (1997). *The innovators dilemma: When new technologies cause great firms to fail*. Cambridge, MA: Harvard Business Review Press.

Christensen, C. M., Baumann, H., Ruggles, R., & Sadtler, T. M. (2006). Disruptive innovation for social change. *Harvard Business Review, 84*, 94.

Das, G. (2012). *India grows at night: A liberal case for a strong state*. New Delhi: Allen Lane.

Davidson, H. (2014, January 2). Cambodia: Child protection workers call for end to 'orphanage tourism'. The Guardian. Retrieved from http://www.thegaurdian.com.

Davies, S. (2016). Singapore students top in Maths, science and reading in PISA international benchmarking Test. Straits Times. Retrieved from http://www.straitstimes.com.

Haugh, H. (2005). A research agenda for social entrepreneurship. *Social Enterprise Journal, 1*(1), 1–12.

International Monetary Fund. (2016). World economic outlook database-April 2016. Retrieved from the International Monetary Fund website: https://www.imf.org/external/pubs/ft/weo/2007/01/data/index.aspx.

Mader, P. (2015). Mechanisms of a microfinance crisis. In P. Mader (Ed.), *The political economy of microfinance* (pp. 160–194). London: Palgrave Macmillan.

Mitchell, J., & Ashley, C. (2010). *Tourism and poverty: Pathways to prosperity*. London: Earthscan.

Ortmans, L. (2017). FT 2017 top 20 business schools in the Americas and Asia-Pacific. Financial times. Retrieved from http://www.ft.com.

Peel, M. (2017, April 27). Thailand under strain from tourist surge. Financial Times. Retrieved from http://www.ft.com/content/a6ba9aa0-2983-11e7-9ec8-168383da43b7.

Prahalad, C. K. (2004). *The fortune at the bottom of the pyramid: Eradicating poverty through profits*. Upper Saddle River: Wharton School Pub.

Rinne, A. (2017, December 13). What exactly is the sharing economy? World economic forum. Retrieved from: https://www.weforum.org/agenda/2017/12/when-is-sharing-not-really-sharing/.

Sheldon, P. (1994). Information technology and computer systems. In S. Witt & L. Moutinho (Eds.), *Tourism marketing and management handbook* (pp. 126–130). London: Prentice Hall.

Sinclair, M. T., & Stabler, M. (1997). *The economics of tourism*. London: Routledge.

Theuvsen, L. (2004). Vertical integration in the European package tour business. *Annals of Tourism Research, 31*(2), 475–478.

Times Higher Education. (2017). Asia University Rankings 2017. Retrieved from: http://www.timeshighereducation.com.

United Nations Department of Economic and Social Affairs. (2015). World population prospects: The 2015 revision. Retrieved from the United Nations Department of Economic and Social Affairs website: https://esa.un.org/unpd/wpp/publications/files/key_findings_wpp_2015.pdf .

United Nations Development Program. (2015). Human development report 2015 – Rethinking work for human development. Retrieved from the United Nations Development Program website: http://hdr.undp.org/sites/default/files/2015_human_development_report.pdf .

UNWTO. (n.d.). Sustainable Development of Tourism: Definition. Retrieved from http://sdt.unwto.org/content/about-us-5 .

UNWTO. (2016). Tourism highlights 2016. Retrieved from the World Tourism Organization website: http://www.e-unwto.org/doi/pdf/10.18111/9789284418145 .

Zhang, X., Song, H., & Huang, G. Q. (2009). Tourism supply chain management: A new research agenda. *Tourism Management, 30*(3), 366–337.

Chapter 11
Sustainable Food Supply Chain Management: An Integrated Framework and Practical Perspectives

A. D. Nuwan Gunarathne, Deemantha G. Navaratne, Amanda E. Pakianathan, and Yasasi T. Perera N.

11.1 Introduction

Among the many reasons why humans need food, the most important is to obtain the requisite nutrients to lead a healthy life. As emphasized in the World Food Summit in 1996, food security, a condition that "exists when all people, at all times, have physical and economic access to sufficient safe and nutritious food to meet their dietary needs and food preferences for a healthy and active" (Pinstrup-Andersen 2009, p. 5), is needed for the progress of any society. The rise in global population and growth in income which, in turn, increase consumer buying power in many major economies have led to an increase in the demand for food both in fresh and processed form, which reflects a similar trend as in many other industries, such as electronic equipment and textiles (Gold et al. 2017). Despite the improvements in technology and agricultural production methods, millions of people around the world are at the risk of hunger arising from a massive global food crisis (Holt-Giménez 2008; Govindan 2017). For example, Yemen, North Nigeria, South Sudan, and Somalia have been identified as "countries at risk of famine" by the Global Report on Food Crises 2017 issued by the Food Security Information Network Secretariat (FSIN) (FSIN 2017). Global issues such as climate change have significant potential to increase food insecurity and malnutrition (World Food Programme 2017). The severity of the food crises was manifested when the world leaders of

A. D. N. Gunarathne (✉)
University of Sri Jayewardenepura, Nugegoda, Sri Lanka

Griffith University, Gold Coast, Australia
e-mail: nuwan@sjp.ac.lk; nuwan.gunarathne@griffithuni.edu.au

D. G. Navaratne · A. E. Pakianathan · Y. T. Perera N.
University of Sri Jayewardenepura, Nugegoda, Sri Lanka
e-mail: deemantha@navaratne.net

H. Qudrat-Ullah (ed.), *Innovative Solutions for Sustainable Supply Chains*, Understanding Complex Systems, https://doi.org/10.1007/978-3-319-94322-0_11

more than 170 countries with the United Nations Development Programme (UNDP) identified "zero hunger" as one of the seventeen Sustainable Development Goals (SDGs) to be realized by 2030 (UNDP 2017b). See Fig. 11.1 for more details on how climate change impacts agricultural systems in the African continent. In some countries, when people die of starvation, in some other countries, there is a significant waste of food and people are suffering from obesity. This represents an imbalance in the global food supply chain. Confirming this view, the Food and Agriculture Organization (FAO) (FAO 2013) highlights that the world is producing more than

Africa has been experiencing the impacts of climate change for many decades. In Africa the adverse effects of climate change are felt in many areas such as health, food security, agriculture productivity, water availability and quality of living. Many of the countries in the African continent are highly dependent on agriculture and these countries face a high risk exposure to climate change. For example, according to the Climate Change Exposure Index (CCEI) of Verisk Maplecroft, agriculture represents more than 30% of the economy of many African countries such as Sierra Leone, Liberia, Central African Republic, Guinea-Bissau, Burundi and Rwanda, all of which are rated as 'extreme risk' category for the climate change. Further, the African continent is home to 17 of the 20 countries of the world that are most economically dependent on agriculture industry. Further these countries are the least prepared to face the consequences of the disruptions to harvests. Hence the impact of climate change on the agricultural systems in Africa are severe than any other parts of the world.

Climate change impacts on Africa have been witnessed in many areas. It has been identified that the disappearance of the glaciers on Mount Kilimanjaro is also due to the changes in climate. Since these glaciers have been a water source for many rivers, drying up of the glaciers also reduces the water levels of the rivers creating scarcity of water available for drinking and agriculture. In some parts of the continent, the temperature has risen leading to poor agricultural productivity which in turn demands an increase in land used for agriculture to meet the continent's growing demand for food. This rise in temperate with droughts and flooding in some parts of the region has led to a drastic reduction in crop and livestock productivity. In sub-Saharan Africa, mobile pastoralists (people who deal with livestock) have already experienced the consequences of climate change and are now facing the reduction in meat production and loss of livelihood. Due to the threats to the agricultural systems, the African continent is already facing many development challenges which have led to the region being vulnerable to many social and economic problems.

Social uprising such as Arab spring and many other civil conflicts in the region have arisen due to the increasing food prices. In addition to these present problems, there are many climate-related future challenges anticipated in the region. It has been predicted that rain-fed agriculture systems could drop by half of its current production by 2020 and crops like maize and wheat, two major staples, will disappear or fall significantly in the next decades to come. Many Africans argue that the continent is least responsible to anthropogenic climate change (i.e. the climate change due to human activities) and needs the most support to prepare for the future climatic change challenges. With these climate change related issues, the achievement of sustainable development goals by 2030 in this region is becoming a daunting target.

Fig. 11.1 Impact of climate change on agricultural systems in Africa. (Sources: (The Conversation 2015; Our World 2008; Verisk Maplecroft 2016))

enough food for the global population. There are many reasons why people do not have enough access to the food produced. Among them the issues associated with the global food supply chains are important. On the other hand, the food industry has major implications for sustainability such as fulfillment of basic human needs, generation of employment opportunities and livelihoods, impacts on natural environment through its use of various resources, and dealing with poverty (Humphrey and Memedovic 2006). Hence, the proper management of food supply chains becomes paramount in addressing these global issues. As Lakovou et al. (2015) state:

> Growing environmental, social as well as ethical concerns, and increased awareness of the impact of food production and consumption on the natural environment have led to increased pressures by consumer organizations, policy-makers, and environmental advocacy groups on agri-food companies to manage social and environmental issues across their supply chains (SCs) from "farm-to-the-fork." (p. 1)

In the context of the importance of food supply chains, the purpose of this chapter is to provide an overview of sustainable food supply chain management and to propose an integrated framework for it.

With the above objective, the rest of the chapter is organized as follows: Section 11.2 provides an overview of sustainable development and the importance of food supply chain management to achieve sustainable development goals. Section 11.3 describes the global food crisis and its severity at a global level. Section 11.4 provides an overview of supply chain management with particular emphasis on sustainable food supply chains. Building on the last three sections, Sec. 11.5 presents an integrated framework for the sustainable food supply chain by extending the framework of the Food and Agricultural Framework (FAO) (FAO 2014). The last section provides the conclusions and the contribution of the chapter. In every section, the chapter provides carefully selected mini practical cases to bring in more insights into the points/arguments mentioned therein.

11.2 Food Supply Chains for Sustainable Development

The purpose of this section is to provide an overview of the notion of sustainable development and the importance of food supply chain management to achieve sustainable development goals.

The term "sustainable development" has now become commonplace. The worldwide attention to the notion of sustainable development was directed when the Brundtland Commission in 1987 issued its famous report "Our Common Future." This report defined sustainable development as "the development that meets the needs of the present generation without compromising the ability of future generations to meet their own needs" (United Nations World Commission on Environment and Development 1987). Sustainable development can also be viewed as an approach to moving toward a more sustainable future (Maxwell and Sheate 2006).

This concept of sustainable development highlights the need to maintain intergenerational and intragenerational equity. In this regard, intergenerational equity ensures that adequate resources are available for future generations to meet their needs, while the intragenerational equity highlights the need for fairness in the utilization of resources among present generations (Maggio 1997). However, the problems associated with the global food crisis compel us to reconsider how far the world has been successful in realizing the inter- or intragenerational equity emphasized in sustainable development. For example, at present, there are nearly 800 million people in hunger, the majority of whom are from developing countries (UNDP 2017a). To feed these hungry people and another two billion that would be added to the world population by 2050, the present global food and agricultural systems require drastic changes. As mentioned in Section 11.1, the food supply chains play a crucial role in achieving the second SDG – "end hunger, achieve food security and improved nutrition and promote sustainable agriculture" (UN 2017). However, it should be noted that food supply chains affect many other SDGs as well. In explaining why zero hunger is important, the UN (2016) states:

> Extreme hunger and malnutrition remains a barrier to sustainable development and creates a trap from which people cannot easily escape. Hunger and malnutrition mean less productive individuals, who are more prone to disease and thus often unable to earn more and improve their livelihoods…poor harvesting practices, as well as food wastage have contributed to food scarcity. (p. 1)

With the objective of achieving "zero hunger," the UN has established several programs, institutions, and projects such as the International Fund for Agricultural Development (IFAD), the Food and Agriculture Organization (FAO), and the World Food Programme (WFP).

IFAD, established in 1977 as one of the major outcomes of the 1974 World Food Conference, is a specialized United Nations agency and an international financial institution. IFAD's programs and projects support the investments in rural people and small- and medium-sized enterprises to reduce poverty and hunger with a community-driven bottom-up approach (IFAD 2017). Refer Fig. 11.2 that explains how IFAD supports the development of agribusiness in Sri Lanka.

IFAD has granted USD 32.9 million for "National Agribusiness Development Programme" for two sub projects, a) to assist smallholder farmers by marketing chain development and linkages, and b) to provide micro-financing and training of youth. In assisting the small holder farmers, the project aims to increase production, productivity and improve the quality and value addition of farmer produce, and with the final objective of increasing farmer income. By 2016 under the 4P (Public-private-producer Partnerships) programme nearly 15,000 households have benefited. An important feature of this subproject is the agreement with private sector and the farmer/producer to purchase 100% of the available produce at a pre-agreed market price to ensure these small-scale farmers have a market for their produce.

Fig. 11.2 IFAD's National Agribusiness Development Programme in Sri Lanka. (Source: (IFAD 2016))

Another institution under the UN is the Food and Agriculture Organization (FAO). Currently FAO has five strategic objectives: (a) help eliminate hunger, food insecurity, and malnutrition, (b) make agriculture, forestry, and fisheries more productive and sustainable, (c) reduce rural poverty, (d) enable inclusive and efficient agricultural and food systems, and (e) increase the resilience of livelihoods from disasters (FAO 2013). Although almost all of these strategic objectives impact sustainable food supply chains, its fourth objective, "enable inclusive and efficient agricultural and food systems," directly deals with food supply chains. In this regard, FAO takes several initiatives. They are to (a) improve the inclusiveness and efficiency of the food system, (b) help strengthen public-private collaboration to improve smallholder agriculture, and (c) improve the inclusiveness and efficiency of markets.

Despite the efforts of these institutions and programs which have contributed to some progress, there are still many challenges in achieving the goal of zero hunger. The next section of this chapter is devoted to explaining the food crisis on a global scale.

11.3 Global Food Crisis

This section presents the global food crisis and its severity at a global level. It first provides evidence of the global food crisis followed by the causes that give rise to this crisis. Next the consequences of the food crisis are highlighted. The chapter provides insights into the severity of the global food crisis through selected mini cases to further enhance knowledge on this global challenge.

11.3.1 Evidence of a Global Food Crisis

The history of mankind is full of various food crises. In the recent decades, the global food crisis has affected communities around the world twice, once in 1972–1974 and again in 2007–2008 (Heady and Fan 2010). The global food crisis can be broadly referred as the sustained rise of global food prices to a level where communities find it difficult to fulfill their daily nutrition needs leading to global food insecurity (Heady 2013, pp. 1–2). The Institute of Hazard, Risk and Resilience (IHRR) (IHRR 2013) states that malnourishment is only a single side to this multi-faceted global issue, and among much else, poverty is a large-scale social dilemma that is exacerbated by the global food crisis.

Global staples prices increased by 85% from April 2007 to April 2008 with wheat prices doubling, maize prices increasing by 67%, rice prices by 160%, and vegetable oil seeds prices doubling within months (Panitchpakdi 2010). According to FAO (2011), the malnourished population in the world exceeded 900 million in 2007, much of this increase being in developing regions in Asia, the Pacific, and sub-Saharan Africa. According to UNICEF (2015), about 21,000 people die due to starvation daily, i.e., one person every 4 s, usually a child under the age of 5, dies of starvation. Concurrently, one third of the edible parts of food produced for human consumption gets lost or wasted through the global food supply chain annually (FAO 2011).

Evidence of a global food crisis is most apparent in low-income earning, least developed, and developing countries, where an average of 50–80% of personal income is spent on daily food requirements (FAO 2011). FAO (2011) further states that the majority of these least developed and developing countries are relying upon net imports of staples rather than being self-sufficient with food harvests. Thus, such economies are highly affected by exchange rate movements. However, larger developing economies and developed countries are not entirely excluded from the negative consequences of rising food prices, poverty tensions, and malnourishment (Panitchpakdi 2010; IHRR 2013). According to the self-assessed food security research by Heady (2013), food insecurity reported the highest in sub-Saharan Africa at 58.3%, which is by far the financially poorest region in the world. South America reported a mean food insecurity rate of 36% followed by Central America and the Caribbean at 34.5% and South Asia at 31%, while OECD countries reported a food insecurity rate of 8.3% (Heady 2013). These figures strongly supplement that evidence of a global food crisis varies on a regional basis, and the impact is highly dependent on the level of development of each region. The following is a mini case emphasizing the severity of malnourishment in Somalia, according to research by the European Union in collaboration with FAO and WFP (see Fig. 11.3).

As indicated above, the global food crisis is caused by an array of factors among which increase in global food prices and poverty stand at the very top. As evidenced in Somalia, population displacements due to armed conflicts and extreme climate events have also caused food insecurity. The section will next focus on various causes of the global food crisis.

11.3.2 Causes of the Global Food Crisis

Among the various studies, Trostle (2008) provides an interesting classification of the factors behind the global food crisis: long-term, medium-term, and short-term factors. Accordingly, strong growth in demand and slowing agricultural production are sample cases of long-term factors leading to the food crisis, while medium-term factors could be exchange rate impact and rising energy prices; and short-term factors are adverse weather and trade shocks. Since these factors are interrelated, a clear demarcation is difficult but used only for analysis. Though armed conflicts have not been a key factor affecting food crisis events experienced so far, this could also form a potential cause of future global food crises as it continues to impact the food security of millions of people in conflict areas. Thus, the following discussion is structured around these factors contributing to the world food crisis.

It is estimated that over 200,000 children in Somalia are suffering from acute malnutrition while closer to 18.5% are severely malnourished. The nutrition survey conducted by the Food Security and Nutrition Analysis Unit (FSNAU) is Somalia in 2015 indicated a median global acute malnutrition (GAM) rate of 13.6% and a median severe acute malnutrition (SAM) rate of 2.3% at national level, with prevalence exceeding the emergency threshold of 15% in livelihoods of the South-Central Region of Somalia (North and South Gedo, Mataban, Beletweyene, Coastal Deeh and Cowpea Belt districts).

Critical levels of GAM prevalence (≥15%) were also recorded in two out of six urban areas surveyed during 2015 (18.4% in Bari and 15.7% in Nugal) while Serious GAM prevalence (above 10%) was noted in urban Mogadishu (10.5%) and Alert (5-9.9%) in urban Sool and urban Kismayo. The overall stunting rate in Somalia was 12%, although there were major differences between zones: 15% in Southern and Central areas, and 1% in the North-East. Out of thirteen internally displaced people settlements surveyed during 2015, five showed critical levels of GAM: Dhobley, Baidoa, Dolow Garowe and Galkayo. It is of concern that acute malnutrition levels in three of these settlements (Dolow, Garowe and Galkayo) have remained at critical levels over the past two years. The nutrition situation in Dollow settlement has also deteriorated since December 2014 with an increase in the levels of GAM, a near doubling of the crude death rate as well as increases in the under-five death rate and morbidity levels.

The nutrition situation in the drought affected areas of Northwest agro pastoral and Gubal Pastoral livelihoods as observed during the survey was expected to deteriorate from 'Serious' to 'Critical' levels of acute malnutrition as the drought conditions were expected to worsen until rains are fully established in October 2015.

Fig. 11.3 Nutrition security situation of Somalia. (Source: (Nkunzimana et al. 2016))

11.3.2.1 Strong Growth in Demand

Demand and supply forces may impose collective pressures on global food prices. Many studies suggest that steady growth in demand, particularly in China and India, which collectively exceeds a population of 2 billion, had caused the hike in global food prices (Panitchpakdi 2010; Jaspars and Wiggins 2009). On the contrary, Heady and Fan (2010) suggest that both China and India are less dependent on staple imports and are becoming more self-sufficient. However, growth of these economies indirectly results in global food price hikes through increased oil and energy prices resulting from increased demand (Heady and Fan 2010). However cereal import demand had grown globally especially from Spain and Mexico in early 2000, which may have affected the food crisis in 2007–2008. This also suggests that the impact of growth factors building gradually over a long period of time will ultimately burst, when the system is unable to fulfill demand in full.

11.3.2.2 Productivity Decline and Falling Supply

Global food production is growing at a slower rate than population growth, resulting in an imbalance in demand and supply. This is reinforced by the trends in production per capita where global cereal production per capita was about 6% lower in 2000 than in the 1980s (Heady and Fan 2010, pp. 20–23). Heady and Fan (2010) further argue that climate changes and unsustainable farming methods are affecting long-term growth of production. Low research and development, land degradation, and constant use of harmful pesticides have worsened the supply side impact on the global food crisis (Panitchpakdi 2010; Heady and Fan 2010). These factors further result in a reduction in the output throughout the food supply chain. That is, during harvesting, food losses occur due to mechanical damage or spillage, post-harvest losses in terms of storage, poor handling, and inappropriate processing, packaging, and distribution (Kummu et al. 2012). Further, the constant decline in the population working in agriculture and farming land had vastly affected the reduction in global supply (Panitchpakdi 2010; Heady and Fan 2010).

Considering the factors causing the food crisis, productivity decline and fall in supply stand out as the factors which could be controlled with proper policy implementation. It could be a focal point in assessing preventive action for future crises.

11.3.2.3 Impact of Fuel and Energy Prices

During the 1972–1974 food crises, the rise in oil prices has been a major contributor. Agricultural production is usually dependent on oil for its energy needs, and the indirect impact of the oil price hike on agriculture production comes from fertilizer and transport. Energy costs constitute about 90% of the cost of fertilizer production, and bulk production needs to reach consumers across regions. Hence increase in oil prices indeed impact the cost of transport and fertilizer, resulting in an increase in food prices (Heady and Fan 2010; Panitchpakdi 2010; Jaspars and Wiggins 2009). The rise in oil prices has indirectly resulted in a decline of the maize supply due to diversion of the maize crop from food to bio-oil.

11.3.2.4 Decline in the Exchange Rate of the US Dollar

There is empirical evidence that patterns of commodity price changes and nominal exchange rate movements of USD have been similar since 1970 (Heady and Fan 2010). When the dollar is weakening, commodity prices have risen including oil prices. The United States, the largest exporter of staples (excluding rice), is the price setter in the international markets (Heady and Fan 2010; Heady 2013). To dampen the impact of exchange rates on real income, and also as a response to increased demand, commodity prices tend to rise when the dollar is weakening. Low interest rates and speculations in financial markets have further aggravated the impact on commodity prices (Heady and Fan 2010).

11.3.2.5 Trade Shocks due to Export Controls and Climate Changes

During 2007, an artificial trade shock was built in international rice markets due to export restrictions imposed by governments to stabilize prices within the country (Heady and Fan 2010). Simultaneously, wheat production was adversely affected by weather shocks in Australia and Africa. They further state that the impact of these short-term export controls, demand surges, and bad weather is more important than that of long-term causes discussed above.

11.3.2.6 Armed Conflicts

Armed conflicts continuing in Africa and the Middle East had interrupted the flow of agricultural and other produce, depriving people living in such areas of their daily nutrition needs. Conflicts result in disrupting the food system through plundering crops and livestock, loss of assets and income, and population displacement (FSIN 2017). Though this has not been a cause of a major global food crisis so far, it has certainly worsened the impact and consequences of such crisis, putting many lives in danger.

11.3.2.7 Poverty and Unequal Access to Food

Despite many efforts to implement solutions for all the above causes, malnourishment and food insecurity still persist with poorer regions showing no improvement and richer countries constantly wasting edible food mass. IHRR (2013) argues that the reason for food insecurity persisting is that the roots of food insecurity lie deep in society in terms of poverty and unequal distribution of resources which have been overlooked and tolerated by past, present, and probably future generations. While many other researchers see poverty as a consequence of the food crisis, IHRR shows that it is the root cause of the food crisis and every other cause discussed are just symptoms.

Irrespective of the cause, food crises result in many adverse consequences, which are discussed next.

11.3.3 Consequences of the Global Food Crisis

The global food crisis and food insecurity have affected many lives in the world to varying extents, and many studies have found that there are regional differences in the severity of impact of the global food crisis (Heady 2013; Jaspars and Wiggins 2009; IHRR 2013). This section now focuses on a few of many consequences of the global food crisis by highlighting the social, financial, and environmental impact in a global context.

11.3.3.1 Social Impact of Global Food Crisis

The global food crisis has endangered many lives posing numerous behavioral changes in the lifestyles of people who are severely affected, especially in the African region. Panitchpakdi (2010) argues that the global food crisis directly undermines one of the most fundamental human rights, which is freedom from hunger and malnutrition. Responding to a constant rise in food and oil prices, many have acted to fulfill their needs at a bare minimum level. The following mini case adopted from Jaspars and Wiggins (2009) provides evidence of the changes in lifestyles of people severely affected by a food crisis (see Fig. 11.4). It has been further noted that women often act as shock absorbers of household food insecurity by reducing their consumption to leave more food for others, especially for children (Heady and Fan 2010). Reduced availability or affordability of food compromises health, education, maternal well-being, sanitation, and many other social indicators such as earning capacity and satisfaction.

11.3.3.2 Financial Impact of Global Food Crisis

Increased cost of production and insufficient infrastructure facilities in the least developed and developing nations have resulted in a drastic drop in the population engaged in agricultural production. Supplemented by subsidies given by developed countries, developing nations that have traditionally been net food exporters have become net food importers in the past 20 years. Hence countries much dependent on net food imports are highly affected by changes in exchange rates imposing greater threats to foreign reserves, balance of payments, interest rates, and inflation rates of the respective countries (Heady and Fan 2010; Panitchpakdi 2010).

It is also worth noting that about one third of edible food produced for human consumption gets wasted or lost throughout the global food supply chain (FAO 2011; IHRR 2013). The amount of resource inputs to generate this waste and its contribution to greenhouse gas emission intensify the impact of food waste and urge global communities to pay more attention to ensuring sustainability of the food supply chains and to minimize food waste by proper management and control.

> *Recent field studies have confirmed that increases in food prices can severely affect urban, displaced, pastoralist and smallholder farming communities. The people affected by the food price increases responded by reducing the quantity and diversity of their diet while cutting down their expenditure on healthcare, education and other basic goods. Somalia provides a classic example on how people responded to an increase in the price of imported rice, the main staple food in Northern and Central Somalia. The Somalian responded by reducing non-food purchases and buying less of some food items, turning to cheaper foods (locally produced sorghum) and eating lower-quality cereals. Children have been taken out of school and purchases of medicine have declined.*

Fig. 11.4 Responses to global food crisis. (Source: (Jaspars and Wiggins 2009))

11.3.3.3 Environmental Impact of Global Food Crisis

There are various environmental consequences of the global food crisis. From a supply side perspective, degradation of land used for agricultural purposes is one such impact of unsustainable agricultural practices. Soil erosion and water pollution due to extensive use of pesticides have affected production capacity as well as biodiversity (Papargyropoulou et al. 2014). Further, the environmental impact of food waste intensifies the environmental impact of the global food crisis. This is in terms of the emission of carbon dioxide and greenhouse gases resulting in adverse climate changes affecting the future production of agriculture.

While there is starvation and food insecurity in most parts of the world, there is also a food surplus and varied levels of food waste taking place, mostly in developed countries. Solutions to these complex global issues are predominantly focused around supply side pressures on production as such factors are more controllable in the short to medium term. Thus, establishing a sustainable food supply chain should dampen the severity of these adverse consequences while providing a sustainable solution for the future, which will be the focus of the following section of the chapter.

11.4 Sustainable Food Supply Chain Management

This section provides a sound theoretical background of sustainable food supply chain management. In doing so, this section first addresses the theoretical background of supply chain management, then sustainable supply chain management, and finally sustainable food supply chain management.

11.4.1 Supply Chain Management

In recent years, there has been heightened attention to supply chain management (Cooper et al. 1997) driven by many factors such as global sourcing, time, and quality-based competition and flexibility required to face environmental uncertainties (Mentzer et al. 2001). Given this popularity, supply chains have been the subject of studies by many scholars. In simple terms, supply chain can be defined as a set of firms that are linked to pass materials forward (La Londe and Masters 1994). Christopher (2011) provides a more descriptive definition of supply chain, where he defines it as a network of organizations involved in the different processes and activities that produce value in the form of products and services delivered to the ultimate consumer, through upstream and downstream linkages. Expanding the above definition, Mentzer et al. (2001, p. 4) define supply chain as three or more entities directly involved in upstream and downstream flows of products, services, finances, and/or information from a source to a customer. Despite the similarity of definitions of supply chain across the literature, there is a disparity regarding the definition on supply chain management (SCM) (Mentzer et al. 2001; La Londe and Masters 1994; Lambert et al. 1998). Mentzer et al. (2001, p. 5) highlight that definitions of supply chain management can be distinctively grouped into three categories: SCM as a management philosophy, SCM as the implementation of a management philosophy, and SCM as a set of management processes.

Cooper and Ellram (1993) highlight SCM as an integrative philosophy oriented toward managing the total flow of a distribution channel from supplier to ultimate user to pinpoint activities such as sourcing, procurement, conversion, and logistics in coordination with the channel partners. As a philosophy, SCM adopts a systemic approach, which views the supply chain as a whole oriented toward common objectives, instead of a set of disintegrated parts performing its own function (Mentzer et al. 2001). SCM as a management concept is used to coordinate traditional fragmented approaches to the buying, storing, and moving of goods (Eastham et al. 2014).

As a set of activities to implement a management philosophy, Mentzer et al. (2001) state that firms within the supply chain must adopt activities, which would enable them to be aligned to the management philosophy. Based on the extant literature, Mentzer et al. (2001, p. 8) present "SCM activities" required to implement a management philosophy: integrated behavior, mutually sharing information, mutually sharing risks and rewards, cooperation, the same goal and the same focus on serving customers, integration of processes, and partners to build and maintain long-term relationships.

Lambert et al. (1998) stress that SCM, as a set of management processes, must overcome functional silos of the firms involved in the supply chain and move to a process approach, where every process is focused on meeting customer needs. They further propose that as a set of management processes, SCM would typically include customer relationship management, customer service management, demand management, order fulfillment, manufacturing flow management, procurement, and product development. Irrespective of the stance toward SCM, sustainability of the SCM has received the attention of many scholars. Having highlighted supply chain and SCM, in general, sustainable SCM is discussed next.

11.4.2 Sustainable Supply Chain Management

The concepts of sustainability (or sustainable development) and sustainable supply chain management (SSCM) have gained attention over the past decades (Ahi and Searcy 2013; Carter and Easton 2011). The sustainable development definition of the Brundtland Commission addresses the concept of sustainability at a much broader level, making it difficult for organizations to apply practically by distinctly addressing future and present needs, identifying technologies and resources required to meet such needs, and balancing organizational responsibilities among multiple stakeholders (Carter and Rogers 2008). On the other hand, the concept of the triple bottom line developed by Elkington helps to operationalize this comprehensive definition of sustainability (Seuring and Muller 2008). Carter and Rogers (2008) stress that the triple bottom line advocates simultaneously considering and balancing the three dimensions of sustainability, i.e., economic, environmental, and social dimensions, for adapting sustainable practices at a micro level.

The impact of the notion of sustainable development and the triple bottom-line approach has been felt across many disciplines, including SCM. Accordingly, SSCM is defined as "the management of material, information and capital flows as well as cooperation among companies along the supply chain while taking goals from all three dimensions of sustainable development, i.e., economic, environmental and social" (Seuring and Muller 2008, p. 1701). Similarly, Carter and Rogers (2008, p. 368) define it as "the strategic, transparent integration and achievement of an organization's social, environmental, and economic goals in the systemic coordination of key inter-organizational business processes for improving the long-term economic performance of the individual company and its supply chains." Furthermore, they also identify four supporting facilitators of SSCM: strategy, risk management, organizational culture, and transparency. In simple terms, it is where supply chains are managed by considering the three dimensions of sustainability – economic, environmental, and social (Ciliberti et al. 2008, p. 1580). Also, Tseng et al. (2015) suggest that supply chain managers need to integrate economic, environmental, and social considerations into their daily decision-making process to achieve sustainable organizational performance.

However, the three dimensions of the triple bottom line were not always included in the definition of SSCM. The definitions put forward by Badurdeen et al. (2009, p. 57) and Haake and Seuring (2009, p. 285) focus on only the environmental and social aspects of sustainability, while Wittstruck and Teuteberg (2011) incorporate ethics as a part of the social dimension of sustainability. Ahi and Searcy (2013), based on a comparative literature analysis, conclude that the definition put forward by Carter and Rogers (2008) is the most comprehensive as it addresses five out of the seven characteristics of SSCM: economic focus, environmental focus, social focus, stakeholder focus, volunteer focus, resilience focus, and long-term focus.

Being more critical on the current status of SSCM, Pagell and Wu (2009, p. 38) stress that a "truly sustainable" supply chain "would at worst do no harm to natural or social systems while still producing a profit over an extended period of time." However, the authors admit that no supply chain has ever achieved such true sustainability. Instead, most supply chains attempt to be more sustainable than others in their industry to continue in business for far longer than their competitors.

On another strand of SSCM, Abukhader and Jönson (2004) advocate three main "themes": reverse logistics, assessment of emissions, and "greening" supply chains. Reverse logistics is a logistics area which begins "from the point of consumption to the point of origin" focused on proper disposal or recapturing value. In other words, reverse logistics encompasses all activities of SCM, but it flows in the opposite direction (Grant et al. 2015). The case of QuickSnap by Fujifilm is illustrated below to better explain the concept of reverse logistics that can mitigate the negative environmental impacts of products (see Fig. 11.5). The second theme on SSCM,

Fujifilm launched QuickSnap, single use cameras in 1986 after careful market research. QuickSnap was soon a popular consumer convenience product and 1 million cameras were sold in the first six months. However, with the growing interest in area of sustainability several stakeholder groups began to attack the product's disposable nature. With consumers beginning to refer to QuickSnap as 'disposables' or 'throwaways', the media highlighted the concerns of the environmental groups regarding the wastefulness and damage to the environment. This resulted in a negative impact on company's brand image and sales.

As a response to such pressures, Fujifilm initiated a voluntary take back programme and started to recycle the collected cameras by utilizing a highly advanced and original recycling system. Fujifilm further altered the product design to use various techniques of product recovery management such that it cannot be reused or resold. Soon Fujifilm was able to achieve 100 per cent recycling rate with its 'inverse manufacturing system' and was able to establish one of the first, fully integrated closed loop or reverse logistics systems for its product, negating the poor environmental image of QuickSnap.

Fig. 11.5 Reverse logistics: the case of QuickSnap. (Source: (Grant et al. 2015))

emissions assessment, involves the assessment of the impact of logistics on the environment. This deals with evaluating the environmental implications of a logistics/supply chain infrastructure. The final theme, greening supply chains, is the reverse of the latter. It focuses on the impact of environment on logistics and aims to identify how implementing environmental measures would influence the logistics/supply chain infrastructure. However, van Hoek (1999) argues that the focus should move from reverse logistics to greening of supply chains as reverse logistics could fail to focus on the entire supply chain. In adopting such greening of supply chains, three green approaches have been advocated by van Hoek (1999). They are reactive, proactive, and value-seeking approaches. The reactive/laid-back approach is when entities satisfy environmental legislations at a basic level as possible by committing minimal resources. This approach could be adopted as a strategy of legitimization. However, under the proactive approach, companies anticipate new environmental regulations and concerns with a modest resource commitment to design green products and initiate recycling. The value-seeking approach is a much broader and comprehensive approach in which companies integrate environmental activities into their business strategy with a strong environmental commitment and operate the firm and its partners in the supply chain to reduce the impact on the environment as a strategic initiative.

Although SSCM emphasizes three aspects, i.e., economic, environmental, and social, the extant literature shows an unequal development of them. The literature highlights that higher prominence has been given to the environmental consequences of supply chain activities while attention to the impact and influence of supply chain activities on the wider society is limited. Confirming this view, Abbasi (2017) highlights that though the concept of sustainability encompasses the three pillars of economic development, environmental protection, and social development each with equal importance, the latter is not emphasized enough both theoretically and practically. Abbasi stresses the importance of social development by highlighting that surrounding societies of the supply chain affect and is also affected by the supply chain. Given below is the case study of AMUL, one of the globally recognized dairy brands, illustrating how a traditional supply chain was replaced by a socially sustainable approach (see Fig. 11.6). The case also provides evidence for Abbasi's claim of interdependence of both surrounding societies and the supply chain.

11.4.3 Sustainable Food Supply Chain Management

As highlighted in the Background Section, in order to address the global food crisis, effective management of food supply chain is important. Hence this subsection provides an overview of sustainable food supply chain management. Aramyan et al. (2007) identify several characteristics, which differentiate food supply chains from other types of supply chains. These could be broadly grouped into three main aspects. They are:

- Production being highly based on biological processes.
- Distinctive features of the supply chain such as perishability, seasonality, and higher safety concerns.
- Growing consumer awareness of the environmental impacts of the production process and the nutritional facts of the products.

Studies on sustainable food supply chains are an emerging area. Hence, clear and commonly used definitions of the area are sparse. However, the Department for Environment, Food and Rural Affairs of the United Kingdom (DEFRA) together with the Sustainable Development Commission of the United Kingdom provides a comprehensive view of sustainable food supply chains as supply chains that:

- Produce safe and healthy products in response to market demands and ensure that all consumers have access to nutritious food and to accurate information about food products.
- Support the viability and diversity of rural and urban economies and communities.
- Enable viable livelihoods to be made from sustainable land management, both through the market and through payments for public benefits.
- Respect and operate within the biological limits of natural resources (especially soil, water, and biodiversity).
- Achieve consistently high standards of environmental performance by reducing energy consumption, by minimizing resource inputs, and by using renewable energy wherever possible.
- Ensure a safe and hygienic working environment and high social welfare and training for all employees involved in the food chain.
- Achieve consistently high standards of animal health and welfare.

AMUL is a dairy cooperative, in Western India that has become world's largest producer of milk through its innovative practices. AMUL sets an example of how to develop a network of firms in an emerging economy, which has fragmented supply bases and under developed markets, by creating value for suppliers as well as the customers. In late 1940s, India was ranked nowhere amongst milk producing countries in the world. In such a point of time, the Polson Dairy dominated the dairy industry of Western India. The government awarded monopoly rights to Polson Dairy to collect milk from the city of Anand in western India and supply to Mumbai city. Polson Dairy followed a traditional supply chain where it sourced milk from fragmented local farmers. Instead of working in cooperation with the supplier base, Polson Dairy used its power of being the only dairy in the city of Anand and focused primarily on maximising its profits by attempting to source at the lowest cost possible.

Wastage in the supply chain was very high. The agents of Polson Dairy decided the prices and the off-take from the farmers by the season. As milk is a commodity that must be collected twice a day from each cow/buffalo, in winter when the demand for milk is low, the farmer was either left with surplus unsold milk or had to sell it at very low prices. Furthermore, farmers had to travel long distances to deliver milk to Polson Dairy in Anand – often milk went sour. As a response to the exploitation of milk producers by Polson Dairy, the nationalist leaders advised farmers to form a cooperative and began to supply directly to the Mumbai Milk Scheme instead of selling it through Polson. This led to the establishment of Kaira District Cooperative Milk Producers' Union Limited in 1946.

The Kaira District Cooperative was established to collect and process milk in the whole district of Kaira in Gujarat, rather than just in Anand city. Milk collection was decentralized, as most producers were small-scale farmers who could deliver 1-2 litres of milk per day. Village level cooperatives were established to organize the small-scale milk producers in each of the villages. Wastage through the supply chain was minimal as the member farmers were assured with a guaranteed purchase of all the milk that they produce at pre-determined prices and the cooperative collected the milk directly from the producers' doorsteps. The cooperatives successfully eliminated the need for middlemen, reduced transaction costs for farmers and were able to directly sell high-quality products at reasonable prices to consumers.

The first modern dairy of the Kaira Union was established at Anand (which popularly came to be known as AMUL dairy after its brand name). The new plant had the capacity to pasteurise 300,000 pounds of milk per day, manufacture 10,000 pounds of butter per day, 12,500 pounds of milk powder per day and 1,200 pounds of casein per day. Indigenous R&D and technology development at the Cooperative had led to the successful production of skimmed milk powder from buffalo milk – the first time on a commercial scale anywhere in the world.

Now India has emerged as one of the largest milk producing country in the world with Gujarat emerging as the most successful State in terms of milk and milk product production through its cooperative dairy movement. The Kaira District Cooperative Milk Producers' Union Limited has become the focal point of dairy development in the entire region with a network of 2.12 million milk producers (called farmers) who are organized in 10,411 milk collection independent cooperatives (called Village Societies). AMUL has emerged as one of the most globally recognized brands, ahead of many international brands.

Fig. 11.6 From a traditional to a socially sustainable supply chain – the story of AMUL. (Source: (Chandra and Tirupati 2003))

- Sustain the resources available for growing food and supplying other public benefits over time, except where alternative land uses are essential to meet other needs of society (DEFRA 2002).

Though the Forum for the Future (2014) does not attempt to define the concept, they stress that to deliver a sustainable food supply chain, three areas for action must be addressed: rethinking value creation, reconnecting people with food, and restoring ecosystems' resilience. Achieving sustainability within food supply chains is challenging due to the barriers inherent in the process. Climate change, water scarcity, demographic changes, land use, and food waste are some key barriers identified by the Forum for the Future (2014). They further identify that institutions and society as a whole are giving wider importance to the food system approach as the "only way" to address such key challenges. The food system approach advocates that all activities relating to the food supply chains such as growing, processing, distributing, and manufacturing should be well connected, linked, and focused to reduce food waste throughout the supply chain. In other words, all the food supply chain parties including farmers, food manufacturers, wholesalers and distributors, brokers, and retailers should be well connected to reduce food waste and make food supply chains sustainable (Fredriksson and Liljestrand 2015).

As highlighted by Bourlakis and Weightman (2004), food supply chains have evolved in focus, from a series of shorter, independent transfers toward more integrated and coherent relationships between producers, manufacturers, and retailers. In the context of a deepening food crisis, the complexity of food supply chains creates a need for an integrated framework for sustainable food supply chain management, which is the focus of the next section.

11.5 Integrated Framework for Sustainable Food Supply Chain Management

Building on the last sections, this section presents an integrated framework for the sustainable food supply chain. We extend the framework of the FAO (2014) by incorporating some guiding principles such as change management, knowledge management, stakeholder engagement, and continuous process improvement (PDCA approach) which will be essential for the sustenance of the model. This will also provide the framework with the ability to be agile and face the dynamic and cyclical nature of sustainable food supply chains. Following a similar pattern as in the previous sections, this section too will provide practical mini cases selected from real-world organizations on how the elements in the integrated framework are operationalized.

11.5.1 Framework of the Food and Agricultural Framework (FAO)

The framework of the FAO has been developed on the value chain concept (see Fig. 11.7 for an overview of value chain concept). The framework represents how the behavior and performance of agri-related stakeholders are determined in a complex environment. It is comprised of mainly two components: extending the value chain and an enabling environment.

The extended value chain is comprised of *core value chain* and *support providers*. At the heart of the core value chain is the upstream and downstream *value chain actors*. These actors can be from private and/or public sectors and are distinct based on the size of operations, technology used, distribution channels used and the end-market they cater to. Following the activities in a traditional supply chain, these value chain actors perform *four core functions*: production, aggregation, processing, and distribution. For developing countries, the aggregation function is especially challenging since they need to efficiently aggregate and store small volumes of produce from a large number of dispersed smallholder producers. This function could be handled by producer groups, intermediaries specializing in aggregation, food producers, or wholesalers and retailers. *Governance structure* plays a key role in the core value chain, and it represents two types of linkages: (a) horizontal linkages (between actors at a stage) and (b) vertical linkages (within the overall chain). These linkages can be related to areas such as information exchange, price determination, standards, and payment mechanisms.

When the *support providers* are added to the core value chain, *an extended value chain* is created. These support providers do not take ownership of the product but facilitate the value creation process. There are three types of support providers. They are (a) physical input providers (such as seeds and packaging material), (b) non-financial services providers (such as field spraying, storage, transport, laboratory

Value chain is one of the commonly used concepts in business management literature and has different conceptualizations. Usually value chain is a meso level analysis which falls between the macro level of the national economy and micro level of the individual firms. There are several concepts of value chain that have been developed over time to address the limitations of the older concepts. The value chain of Michel E Porter is one such widely recognized value chain concept. Porter (1985), in his famous book, "Competitive Advantage" presents his conceptualization of value chain, which he defines as a set of activities an organization carries out to create value for its customers. He classifies these activities of a business organization as primary and support activities. Primary activities include inbound logistics, operations, outbound logistics, marketing and sales and service while procurement, human resource management, technological development and infrastructure are identified as support activities. Both primary and support activities are necessary for the effective functioning of an organization.

Fig. 11.7 Value chain concept. (Source: (FAO 2014))

testing, and market research), and (c) financial services providers (such as banks, financial institutions, and other players).

The ultimate value of the agricultural produce generated via the value chain is sold to consumers who are from national or international markets. The effects of the consumer purchases trickle down to value chain actors and support providers. Value chain actors, support providers, and consumers operate in an enabling environment where natural and societal elements can be distinguished. Natural elements include natural resources such as soil, air, water, biodiversity, etc. Societal elements include sociocultural elements (e.g., consumer preference and religious requirements), institutional elements (e.g., regulations, laws, and policies), and organizational elements (e.g., professional associations and educational facilities) and infrastructural elements (transport and energy systems and communication networks). The enabling environment can be further demarcated as national and international environments.

The framework also emphasizes that the sustainability of the value chain has to be decided on along three dimensions: economic, social, and environmental. From an economic dimension, the activities at individual actor level and support provider level should be commercially viable. From a social dimension, the distribution of benefits and costs of increased value creation should be socially and culturally acceptable. From an environmental dimension, there should be little or no negative impact and/or maximum positive impacts on the environment by the actors and support providers in their value-generating endeavors.

Although this framework is comprehensive in nature, there are some other aspects that should be emphasized to make food supply chains truly sustainable. Sustainability is dynamic due to its cyclic and path-dependent nature. This means the sustainable performance of one period strongly affects the next, which leads us to the downsides of the model discussed above. This framework thus fails to focus on some of the essential elements needed for the sustenance of the system due to the dynamic nature of food supply chains. Hence, in our proposed framework, we extend this framework by incorporating the guiding principles necessary for the sustenance of the system.

11.5.2 Proposed Extended Framework for Sustainable Food Supply Chain Management

A vital element of the proposed model is its guiding principles (refer Fig. 11.8). The aim of these guiding principles is primarily twofold: (a) to address the dynamic nature of sustainable food supply chains by adopting a continuous improvement cycle and (b) to provide a broader stakeholder-oriented perspective for the implementation and continuous sustenance of the model. We believe that without these guiding principles, no framework will be truly sustainable in the long run in ever-changing business environments.

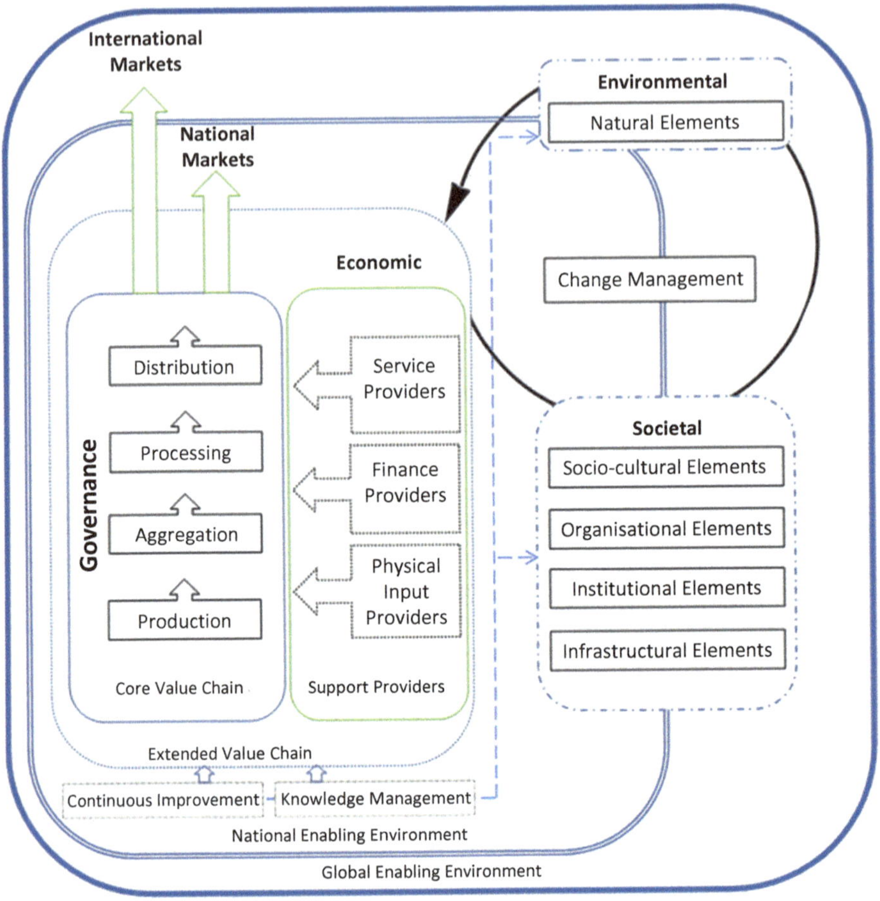

Fig. 11.8 Proposed extended framework for sustainable food value chain management. (Source: Adapted from FAO (2014))

To ensure successful operationalization and sustenance of a sustainable food supply chain, a sound continuous improvement cycle is required. This assists in the continuous evolution to meet future demands with respect to the sustainability of food supply chains. The suggested approach, PDCA cycle (Plan-Do-Check-Act), is a well-known model in process improvement (Johnson 2002). Hence, there should be continuous planning, implementation, checking, and improvement at each stage of the food supply chain, i.e., each of the core functions. More than being a simple tool, PDCA cycle/principles highlight the embodiment of continuous improvements in the culture of an organization (Sokovic et al. 2010). Due to the dynamism in the food supply chain, it is necessary to continuously improve the core functions in the supply chain. Not only the core functions but also the support providers should be part of this PDCA cycle. In other words, the continuous improvement cycle should be an integral part of the extended value chain.

Another guiding principle for sustainable food supply chains is knowledge management (Beske 2012; Beske et al. 2014). Knowledge management ensures that the partners of the extended supply chain create internal knowledge, acquire external knowledge, store knowledge as well as update and share knowledge internally and externally (Alavi and Leidner 2001; Teece 1998). According to the framework of FAO, this aspect is part of the governance structure, but it should be extended to embrace all the actors in the extended value chain. It should also be noted that the relationship and need of continuous improvement and knowledge management to the economic factors are direct. However, there can be an indirect relationship to the environmental and societal dimensions. The dotted arrow to societal and environmental dimensions represents this indirect relationship in the model.

When converting traditional supply chains into sustainable supply chains, change management throughout the supply chain becomes crucial. Hence, it is essential to continually renew the direction, structure, and capabilities of the extended value chain to serve the ever-changing needs of external and internal enabling environment (Moran and Brightman 2001). This requires changing the current practices to make them sustainable in the spheres of economic, social, and natural dimensions in line with the expectations of stakeholders. A shared vision and a common direction are fundamental. This is followed by measures to change past practices by unfreezing the present to move to a new level and refreezing this new level. This model of change recognizes the need to discard prevailing sole economic focused, structures, processes, and culture before successfully adopting new sustainability approaches (Bamford and Forrester 2003). This will be followed by crafting a sustainability implementation plan of the food supply chain and developing enabling structures. Finally, it is reinforcing and institutionalizing change that facilitate monitoring and improving strategies in remedying issues arising in the process of changing toward sustainability.

Due to the wide spectrum this framework covers, the authors do not find a single organization that demonstrates all the elements of this framework. However, given below is the case of Coles Supermarkets in Australia which have incorporated some of the concepts in its supply chain in their day-to-day business operations (see Fig. 11.9). The practices followed by Coles provide a classic example of how to build an extended sustainable food supply chain by considering economic, environmental, and social dimensions. The approach followed by Coles considers four aspects of sustainability in their supply chain. Its "people and community" aspect considers the employees and societal elements, while the "environment" aspect directly deals with the natural environment in the framework. "Responsible and ethical sourcing" together with "suppliers" is associated with the support providers and societal elements including institutional, sociocultural, and organizational elements.

The benefits of this framework are as follows: First, the conceptual analysis presented in the framework can serve as a basis for practitioner to comprehend the complexity in the food supply chain to develop innovative inter- and intraorganizational business processes (Gold et al. 2017). As such, the framework overcomes the fragmented nature of any of the process models of (food) supply chain management

Coles which is a leading food retailer in Australia, who is concerned with the responsibility towards their stakeholders in many areas of their business operations. Their focus is to continue to work towards a sustainable future while supporting Aussie farmers and the local community. Their sustainability strategy addresses many aspects in the framework to cover people and community, environment, suppliers while following ethical and responsible sourcing practices. The way they practise corporate social responsibility and work towards sustainability is as follows:

People and community:
At the heart of the Coles community are its dedicated team members. Coles employs over 100,000 Australians across the country and are passionate about the people and proud to lend a helping hand to the many organisations that their team members are involved in. Coles proudly supports a number of charity partners throughout Australia and work with team members and customers to raise funds and awareness of these important causes. Last financial year, Coles direct support totalled $43.9 million and an additional $7.9 million was contributed by customers, team members and suppliers. Coles follows the London Benchmark Group framework to measure its community investment, which is internationally recognised and used around the world by more than 300 companies.

Environment:
Coles is constantly exploring and investing in important environmental projects and partnerships to reduce the environmental impact. The focus is to improve the efficiency in the supply chain and reduce greenhouse gas emissions across business while focussing on waste reduction and recycling. The emissions, energy use and waste are measured and reported, while ways to minimise the impact of operations on the environment is evaluated. Due to these dedicated efforts, Coles has been able to significantly reduce its energy usage since 2009 despite the expansion in store network.

Responsible and ethical sourcing:
Coles has given priority to identify where food and ingredients come from and constant efforts are put in place with suppliers and industry experts to adopt responsible practices that look after people, animals and the environment. For example, the company provides a range of RSPCA Approved fresh meat options [The RSPCA Approved Farming Scheme has been developed to improve the welfare of farm animals on farm, during transport and at slaughter]. Moreover, Coles offers sea food certified by the Marine Stewardship Council (MSC), Aquaculture Stewardship Council (ASC) and WWF Australia.

Suppliers:
Coles believes Australia produces some of the best food in the world, hence adopted the 'Australian First' sourcing policy. As per this policy Coles sources Australian grown food whenever possible, which currently accounts for around 96% of fruit and vegetables and 100% Coles fresh lamb, pork, chicken, beef, milk and eggs.

Fig. 11.9 Sustainable supply chain management practices at Coles. (Source: (Coles 2017))

that only deals with supply chain partners. Hence, organizations and managers can identify where the present unsustainable practices arise through the lenses of this framework while instigating actions necessary to overcome them. These initiatives can be focused on the core value chain activities or activities obtained through the service providers or interactions with the broader environmental or social dimensions. Next, the incorporation of the guiding principles such as knowledge and change management and continuous improvement stresses the need for managers and academics to look beyond techno-oriented solutions to embrace stakeholder and social interactions when making food supply chains sustainable (Govindan 2017). Finally, a framework of this nature would facilitate the regulators, institutions related to food supply chains, and policy-makers to bring in regulations, standards, and guidelines to make the global food supply chains more sustainable. As highlighted in the framework, any of these regulations or standards will have to embrace the extended value chain partners, local and global consumers, and the dynamics in the environmental and social spheres to make the food supply chains truly sustainable. On the flip side, this framework does not consider the asymmetric power relationships in a global (or to some extent local) food supply chain. For example, as Trienekens (2011) opine, disproportionate power interactions in the global food supply chains favor Western supply chain parties, which could result in unequitable distribution of risks and rewards in the supply chain. Hence, the supply chain partners in developing countries who mainly supply primary commodities can be at a disadvantage, which is not reflected in a static model of this nature.

11.6 Conclusions

In the face of the global importance of food supply chains, the purpose of this chapter was to provide an overview of sustainable food supply chain management and to propose an integrated framework. With the uncontrollable growth of population, the world is currently facing a food crisis. In some parts of the world, millions of people are at risk of hunger, while in certain countries, there is significant food waste and people are suffering from obesity. While there are many reasons for this pathetic situation, it clearly reflects an imbalance in the global food supply chain. Hence, building a sustainable food supply chain is of global importance. With this objective, the chapter provided an overview of sustainable development and the importance of food supply chain management to achieve sustainable development goals while highlighting the severity of the global food crisis. The chapter presented a theoretical background of sustainable food supply chains in the broad context of supply chain management. With the help of the theoretical background, finally the chapter presented an integrated framework for the sustainable food supply chain management by extending the prior work of the FAO (2014). The proposed framework demonstrates the complexity of the food supply chains and highlights the need for continuous improvement in the light of the ever-changing national and international business environment along economic, natural, and social dimensions. An important feature of the chapter is its provision of mini cases with the objective of providing readers with practical insights in each section.

References

Abbasi, M. (2017). Towards socially sustainable supply chains – Themes and challenges. *European Business Review, 29*(3), 261–303.

Abukhader, S., & Jönson, G. (2004). Logistics and the environment: Is it an established subject? *International Journal of Logistics, 7*(2), 137–149.

Ahi, P., & Searcy, C. (2013). A comparative literature analysis of definitions for green and sustainable supply chain management. *Journal of Cleaner Production, 52*, 329–341.

Alavi, M., & Leidner, D. E. (2001). Knowledge management and knowledge management systems: Conceptual foundations and research issues. *MIS Quarterly, 25*(1), 107–136.

Aramyan, L., Oude Lansink, A., van der Vorst, J., & van Kooten, O. (2007). Performance measurement in Agri-food supply chains: A case study. *Supply Chain Management: An International Journal, 12*(4), 304–315.

Badurdeen, F., Iyengar, D., Goldsby, T., Metta, H., Gupta, S., & Jawahir, I. (2009). Extending total life-cycle thinking to sustainable supply chain design. *International Journal of Product Lifecycle Management, 4*(1/2/3), 49–67.

Bamford, D. R., & Forrester, P. L. (2003). Managing planned and emergent change within an operations management environment. *International Journal of Operations and Production Management, 23*(5), 546–564.

Beske, P. (2012). Dynamic capabilities and sustainable supply chain management. *International Journal of Physical Distribution and Logistics Management, 42*(4), 372–387.

Beske, P., Land, A., & Seuring, S. (2014). Sustainable supply chain management practices and dynamic capabilities in the food industry: A critical analysis of the literature. *International Journal of Production Economics, 152*, 131–143.

Bourlakis, M., & Weightman, P. (2004). *Food supply chain management*. Oxford: Blackwell.

Carter, C., & Easton, L. P. (2011). Sustainable supply chain management: Evolution and future directions. *International Journal of Physical Distribution and Logistics Management, 41*(1), 46–62.

Carter, C., & Rogers, D. (2008). A framework of sustainable supply chain management: Moving toward new theory. *International Journal of Physical Distribution and Logistics Management, 38*(5), 360–387.

Chandra, P., & Tirupati, D. (2003). *Business strategies for managing complex supply chains in large emerging economies: The story of AMUL*. Ahmedabad: Indian Institute of Management.

Christopher, M. (2011). *Logistics and supply chain management*. Harlow: Financial Times/ Prentice Hall.

Ciliberti, F., Pontrandolfo, P., & Scozzi, B. (2008). Investigating corporate social responsibility in supply chains: A SME perspective. *Journal of Cleaner Production, 16*(15), 1579–1588.

Cooper, C. M., & Ellram, L. M. (1993). Characteristics of supply chain management and implications for purchasing and logistics strategy. *The International Journal of Logistics Management, 4*(2), 13–24.

Cooper, M., Lambert, D., & Pagh, J. (1997). Supply chain management: More than a new name for logistics. *The International Journal of Logistics Management, 8*(1), 1–14.

Coles. (2017). *Corporate responsibility*. Available via https://www.coles.com.au/corporate-responsibility. Accessed 29 May 2017.

Department for Environment, Food and Rural Affairs (DEFRA). (2002). *The strategy for sustainable farming and food: Facing the future*. London: Defra.

Eastham, J., Sharples, L., & Ball, S. (2014). *Food supply chain management: Issues for hospitality and retail sectors*. Oxford: Butterworth-Heinemann.

Food and Agricultural Framework (FAO). (2014). *Developing sustainable food value chains-guiding principles*. Rome: FAO.

Food and Agricultural Organization (FAO). (2011). *Global food losses and food waste – Extent, causes and prevention*. Rome: FAO.

Food and Agricultural Organization (FAO). (2013). *Our priorities-the FAO strategic objectives*. Rome: FAO.

Food Security Information Network (FSIN). (2017). *Global report on food crises 2017*. FSIN. Available via http://documents.wfp.org/stellent/groups/public/documents/ena/wfp291271.pdf?_ga=2.68548753.189487937.1496587970-882440612.1496587970. Accessed 10 June 2017.

Forum for the Future. (2014). Key challenges for a sustainable food supply. *Forum for The Future*, 1–17.

Fredriksson, A., & Liljestrand, K. (2015). Capturing food logistics: A literature review and research agenda. *International Journal of Logistics Research and Applications, 18*(1), 16–34.

Gold, S., Kunz, N., & Reiner, G. (2017). Sustainable global Agrifood supply chains: Exploring the barriers. *Journal of Industrial Ecology, 21*(2), 249–260.

Govindan, K. (2017). Sustainable consumption and production in the food supply chain: A conceptual framework. *International Journal of Production Economics, 195*, 419. https://doi.org/10.1016/j.ijpe.2017.03.003.

Grant, D., Trautrims, A., & Wong, C. (2015). *Sustainable logistics and supply chain management* (Revised ed.). London: Kogan Page.

Haake, H., & Seuring, S. (2009). Sustainable procurement of minor items – Exploring limits to sustainability. *Sustainable Development, 17*(5), 284–294.

Heady, D. (2013). The impact of the global food crisis on self-assessed food security. *World Bank Economic Review, 27*(1), 1–27.

Heady, D., & Fan, S. (2010). *Reflections on the global food crisis*. Washington, DC: International Food Policy Research Institute.

Holt-Giménez, E. (2008). *The world food crisis: What is behind it and what we can do*. Available via http://www.worldhunger.org/world-food-crisis/. Accessed 14 Feb 2017.

Humphrey, J., & Memedovic, O. (2006). *Global value chains in the agrifood sector*. Vienna: United Nations Industrial Development Organization.

Institute of Hazard Risk and Resilience (IHRR). (2013). Food crisis? The real problem is poverty. *Global Policy Journal*. Available via http://www.globalpolicyjournal.com/blog/22/07/2013/food-crisis-real-problem-poverty. Accessed 10 May 2017.

International Fund for Agricultural Development (IFAD). (2017). *IFAD at a glance*. Rome: IFAD.

International Fund for Agricultural Development (IFAD). (2016). *Sri Lanka National Agribusiness Development Programme, Supervision report*, IFAD, Asia and the Pacific Region Division, Programme Management Department.

Jaspars, S., & Wiggins, S. (2009). *The global food crisis: An overview*. Available via http://odihpn.org/magazine/the-global-food-crisis-an-overview/. Accessed 10 May 2017.

Johnson, C. N. (2002). *The benefits of PDCA, quality progress*. Available from: http://asq.org/quality-progress/2002/05/problem-solving/the-benefits-of-pdca.html. [12 May 2017].

Kummu, M., De Moel, H., Porkka, M., Siebert, S., Varis, O., & Ward, P. J. (2012). Lost food, wasted resources: Global food supply chain losses and their impacts on freshwater, cropland, and fertiliser use. *Science of the Total Environment, 438*, 477–489.

Lakovou, E., Bochtis, D., Vlachos, D., & Aidonis, D. (2015). Sustainable Agrifood supply chain management. In D. Bochtis, E. Lakovou, D. Vlachos, & D. Aidonis (Eds.), *Supply chain management for sustainable food networks* (pp. 1–72). Sussex: John Wiley and Sons.

La Londe, B., & Masters, J. (1994). Emerging logistics strategies. *International Journal of Physical Distribution and Logistics Management, 24*(7), 35–47.

Lambert, D., Stock, J., & Ellram, L. (1998). *Fundamentals of logistics management*. Boston: Irwin/McGraw-Hill.

Maggio, G. F. (1997). Inter/intra-generational equity: Current applications under international law for promoting the sustainable development of natural resources. *Buffalo Environmental Law Journal, 4*(2), 161–223.

Maxwell, D., & Sheate, W. (2006). Enabling sustainable development through sustainable consumption and production. *International Journal of Environment and Sustainable Development.*, *5*(3), 221–239.
Mentzer, J., DeWitt, W., Keebler, J., Min, S., Nix, N., Smith, C., & Zacharia, Z. (2001). Defining supply chain management. *Journal of Business Logistics, 22*(2), 1–25.
Moran, J. W., & Brightman, B. K. (2001). Leading organizational change. *Career Development International, 6*(2), 111–118.
Nkunzimana, T., Custodio, E., Thomas, A. C., Tefera, N., Perez Hoyos, A., & Kayitakire, F. (2016). *Global analysis of food and nutrition security situation in food crisis hotspots*. Belgium: EUR.
Our World. (2008). *Africa and climate change*. Available via https://ourworld.unu.edu/en/africa-and-climate-change. Accessed 18 May 2017.
Pagell, M., & Wu, Z. (2009). Building a more complete theory of sustainable supply chain management using case studies of 10 exemplars. *Journal of Supply Chain Management, 45*(2), 37–56.
Panitchpakdi, S. (2010). *The global food crisis: Causes and policy response*. Switzerland: UNCTAD.
Papargyropoulou, E., Lozano, R., Steinberger, J. K., Wright, N., & Bin Ujang, Z. (2014). The food waste hierarchy as a framework for the management of food surplus and food waste. *Journal of Cleaner Production, 76*, 106–115.
Pinstrup-Andersen, P. (2009). Food security: Definition and measurement. *Food Security, 1*(1), 5–7.
Porter, M. E. (1985). *Competitive advantage*. New York: The Free Press.
Seuring, S., & Muller, M. (2008). From a literature review to a conceptual framework for sustainable supply chain management. *Journal of Cleaner Production, 16*(15), 1699–1710.
Sokovic, M., Pavletic, D., & Pipan, K. K. (2010). Quality improvement methodologies – PDCA Cycle, RADAR Matrix, DMAIC and DFSSM. *Journal of Achievements in Materials and Manufacturing Engineering, 43*(1), 476–483.
Teece, D. (1998). Capturing value from knowledge assets: The new economy, markets for know-how, and intangible assets. *California Management Review, 40*(3), 55–79.
Trienekens, J. H. (2011). Agricultural value chains in developing countries: A framework for analysis. *International Food and Agribusiness Management Review, 14*(2), 51–82.
Trostle, R. (2008). *Global agricultural supply and demand: Factors contributing to the recent increase in food commodity prices*. Washington, DC: United States Department of Agriculture.
Tseng, M., Lim, M., & Wong, W. P. (2015). Sustainable supply chain management. *Industrial Management and Data Systems., 115*(3), 436–461.
The Conversation. (2015). *Climate change is hitting African farmers the hardest of all.* Available via http://theconversation.com/climate-change-is-hitting-african-farmers-the-hardest-of-all-40845. Accessed 28 April 2017.
UNICEF. (2015). *Eradicate extreme poverty and hunger, millennium development goals*. Available via https://www.unicef.org/mdg/poverty.html. Accessed 13 April 2017.
United Nations (UN). (2016). *Zero hunger: Why it matters?* Available via http://www.un.org/sustainabledevelopment/wp-content/uploads/2016/08/2_Why-it-Matters_ZeroHunger_2p.pdf. Accessed 22 May 2017.
United Nations (UN). (2017). *Goal 2: End hunger, achieve food security and improved nutrition and promote sustainable agriculture*. Available via http://www.un.org/sustainabledevelopment/hunger/. Accessed 16 May 2017.
United Nations Development Program (UNDP). (2017a). G*oal 2 targets*. Available via http://www.undp.org/content/undp/en/home/sustainable-development-goals/goal-2-zero-hunger/targets/. Accessed 06 June 2017.
United Nations Development Program (UNDP). (2017b). *Sustainable development goals*. Available via http://www.undp.org/content/undp/en/home/sustainable-development-goals.html. Accessed 20 Feb 2017.

United Nations World Commission on Environment and Development (UNWCED). (1987). *Report of the world commission on environment and development: Our common future*. Oxford: Oxford University Press.

van Hoek, R. (1999). From reversed logistics to green supply chains. *Supply Chain Management: An International Journal, 4*(3), 129–135.

Verisk Maplecroft. (2016). *Africa's agriculturally dependent nations facing highest costs of climate change*. Available via https://maplecroft.com/portfolio/new-analysis/2016/12/18/study-africas-agriculturally-dependent-nations-facing-highest-costs-climate-change-key-agri-commodities-risk/. Accessed 12 June 2017.

Wittstruck, D., & Teuteberg, F. (2011). Understanding the success factors of sustainable supply chain management: Empirical evidence from the electrics and electronics industry. *Corporate Social Responsibility and Environmental Management, 19*(3), 141–158.

World Food Programme. (2017). *Climate impacts on food security*. Available via https://www.wfp.org/climate-change/climate-impacts. Accessed 10 June 2017.

Chapter 12
Risk Assessment of the Timber Supply Chain in Southern Ontario Using Agent-Based Simulation

Marcin Lewandowski and Ali Asgary

12.1 Introduction

The bioenergy sector has been experiencing significant growth in the last two decades in Ontario, Canada (Krigstin et al. 2016). However, due to the fact that compared to fossil fuels biomass has relatively little energy concentration by weight, in order to be economic biomass has to be sourced locally (McKendry 2002). Because of this constraint, in order to make sound decisions, bioenergy developers and investors have to understand risks associated with local biomass supply chains.

Biomass is typically derived as a by-product of another industry. In this case, we look at wood chips and sawdust, collectively referred to as sawmill residue, generated as a by-product of lumber production. The objective of this paper is to determine the most resilient areas in the Southern Ontario timber supply chain from a perspective of a developer looking to build a potential new biofuel plant consuming sawmill residues as raw material, such as a pellet mill or a cellulosic ethanol plant. Sawmill residue production is directly proportional to the quantity of timber processed by sawmills. On average, approximately 50% of timber weight is converted into sawmill residue, mainly wood chips (Krigstin et al. 2012). Due to this direct relationship between quantity of timber processed and quantity of sawmill residue produced, all things being equal, it is possible to determine sawmill residue supply risk through estimating timber supply risk. In other words, assuming economic variables (such as demand for Ontarian lumber) remain constant, one can derive a risk

M. Lewandowski
Ecostrat Inc, Toronto, Canada
e-mail: marcin@ecostrat.com

A. Asgary (✉)
York University, Toronto, Canada
e-mail: asgary@yorku.ca

H. Qudrat-Ullah (ed.), *Innovative Solutions for Sustainable Supply Chains*, Understanding Complex Systems, https://doi.org/10.1007/978-3-319-94322-0_12

profile for sawmill residue supply by estimating the risk profile of timber supply. Such rationale is applied in this analysis.

12.2 Timber Supply Chain Risks

The main objective of the timber supply chain is to provide timber for different uses by the forest product sector. Providing suitable supply of timber requires dealing with issues ranging from strategic forest management, such as harvesting activities, to operational risks related to harvesting and transportation (D'Amours et al. 2008). Timber supply chains, like all other supply chains, have inherent risks associated with them. Timber supply chains are complex, and their robustness depends on a number of factors interacting together (Vahid 2011). Because timber is harvested on geographically distributed forests, the supply chains are susceptible to various disturbances, such as natural events (e.g., floods, wildfires, extreme weather), infrastructural risks (e.g., road and rail closure), or political events and labor actions (e.g., demonstrations, strike). Since the forest sector depends on raw wood supply, these supply chain risks translate into operational risks for the entire forest industry. Understanding these risks, what factors influence them most would provide for better mitigation strategies (Chauhan et al. 2009).

A supply chain is defined as "a network of organizations that are involved, through upstream and downstream linkages, in different processes and activities that produce value to consumers" (Christopher and Holweg 2011). From modeling perspective, a supply chain can be defined as "a network of autonomous or semiautonomous business entities collectively responsible for procurement, manufacturing and distribution activities associated with one or more families of related products" (Swaminathan et al. 1998).

Risk is most commonly (e.g., Christopher and Peck 2004) defined by the following simplified equation:

$$\text{Risk} = \text{Likelihood of Occurrence} \times \text{Consequences of Disturbance}$$

There are three broad strategies when dealing with supply chain risks: (1) inventory management, building up safety stock; (2) sourcing, developing contingency strategies for specific suppliers or supply chain links; and (3) acceptance, doing nothing as costs of mitigation outweigh benefits (Manners-Bell 2014).

In general, the higher the complexity of supply chains, the higher risk becomes, due to factors such as the globalization of supply chains, increased volatility and variability of demand and supply, increased outsourcing, etc. (Chen et al. 2013), although complexity can also promote resiliency through buildup of network density (Falasca et al. 2008). The risk in supply chains usually originates from uncertainty and volatility of the business environment. Also, disruptive events, such as natural disasters, have a significant effect on supply chains (Chen et al. 2013).

Four main concepts in supply chain risk assessment include (1) vulnerability, (2) robustness, (3) disturbances, and (4) resilience.

Svensson (2000) defines supply chain vulnerability as "the existence of random disturbances that lead to deviations in the supply chain of components and materials from normal, expected or planned schedules or activities, all of which cause negative effects or consequences for the involved manufacturer and its sub-contractors" (p.732). In line with this definition, Vlajic et al. (2012) identify vulnerability sources as "characteristics of the supply chain or its environment that lead to the occurrence of unexpected events and as such, they are direct or indirect causes of disturbances" (p.179).

In supply chain management theory, robustness and vulnerability are perceived as opposite. The concept of robustness describes a supply chain as a system which has an ability to resist disruptions, retaining the system structure intact. Based on this concept, Vlajic et al. (2012) further refine the definition of supply chain robustness as "the degree to which a supply chain shows an acceptable performance in (each of) its Key Performance Indicators (KPIs) during and after an unexpected event that caused disturbances in one or more logistics processes" (p.188). In other words, "a supply chain is robust with respect to a KPI if the value of that KPI, adequately measured over an observation period, is sustained in a predefined desired range [the Robustness Range], even in the presence of disturbances" (p.188). This definition indicates that supply chain robustness can be quantified. This characteristic allows researchers to model supply chain robustness. For example, supply chain robustness is often modeled by stochastic programming, such as the Monte Carlo methods.

Svensson (2000) defines supply chain disturbance as "a deviation that causes negative consequences for the firm involved in the supply chain" (p.735). Vlajic et al. (2012) are more specific: "supply chain disturbance [is] a minor or major deviation, or failure of one or more logistics processes triggered by unexpected events in the supply chain or its environment resulting in poor performance of the process itself, company and potentially along the supply chain in a given time period" (p.179). Vlajic et al. further write that supply chain disturbances can be characterized by a number of elements, including (a) frequency of occurrence, (b) possibility of detection, (c) impact on supply chain performance, and (d) disturbance cause and size.

Disturbances are significant because of impacts they have on supply chain robustness. Disturbance becomes larger if a supply chain is not flexible and responsive to that disturbance. In other words, disturbances grow in significance if a supply chain cannot adapt to a new situation. Disturbances can range from local, affecting just one linkage in the chain, or system wide, affecting the entire supply chain. For instance, in the context of timber production, harvest failure would be a system-wide disturbance (Svensson 2000; Vlajic et al. 2012).

Damage due to a disturbance can be limited through agility or rapidity of response. A supply chain is agile when each member of the supply chain is agile (Manners-Bell 2004). Agility has two elements: supply chain visibility and supply chain velocity (Christopher and Peck 2004). Supply chain visibility refers to a clear understanding of demand and supply. Supply chain velocity refers to the time lapsed between order from a supplier and delivery and the supply chain's ability to accelerate deliveries after a disturbance, i.e., "how quickly can product delivery be speeded up to respond to a particular event in the supply chain" (Manners-Bell 2004).

Supply chain resilience (SCRES) is defined as the ability of a system to go back to its original or a more desirable new state after being disturbed (Christopher and Peck (2004). The concept of resilience can be regarded as an extension of the concept of resistance traditionally used in disaster management field. Resistance is used as a measure to determine performance of structures, infrastructure, and other elements in reducing losses from a disaster. As such, disaster resistance emphasizes the importance of pre-disaster mitigation. However, the concept of resilience builds upon this notion and includes improvements in the flexibility and performance of a system both during and after a disturbance. Consequently, Falasca et al. (2008) define supply chain resilience as "the ability of a supply chain system to reduce the probabilities of a disruption, to reduce the consequences of those disruptions once they occur, and to reduce the time to recover normal performance" (p. 596).

Based on this definition, Falasca et al. (2008) developed a model which quantifies supply-chain resilience. There are three determinants of resilience within a supply chain. The first determinant is supply chain density, which is defined by the quantity and geographical spacing of nodes within a supply chain. A supply chain is denser if a large number of nodes are clustered closely together. Therefore, in the case of the timber supply chain, if there are clusters of mills together, these regions could suffer significantly if a nearby disturbance occurs.

Secondly, the number of nodes and connections between them is directly related to supply chain resilience. However, as pointed out by Craighead et al. (2007), it could differ whether this relationship is positive or negative. For example, because a lower complexity supply chain by definition has less nodes or less interconnections, a disturbance would have a relatively lower impact on it. That said, if each node is crucial to a supply chain and it is taken out, without sufficient complexity of interconnections, the supply chain is likely to be less resilient. This is likely to be the case in the timber supply chain.

The third determinant of supply chain resilience is somewhat related to the previous one: node criticality. Node criticality is defined as "the relative importance of a given node or set of nodes within a supply chain" (Craighead et al. 2007). In the timber supply chain node, criticality is based on production by Forest Management Units and the consumption levels by sawmills.

Forest product supply chains are complex. Complexity of a supply chain is defined to be the sum of two components: the total number of nodes and the total number of forward, backward, and within-tier material flows (Craighead, et al. 2007). Although private companies are the main actors in these supply chains, the networks are significantly influenced by government agencies, nongovernmental organizations, environmental groups, and community organizations. Macroeconomic variables, such as business cycles, international trade trends, and currency fluctuations, also have an important influence on timber supply chains (Vahid 2011).

Additionally, these supply chains are susceptible to meteorological and other natural events, such as floods, forest fires, and insect outbreaks (D'Amours et al. 2008).

12.3 Forest Sector Supply Chain in Ontario

In Ontario, the availability of timber is determined by the Ontario Ministry of Natural Resources and Forestry (OMNRF). This is due to the fact that the vast majority of forests in Ontario are owned by the Crown, that is, are publicly owned. The OMNRF serves as a management agency of Crown forests and therefore oversees the planning of timber harvests. Consequently, the OMNRF maintains a dataset with forest inventory information, called the forest resource inventory (FRI). This dataset is later used to prescribe future harvests by modeling forest growth and structure, thereby ensuring that the harvests result in healthy forests. The models' outputs are later used in the preparation of forest management plants (Krigstin et al. 2016).

The Crown forest in Ontario is divided into Forest Management Units (FMUs). In total, there are 41 FMUs in the province. To limit the scope to the most active region in the province – Southern Ontario – only 11 of the 41 FMUs are analyzed in this project.

Each FMU is managed by an agency composed of local stakeholders, in most cases companies who have rights to harvest on that specific unit. These agencies are responsible for preparing Forest Management Plans (FMPs), which typically lay down detailed plan for harvesting operations over a 20-year period. The FMPs are updated every 10 years, with more frequent updates if necessary (OMNRF 2016).

The OMNRF controls timber allotments to optimally utilize the forest resources to maximize economic, social, and environmental aspects of the forest resources. To ensure optimal use of forest resources, OMNRF allocates specific quantities of timber to be cut by mills in Ontario through a mechanism called Timber Supply Agreement, or alternatively called "commitment." These commitments are based on Forest Management Plans and are therefore assigned for a 20-year period, with a review process every 5 years (Wallace 1992).

In addition to long-term timber allocation, OMNRF can allocate temporary timber licenses when such are needed to maintain continuity of supply to a particular mill. The mechanism is called an Interim Management Plan and is conducted on an annual basis, until the additional allotments are included in the main plan (Wallace 1992).

12.4 Study Area

The simulation focuses on Southern Ontario, Canada. In particular, there are 11 Forest Management Units under investigation here (Map 12.1). These Forest Management Units have been chosen because traditionally Southern Ontario was the epicenter of the Ontarian forest industry and currently has lots of potential for the wood pellet industry (Krigstin et al. 2016).

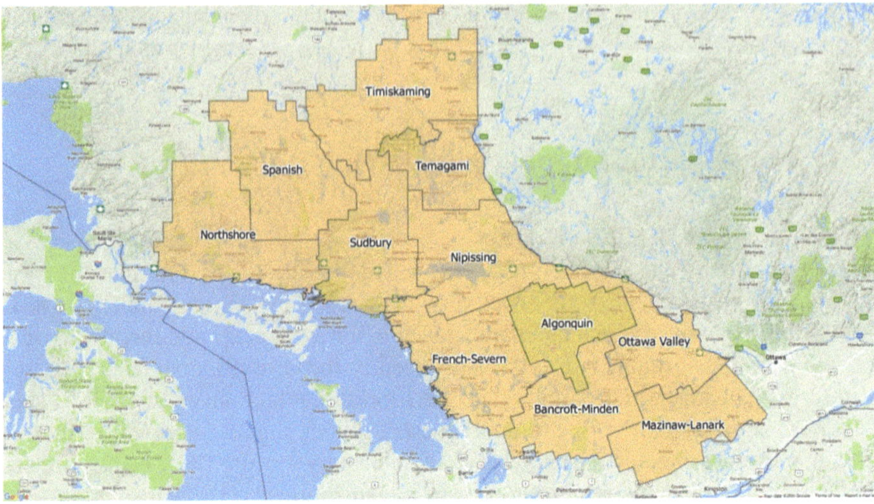

Map 12.1 Forest management units in the study region

Table 12.1 Forest management units analyzed in the simulation

Forest management unit	Commitments (cubic meters/year)
Algonquin	937,735
Bancroft-Minden	108,050
French-Severn	192,700
Nipissing	400,900
Northshore	367,500
Ottawa Valley	120,800
Spanish	1,014,300
Timiskaming	907,300
Temagami	94,000
Mazinaw-Lanark	73,400
Sudbury	358,900

Source: Ontario Ministry of Natural Resources (2016)

Table 12.1 indicates Forest Management Units analyzed in this simulation, together with corresponding quantities committed to sawmills. Map 12.2 shows all sawmills to which wood fiber from within the study area is committed to.

The simulation was set up in AnyLogic 7.3.5, which allows for spatial representation of a supply chain. Each Forest Management Unit is represented by a polygon, for which shapefiles were acquired from the Ontario Ministry of Natural Resources and Forestry (OMNRF). The point from which timber is picked up by trucks is determined by the nearest city to the center of the forest. That said, the model is programmed in a way that the user can define exact locations from which timber needs to be picked up. For example, with additional work and access to files describ-

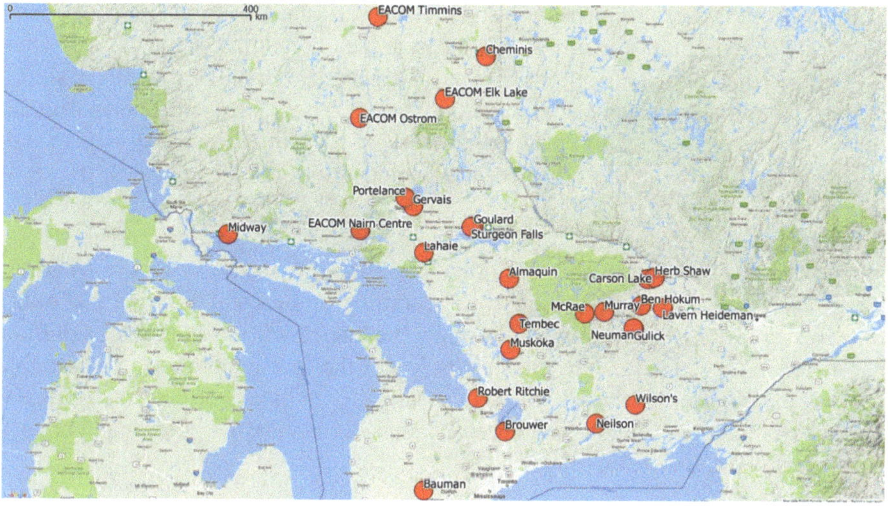

Map 12.2 Sawmills in the study area

ing harvesting plans in the Forest Management Plans (OMNRF 2016), it is possible to simulate actual harvesting activities as planned by forest companies.

12.5 Model Structure

In this model, there are three types of agents categorized into two categories: stationary and dynamic. Stationary agents have one specific geographic location. Dynamic agents move, following established routes between stationary agents. There are two types of stationary agents: forests and sawmills. The dynamic agent is represented by a truck. The truck agent delivers timber from the forest agent to the sawmill agent.

Forests are agents from which timber flow originates. In other words, the supply chain starts off in forest agents. Each forest agent features a predetermined quantity of timber available for consumption each year. These quantities are defined by OMNRF, and although they may slightly change over time, for the purposes of this model, they are kept constant. It is important to note however that the quantities designated by OMNRF indicate maximum timber available, meaning that sawmill agents can consume less timber than designated. Forest agents interact with truck agents. A truck agent arrives at a forest agent, loads 32 metric tons of timber, and leaves the forest agent. The 32-metric ton quantity is then subtracted from the annual timber availability predetermined for the sawmill agent procuring timber from the forest agent.

The sawmill agent is a stationary agent represented by a point. The sawmill agent procures timber from a forest zone agent. Locations and names of sawmills were acquired from Ontario Ministry of Natural Resources and Forestry (2016). Because the OMNRF data may sometimes be outdated, the operational status of each sawmill was verified directly by phone. In total, there are 29 sawmill agents in the simulation. Sawmill agent interacts with the truck agent. The truck agent arrives with 32 metric tons of timber. It unloads the timber for 25 min (Liew 2016 – personal comm.). The 32 metric tons of timber is then added to timber inventory at the sawmill agent.

Truck is a dynamic agent. It interacts with and moves between forest and sawmill agents. Forest sends a signal to truck that timber is ready for pickup. A truck leaves a sawmill, travelling along a road according to the speed limit on that road. It arrives at a forest zone and waits 30 min to be loaded with 32 metric tons of timber material. Once loaded, the truck agent leaves the forest and travels back to the sawmill it came from. When it reaches the sawmill, it unloads timber into the sawmill's inventory; the unloading takes 25 min (Liew 2016 – personal comm.) (Figs. 12.1, 12.2, and 12.3).

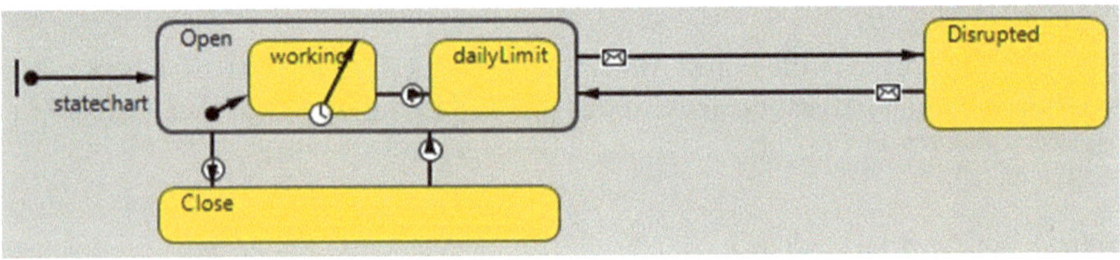

Fig. 12.1 Statechart for the forest agent

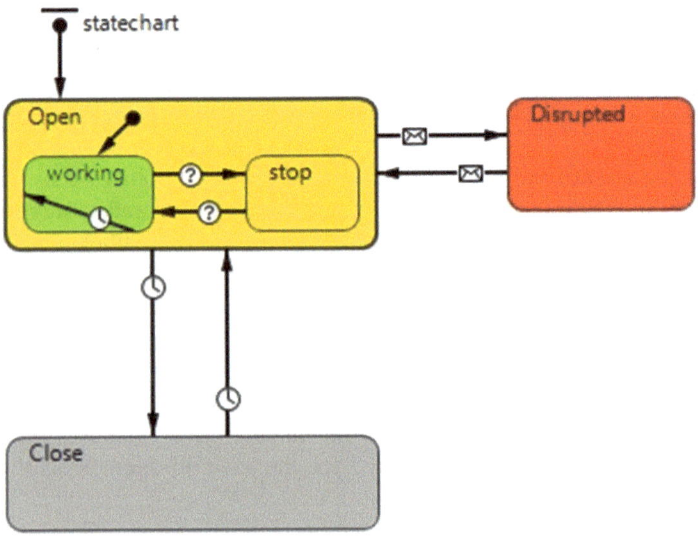

Fig. 12.2 Statechart for the sawmill agent

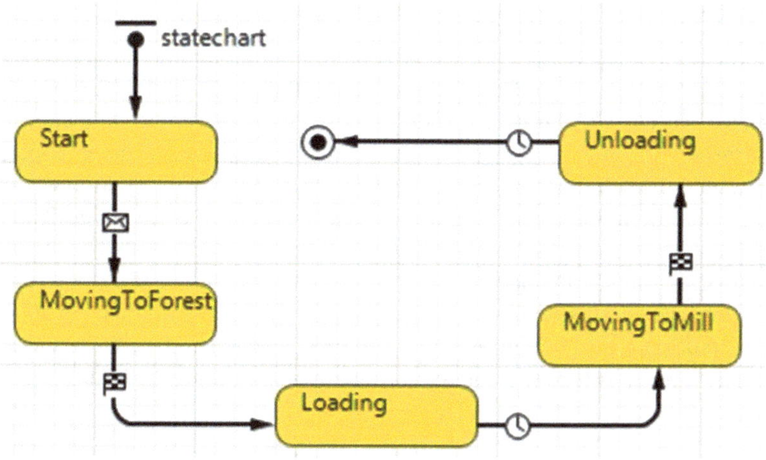

Fig. 12.3 Statechart for the truck agent

12.6 Disturbance

The objective of this research is to simulate disturbances and gain insight into their effects on the supply chain. For the simplicity sake, only one type of disturbance is modeled in this simulation: a political disturbance. An example of a political disturbance would be a campaign waged by an environmental group which stops logging operations in one administrative unit. The political disturbance was specifically chosen because such disturbance would cause a removal of one entire Forest Management Unit (or forest agent) from the simulation. This means that each of the 11 forests can be taken out easily, and a comparative analysis between impacts of each forest's absence on the timber inventories can be measured within the timeframe of this project.

This is just one example of a disturbance that could be modeled by the simulation. Other disturbances, such as forest fires, floods, road blockades, etc., can be easily modeled, as long as their estimated spatial extent is known.

Based on the scenario above, the objective of these simulations is to answer the following:

1. What is the minimum sawmill inventory level to keep the supply chain resilient to forest disturbances?
2. How resilient is the timber supply chain to forest disturbances, at each inventory level? (10-, 20-, 30-day inventory levels simulated).
3. Which forests are most significant to the supply chain, i.e., which forests, when disturbed, cause most negative reaction in the supply chain?
4. What is the risk a specific sawmill will close down during the 2-week forest disruption, at each inventory level?
5. What are the most risky areas to site a biofuel plant in, in terms of sawmill closure and wood fiber availability risk, based on the forest closure scenario?

12.7 Results

Table 12.2 indicates the number of times a sawmill closed, i.e., the number of times its inventory reaches 0. The table is ordered by the number of times a sawmill closes. For example, Sturgeon Falls experiences most closures, and Tembec experiences least closures. It is important to note that five sawmills, starting with inventory for 10 days, shut down operations in almost all instances – 10 or 11 times. This result simply indicates that at the 10-day inventory level, these sawmills cannot operate, even when all forests are open. In the real world, therefore, the sawmills likely import timber from outside the study area or trade timber with other sawmills.

Other than those five sawmills, the supply chain seems to be robust. Gulick and Portelance Sawmills close twice at the starting inventory levels of 10 days. However,

Table 12.2 Number of sawmill shutdowns by simulated starting inventory days. Sawmills are ordered by susceptibility to closures due to forest disturbances

Mill name	Inv10	Inv20	Inv30
Sturgeon falls	11	11	1
Wilson's	11	11	1
Robert Ritchie	11	1	1
Bauman	10	1	1
Brouwer	10	1	1
Gulick	2	0	0
Portelance	2	0	0
Almaquin	1	1	0
Carson Lake	1	1	0
EACOM Elk Lake	1	1	0
EACOM Ostrom	1	1	0
EACOM Timmins	1	1	0
Lahaie	1	1	0
Midway	1	1	0
Muskoka	1	1	0
Neilson	1	1	0
Ben Hokum	1	0	0
Cheminis	1	0	0
Dament & Charles	1	0	0
EACOM Nairn Centre	1	0	0
Gervais	1	0	0
Goulard	1	0	0
Herb Shaw	1	0	0
Lavern Heideman	1	0	0
McRae	1	0	0
Murray	1	0	0
Neuman	1	0	0
Tembec	1	0	0

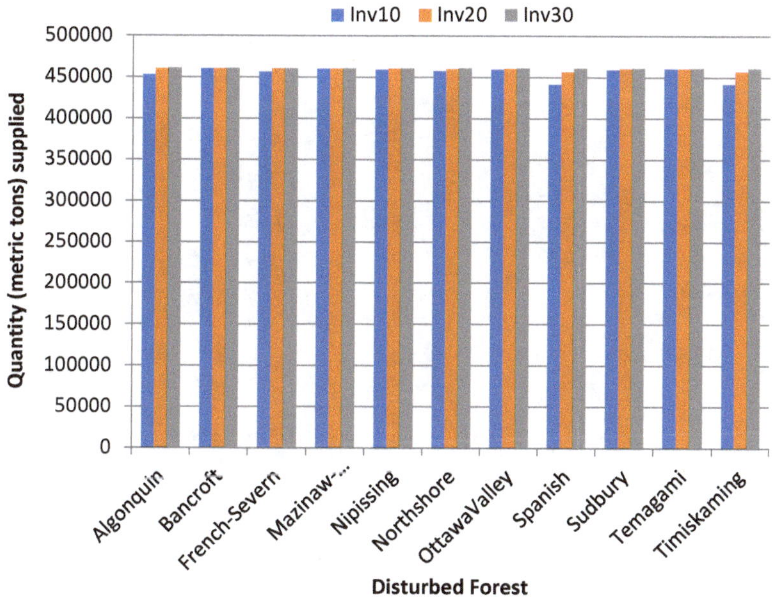

Fig. 12.4 Quantity of wood delivered to all sawmills by starting inventory level and forest disturbed

once the starting inventory level increases to 20 days, they do not close at all. The remaining 21 sawmills close only once. Nine of them close at both 10- and 20-day starting inventory levels. Twelve close only at 10-day starting inventory levels.

The conclusion that the timber supply chain in Ontario is robust, from the perspective of this simulation, is further visible in Figs 12.4 and 12.5. The figures indicate the total quantity of wood supplied to sawmills, for each disturbance scenario, and at each starting inventory level. It is apparent from Fig. 12.4 that the 2-week disturbance during a 6-week timeframe does not significantly affect the supply chain as a whole. Forests whose disturbance has the most significant effect on the total quantity of wood supplied are Spanish, Timiskaming, and Algonquin.

Figure 12.5 better illustrates the relative importance of each forest to the supply chain. The vertical axis was shortened to make differences in the quantities of supplied wood better visible. Here we see that, in comparison to other forests, Spanish and Temagami affect the supply chain the most. However, this effect is only visible at the 10-day starting inventory level. Once the starting inventory level increases to 20 days, the disturbances are barely noticeable. It is safe to conclude then that the timber supply chain in Southern Ontario is resilient to 2-week forest closures, provided that the inventories are kept at 20 days.

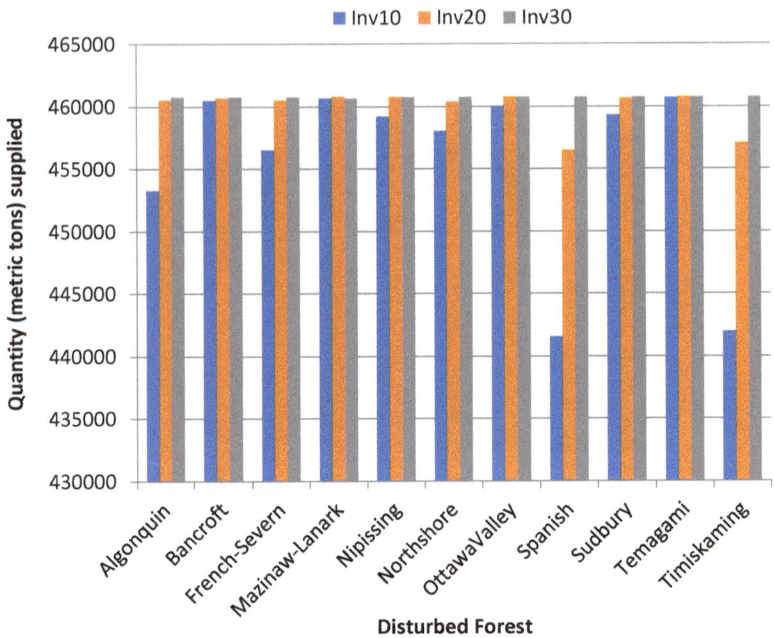

Fig. 12.5 Quantity of wood delivered to all sawmills by starting inventory level and forest disturbed. Vertical axis starts at 430,000 MT to better illustrate differences in the quantity of wood supplied

12.8 Most Resilient Locations

When siting a new biofuel plant, a developer does not look at a supply chain as a whole but rather at each region separately. It is therefore important to analyze simulation results in a spatially based environment. In this case, the results were analyzed in a geographic information system (GIS) environment using QGIS 2.14.3 Essen.

Due to the limitations of this simulation, such as limited spatial extent of the supply chain and lack of accounting of timber imports into the system, risk of disturbance in wood fiber supply cannot be measured objectively. Therefore, the risk of wood fiber supply loss is presented as *relative* risk here. There are two separate methods the risk is measured using results from the simulation.

1. All quantities of loss of wood supply were aggregated for each sawmill (for each starting inventory level). The resulting quantity indicates how much wood (in total) would be lost if all forests were disturbed once.
2. The quantities calculated in (1) are expressed as proportion of the total quantity of wood consumed if no forests were disturbed. For example, a 0.25 represents a proportion of total consumption that is not supplied to the sawmill due to disturbances. Just as in (1), the data are aggregated for each sawmill, representing proportion of wood lost if all forests were disturbed once.

It is important to note that the quantities presented in these maps are meaningless on their own. That is, it is inconceivable that each forest would close down precisely once and exactly for 2 weeks. These metrics are meant to indicate *relative risk* posed in any given region. For instance, a cluster of large red dots in a particular region indicates that Sawmills in this region may not be resilient to forest disturbances, and therefore this region poses high risk of loss of wood fiber supply – *in relation to other regions*. One cannot conclude, for instance, that there is a specific probability that a particular sawmill will lose exactly 10,000 metric tons of fiber supply.

Map 12.3 indicates each sawmill's relative risk of wood fiber supply loss, as simulated for 10-day starting inventory levels. The results indicate that the least resilient region in the study area is the northern part of it, with two sawmills showing a high degree of risk. However, this could be indicative of the fact that in the real world these sawmills likely draw wood from forests other than the ones analyzed in this study – which would explain why the two sawmills with highest relative risk are located close to the northern border of the study region. Another explanation for the results is that the two sawmills showing the highest relative risk, in terms of quantity of wood fiber loss, are large operations for which even slightest disruption of supply translates into large fiber losses.

As initially shown in Table 12.2, and now indicated on Map 12.3, at 10-day starting inventory levels, all sawmills experience loss of fiber supply. This is in comparison to sawmills simulated at 20-day starting inventory levels, as shown on Map 12.4. The area south of Algonquin Forest, represented by a cluster of sawmills, may seem as relatively non-resilient region if 10-day starting inventories are considered. However, considering 20-day starting inventory output (Map 12.4), this region appears relatively robust, with eight out of ten sawmills not experiencing any interruption in fiber supply.

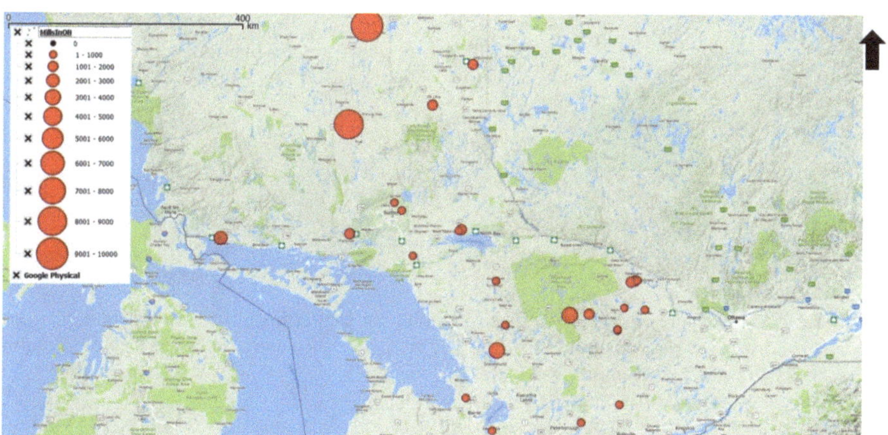

Map 12.3 First measure of relative risk of wood fiber supply loss, based on the total quantity of wood supply loss after each of the 11 forests had been disturbed, and results aggregated. Results based on 10-day starting inventory levels

Map 12.4 First measure of relative risk of wood fiber supply loss, based on the total quantity of wood supply loss after each of the 11 forests had been disturbed, and results aggregated. Results based on 20-day starting inventory levels

Map 12.5 First measure of the relative risk of wood fiber supply loss, based on the total quantity of wood supply loss after each of the 11 forests had been disturbed, and results aggregated. Results based on 30-day starting inventory levels

Looking at the maps of relative fiber supply loss based on 30-day starting inventory levels (Map 12.5), it is apparent that the loss of wood fiber to the supply chain is negligible. In other words, if all sawmills start off with 30-day inventory, a 2-week disturbance to any (or all) forests does not affect the supply of wood fiber. Therefore, maintaining inventory levels at 30 days makes the entire supply chain robust to 2-week forest closures.

12.9 Conclusion

A model was developed here simulating the timber supply chain in Southern Ontario. The objective of the simulation was to show the applicability of computer simulation methods in determining the most resilient areas from a perspective of a developer looking to build a new biofuel plant. In this case, a scenario was tested whereby each Forest Management Unit in the study area is closed (due to a political disturbance) once for a period of 2 weeks. The results indicated which sawmills, and by extension which areas, are most resilient to the 2-week disturbance. Two different methods were applied to measure relative risk of wood fiber loss, one based on the total quantity of wood supply loss and the other based on percent of total wood supply loss. Additionally, each simulation was performed based on three starting timber inventory levels: 10, 20, and 30 days.

Simulation results show that in general the wood supply chain in Southern Ontario is robust under a scenario of a 2-week timber supply disturbance. Aggregated results for all simulation runs, and separated by starting inventory levels, show that during the 2-week disturbance of any Forest Management Unit, the vast majority of sawmills remain open; those that shut down due to the lack of timber inventory are the very small operations. The total quantity and the percentage of wood loss due to disturbances are further visualized on GIS-rendered maps. The maps specify areas where the supply chain is least resilient to the disturbance by quantifying relative risk, based on the total quantity and the percentage of wood loss. Therefore, the maps serve as indicators of best potential locations for a biofuel plant, from a perspective of risk posed by a theoretical 2-week closure of a Forest Management Unit.

The biofuel industry faces many challenges, of which raw material supply risk is one of the most significant. The experiment presented here provides an example of the applicability of using an agent-based simulation in biofuel industry development decision-making process. By simulating the entire supply chain, a decision-maker can understand areas of highest risk of, for example, supplier closure due harvesting disturbance. Understanding risk has a potential to lower risk perception, promoting investment into biofuel industry, and lowering investment debt. With lower investment debt, the likelihood of biofuel plant success increases (Solomon 2016 – personal comm.). The potential of agent-based simulation in the bio-energy industry is therefore significant.

References

Chauhan, S. S., Frayret, J.-M., & LeBel, L. (2009). Multi-commodity supply network planning in the Forest supply chain. *European Journal of Operational Research, 196*, 688–696.

Chen, X., Ong, Y., Tan, P., Zhang, N., & Li, N. (2013). Agent-based modling and simulation for supply chain risk management – A survey of the State-of-the-Art. 2013 IEEE International Conference on Systems, Man, and Cybernetics, 1294–1299.

Christopher, M., & Holweg, M. (2011). "Supply Chain 2.0": Managing supply chains in the era of turbulence. *International Journal of Physical Distribution & Logistics Management, 41*(1), 63–82.

Craighead, C. W., Blackhurst, J., Rungtusanathan, M. J., & Handfield, R. B. (2007). The severity of supply chain disruptions: Design characteristics and mitigation capabilities. *Decision Systems, 38*(1), 131–156.
D'Amours, S., Ronnqvist, M., & Weintraub, A. (2008). Using operational research for supply chain planning in the forest products industry. Available at: http://repositorio.uchile.cl/bitstream/handle/2250/125029/D%E2Amours_Sophie.pdf?sequence=1
Falasca, M., Zobel, C.W., & Cook, D., (2008). A decision support framework to assess supply chain resilience. Paper presented at international ISCRAM conference, May, Washington, DC.
Krigstin, S., Hayashi, K., Tchorzewski, J., & Wetzel, S. (2012). Current inventory and modeling of Forest Residues in Eastern Canada. *The Forestry Chronicle, 88*(5), 626–635.
Krigstin, S., Wetzel, S., Mabee, W., & Stadnyk, S. (2016). Can Woody biomass support a pellet industry in Southeastern Ontario: A case study. *The Forestry Chronicle, 92*(2), 189–199.
Manners-Bell, J. (2014). *Supply chain risk: Understanding emerging threats to global supply chains*. London: Kogan Page.
McKendry, P. (2002). Energy production from biomass (Part 1). Overview of biomass. *Bioresource Technology, 86*(1), 37–46.
OMNRF (Ontario Ministry of Natural Resources and Forestry) (2016). *Forest Management Planning*. Retrieved from: https://www.ontario.ca/page/forest-management-planning
Christopher, M., & Peck, H. (2004). Building the resilient supply chain. *International Journal of Logistics Management, 15*(2), 1–13.
Vahid, S. (2011). An agent-based supply chain model for strategic analysis in forestry. A thesis submitted in partial fulfillment of the requirements for the degree of Doctor of Philosophy, The University of British Columbia, The Faculty of Graduate Studies.
Wallace, J.-L. (1992). Timber Administration in the Province of Ontario. Available at: http://e-collection.library.ethz.ch/eserv/eth:2558/eth-2558-01.pdf#page=266
Svensson, G. (2000). A Conceptual Framework for the Analysis of Vulnerability in Supply Chains. *International Journal of Physical Distribution & Logistics Management, 30*(9), 731–189.
Swaminathan, J. M., Smith, S. F., Sadeh, N. M. (1998). Modeling supply chain dynamics: A multiagent approach. *Decision Sciences, 29*(3), 607–628.
Vlajic, J. V., Vand der Vorst, J. G. A. J., Haijema, R. (2012). A framework for designing robust food supply chains. *International Journal of Production Economics, 137*(1), 176–189.

Part IV
Finale: Sustainable Supply Chains Continuum

Chapter 13
Finale: Sustainable Supply Chain Continuum

Hassan Qudrat-Ullah

13.1 Roadmap for Improved Decision-Making in the Sustainable Supply Chains

At the outset of this book, we set the objective to aid managerial decision-makers, in the domain of supply chain, to make improved decisions by embracing unique perspectives and various model-based approaches. Here we present some key findings and insights based on these the perspective described and model-based solutions applied to address various issues related to the design, development, growth, and profitability of sustainable supply chains.

R&D in supply chains To increase innovation in logistics and the supply chain industry, it is critical to address the gaps at each stage of the innovation process: idea generation ➔ R&D ➔ technology implementation ➔ adoption by industry. To increase the speed of innovation in and across the supply chains, there are three key factors in the development process: (i) assessment of industry need/requirement at R&D proposal stage, (ii) provision of technical support to system integrator to enable R&D technology transfer to industry adoption, and (iii) simplification of the technical and administrative inquiries among the stakeholders of R&D projects (Chap. 2 by Ho and Chuah (2018)).

Supplier selection Supplier selection is a critical dimension for the successful design, development, and growth of sustainable supply chains. Regarding supplier selection, the key insights are:

H. Qudrat-Ullah (✉)
College of Industrial Management, King Fahd University of Petroleum Engineering and Minerals, Dhahra, Saudi Arabia
e-mail: hassan.qudratullah@kfupm.edu.sa

H. Qudrat-Ullah (ed.), *Innovative Solutions for Sustainable Supply Chains*, Understanding Complex Systems, https://doi.org/10.1007/978-3-319-94322-0_13

- Economic criteria did not change much since 1966.
- Quality, delivery, and cost remain the most popular economic criteria.
- Environmental management systems in terms of qualitative criteria are the most popular.
- Environmental costs (pollutant effects) in terms of quantitative criteria are the most popular.
- In terms of social criteria, "labor practices and decent work" are the most popular.
- In order to cope with uncertainty, the use of fuzzy set theory (as a method) is popular across sustainability dimensions (Chap. 3 by Tuczek and Wakolbinger (2018)).

Green supply chains To achieve and sustain a green supply chain, the role of internal environmental management and the firm's top management commitment is critical; instead of acquiring new technology, the majority of supply chains can embrace green innovations by simply improving their existing business operations; collaboration with upstream and downstream supply chain stakeholders facilitates the achievement of green supply chains; and being a green supply chain leads to improved triple bottom line (Chap. 4 by Burki (2018)).

Design modularity for supply chains To consider design modularity in the products of supply chains partners, here are some insights:

- When product return flows are considered in the decision-making process of supply chains, the reactive strategy turns out to be economically beneficial for high return rates, while the anticipatory strategy appears beneficial for relatively low return rates.
- In sustainable supply chains, the relative advantage of the reactive strategy might be weakened when a high product return rate is achieved and consumers' preference for modular products is low.
- From a manufacturer's viewpoint, consumers' willingness to pay for new or remanufactured, integral, or modular products and the chosen strategy are key determinants for the optimal modularity level of a product.
- Coordination between consumers and manufacturers would be indispensable for pursuing the circular economy through modular product designs (Chap. 5 by Nowak et al. (2018)).

Effects of model-based training To improve decision-making capability and develop insights for the management of supply chains, a complex dynamic task, simulation-based training is essential. The key results of a system dynamics model-based training are (i) experimentation with the model suggests different policy interventions to reduce the waste level, lower the cost of failures, increase the motivation level, and enhance the product quality index and supply chain performance, (ii) there is a need to design policies on the basis of system structures for long run and sustainable results, and (iii) training enhances management perceptions of the issues which seem to be simple but are not so, and policies based on mere intuition may appear as the disposable initiatives in terms of sustainable improvement (Chap. 6 by Yosuf and Azhar (2018)).

Sustainable procurement operation Achieving a sustainable procurement operation in today's supply chain is a complex, dynamic task. To achieve a sustainable operation, business firms have to (i) deal with the capabilities' trap and (ii) isolate strategy from operations by dedicating resources which are entirely focused on strategic procurement coupled with industry experts (external support) to help baseline spend and procurement processes to set the tone and deliver a roadmap for improvement (Chap. 7 by Barrad et al. (2018)).

Wind power industry supply dynamics in Brazil Based on the application of system dynamics approach, key insights pertaining to wind-power expansion are:

- Despite current wind-power industry growth in Brazil, the northeast and north regions will be facing a strong reduction in the energy supply as a result of time delays in the construction of transmission lines.
- The investment costs in wind power should decrease with expansion of transmission systems, making wind power more attractive for other regions in the near future.
- The synchrony in the supply chain may be achieved with coordinated investment policy together with reductions in delivery times and the maintenance time of wind farms (Chap. 8 by Herrera et al. 2018)).

Essential consumer goods supply dynamics in Europe Based on an innovative application of system dynamics approach, here are some interesting findings: (i) any evaluation of essential good supply should include the comparison and sensitivity analysis of all primary and secondary causes of medicines shortages, and (ii) the focus should be on the determination of system leverage points or the identification of model segments most crucial to supply reliability (Chap. 9 by Grossor (2018)).

Tourism industry supply chain dynamics Building on an actual case study, here are some interesting and innovative insights applicable across the sustainable supply chains in many domains including tourism:

- A redesigned and flexible tourism supply chain can reduce fixed cost, thus allowing a push system to be replaced with a pull system.
- Scouting, training, and developing local hosts in the tourism supply chain are of great importance for building an authentic enjoyment for both tourists and locals.
- Building networks with local NGOs improves the performance of tourism supply chain.
- Adaptation to digital marking improves communication and coordination among geographically dispersed teams of tourism supply chain (Chap. 10 by Gunarathne et al. (2018)).

Sustainable food supply chain management Building a sustainable food supply chain in regions across the world is a challenge. Based on a theoretical model and developed framework, some insights are presented: (i) management of a sustainable food supply chain is complex task, and (ii) there is a need for continuous improvement in light of the ever-changing national and international business environment along economic, natural, and social dimensions (Chap. 11 by Wongsurawat and Shrestha (2018)).

Feedstock supply risk An agent-based model is used to better understand and manage supply risk pertaining to the timber supply chain in Southern Ontario. Here some key results:

- The model-based solution shows that in general the wood supply chain in Southern Ontario is robust under a scenario of a 2-week timber supply disturbance.
- By simulating the entire supply chain of a feedstock, a decision-maker can understand areas of highest risk of supplier closure due to harvesting disturbance.
- A better understanding of feedstock risk has a potential to lower risk perception, promoting investment into the biofuel industry and lowering investment debt (Chap. 12 by Lewandowsk and Asgary (2018)).

Finale of finale Please note that we are not claiming that reading this book will make you an expert on building sustainable supply chains or decision-makers in supply chains have to become an expert on the presented perspectives and the modeling approaches such as system dynamics and agent-based modeling (great, if they are) but rather this edited volume is to motivate you to take actions in your organization to instill the virtues of sustainable supply chains. To deal with complex, dynamic sustainable supply chain issues (i.e., design and development, profitability, and growth), one can benefit from the state-of-the-art chapter-based contributions.

Overall, as shown in Fig. 13.1, building and growing a sustainable supply chain is an ongoing and iterative process and a continuum. Readers of this book can build their knowledge about the design, development, profitability, and growth of sustainable supply chains by following the similar iterative process going through various perspectives (e.g., Chap. 2 is about R&D perspective) and unique model-

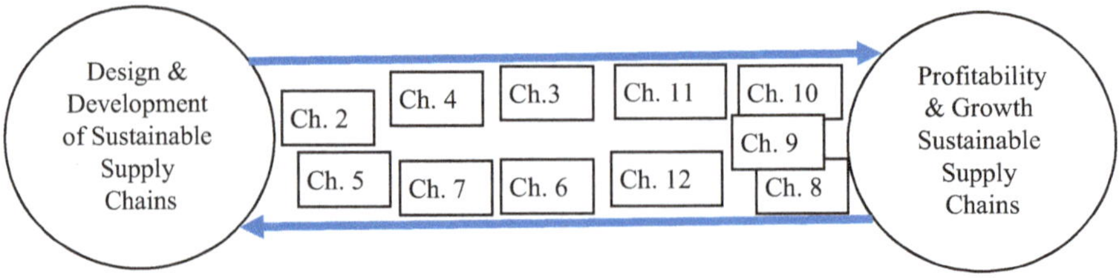

Fig. 13.1 Sustainable supply chain continuum

based applications (e.g., Chap. 8 applies a system dynamics model to understand wind-power industry supply dynamics in Brazil). It should be noted that chapters are placed on a knowledge building continuum: for instance, readers who are more interested in the design and development phase can begin with Chap. 2 and then Chaps. 3, 4, 5, 6, and 7. On the other hand, those readers who want to see how the performance of sustainable supply chains in various industries has improved by adopting model-based innovations can readily benefit from the reading of Chaps. 8, 9, 10, 11, and 12.

13.2 A Way Forward

It is worth mentioning that all the contributed chapters (Chap. 2 through Chap. 12) have presented some fruitful avenues for future research pertaining to the design, development, profitability, and growth of sustainable supply chains. For instance, in terms of perspectives:

- An R&D perspective by technologists in the logistics and supply chain industry in Hong Kong is analyzed in Chap. 2. Future research can explore to what extent these insights and results are applicable to the supply chains of (i) other domains (e.g., healthcare industry) and (ii) other geographical locations. Future research can also investigate the impact of private investments on the scale of innovations in logistics and supply and other industries.
- A comprehensive review of literature on supplier selection approaches and practices is presented in Chap. 3. Authors of this chapter, Tuczek and Wakolbinger (2018), have themselves identified, "integrating the assessment of R&D capabilities and consequently potential synergy effects between buyers and suppliers into supplier selection approaches" as a future research area.
- Chapter 4 presented a unique model regarding green innovations and green practices in supply chains. In this context, here are some testable prepositions: (i) when businesses integrate green innovations and green practices in their supply chains, they achieve encouraging triple bottom line outcomes; (ii) emerging green practices and tendencies among consumers and supply chain stakeholders are forcing businesses and governments to show a clear commitment toward sustainability; (iii) an increase in the stakeholders' commitment toward green stewardship leads to increased adoption and implementation of green innovations and green practices by the supply chains; and (iv) increased economic and ecological results strengthen firms' attitudes and commitments toward green supply chains.
- The links between anticipatory and reactive supply chain strategies, consumer preferences, as well as a company's optimal product design decision by considering reverse logistics operations are explored in Chap. 5. However, there are several assumptions used in the model that need further research: (i) How does an effective future demand impact the possible gains from the anticipator strategy model? (ii) How does an adaptive product design impact the life-cycle profits

of a modular product of the supply chains? and (ii) Do companies decide freely on the optimal level of modularity?
- The impact of the training on behavior and attitude of the employees the supply chain of the focal firm is investigated in Chap. 6. An important future research question is: Do the policies based on parametric changes in the model remain plausible when model is subjected to structure changes?
- In Chap. 7, the authors, by applying system dynamics approach, developed and applied a simulation model to better understand the dynamics of sustainable procurement function of supply chains. Unfortunately, the developed model is neither exposed to structural validity testing nor to behavioral testing prior to the application of the model. Future researchers can subject this model to structural as well as behavioral validity tests and then validate the findings of the model.

From Chap. 8 through Chap. 12, there are five unique case-based and modeling-based innovative solutions and applications aimed at improving the performance of sustainable supply chains. Here too, several research questions and testable propositions require future research to gain further insights and decision-making aid for the decision-makers in sustainable supply chains:

- A system dynamics model is used for understanding how to synchronize wind-power expansion and the corresponding transmission facilities to guarantee power supply in the Brazilian market in Chap. 8. Some interesting future research questions are: Do suppliers' R&D capabilities affect reduction in the lead time in the wind-power industry's supply chain? How does the asynchrony of the electricity market affect the prices? How do increased investments in wind-power sector impact the overall supply mix of the region?
- A unique application of system dynamics approach: the case of inexpensive, generic, injectable oncological medication shortages in Europe is presented in Chap. 9. Two important research questions areas are proposed here: (i) Any future evaluation regarding the availability of essential goods across the supply chains should include the comparison and sensitivity analysis of all primary and secondary causes of medicine shortages identified in this model and (ii) impacts of the evaluation based on the determination of system leverage points or the identification of model segments most crucial to supply reliability should be investigated.
- In an in-depth case study, in Chap. 10, the authors provide two assertions: (i) In the tourism industry, creating greater equity in the distribution of benefits from tourism requires a disruptive change to the current industry supply chain and (ii) by arranging genuine exchanges between tourists and locals, travelers can contribute to a better livelihood for less privileged people while learning to appreciate a different culture, a win-win situation for all the stakeholders in the redesigned sustainable supply chain. Future research, by using empirical research methodology, can investigate the validity of these two assertions. Also, will scouting, training, and developing local hosts in the healthcare supply chains lower the cost of operations (as was the case for the tourism industry) needs empirical testing by future researchers in this area.

- Based on several mini-cases and a comprehensive literature review, in Chap. 11, a new framework for better understanding the complexity of the food supply chains is presented. However, the key assertion advanced, "the conceptual analysis presented in the framework can serve as a basis for practitioner to comprehend the complexity in the food supply chain to develop innovative business processes," needs to be tested empirically. Do practitioners in food industry understand the dynamics of food supply chain better when they apply or use this new framework?
- An agent-based model is developed and applied to better understand the feedstock (of timber) supply risk in the case of Southern Ontario's timber supply chain. It would be interesting to test the applicability of this model to apply it to other timber supply areas in Canada and abroad. How does the model behave when it assumes the capacity of the supplier is fixed and relaxed to become dynamic? Future research can address such questions and provide useful insights and innovative solution to both policy makers and practitioners of timber industry to build and sustain their supply chains.

Furthermore, with respect to the developed and validated system dynamics and agent-based models, several research questions can be pursued: Will the agent-based model provide similar outcomes and insights as did the system dynamics-based simulation model (in the case of procurement function of the sustainable supply chain)?

Let me conclude by encouraging you, the reader of this book, to embrace any perspective and/or model-based innovative solution presented here those suites to your need (depending upon your position on the continuum of sustainable supply chain of your business).

Acknowledgment I would like to recognize and thank Anam Qudrat for her insightful comments on the draft version of this chapter.

References

Barrad, S., Valverde, R., & Gagnon, S. (2018). The application of system dynamics for a sustainable procurement operation. In H. Qudrat-Ullah (Ed.), *Innovative solutions for sustainable supply chains*. New York: Springer, USA (in press).

Burki, U. (2018). Green supply chain management, green innovations, and green practices. In H. Qudrat-Ullah (Ed.), *Innovative solutions for sustainable supply chains*. New York: Springer, USA (in press).

Tuczek, F., & Wakolbinger, T. (2018). A meta-analysis of sustainable supplier selection approaches. In H. Qudrat-Ullah (Ed.), *Innovative solutions for sustainable supply chains*. New York: Springer, USA (in press).

Gunarathne, A., Nuwan, D., Navaratne, D. G., Pakianathan, A. E., Perera, N., & Yasasi, T. (2018). Sustainable food supply chain management: An integrated framework and practical perspectives. In H. Qudrat-Ullah (Ed.), *Innovative solutions for sustainable supply chains*. New York: Springer, USA (in press).

Herrera, M., Dyner, I., & Cosenz, F. (2018). The wind-power industry's supply chain in Brazil. In H. Qudrat-Ullah (Ed.), *Innovative solutions for sustainable supply chains*. New York: Springer, USA (in press).

Ho, S. C., & Chuah, K. B. (2018). Determinants of ITF R&D Technology Commercialization in logistics and supply chain industries: R&D technologist perspective. In H. Qudrat-Ullah (Ed.), *Innovative solutions for sustainable supply chains*. New York: Springer, USA (in press).

Yosuf, I., & Azhar, T. (2018). Policy Design for Sustainable Supply Chains through trainings. In H. Qudrat-Ullah (Ed.), *Innovative solutions for sustainable supply chains*. New York: Springer, USA (in press).

Lewandowsk, M., & Ali Asgary, A. (2018). Risk assessment of the timber supply chain in Southern Ontario using agent-based simulation. In H. Qudrat-Ullah (Ed.), *Innovative solutions for sustainable supply chains*. New York: Springer, USA (in press).

Nowak, T., Toyasaki, F., & Wakolbinger, T. (2018). The road towards a circular economy: The role of modular product designs in supply chains. In H. Qudrat-Ullah (Ed.), *Innovative solutions for sustainable supply chains*. New York: Springer, USA (in press).

Grossor, S. (2018). Resilience as basis for sustainability: Shortages in production supply chains for essential consumer goods. In H. Qudrat-Ullah (Ed.), *Innovative solutions for sustainable supply chains*. New York: Springer, USA (in press).

Wongsurawat, W., & Shrestha, V. (2018). Designing a sustainable tourism supply chain-a case study from Asia. In H. Qudrat-Ullah (Ed.), *Innovative solutions for sustainable supply chains*. New York: Springer, USA (in press).

Endorsements

"Finally, the long expected book about sustainable supply chains! Its topic is of great importance. This opus conveys powerful insights for managers and professionals concerned with value chains and networks."

- **Markus Schwaninger**, Professor, Institute of Management, University of St. Gallen, Switzerland.

***

"This book provides unique perspectives and modeling-based solutions for sustainable supply chains. No matter at what stage of sustainability is your organization, this book is an essential read for you to achieve next level of sustainability."

- **BaekSeo Seong**, Professor, Operations & Management, Konkuk Business School, Konkuk University, Seoul, South Korea.

***

"This book offers a unique and an effective combination of experts' perspectives and state-of-art technologies to design innovative solutions for supply chain problems. The book includes a number of case studies that provides useful insights for decision makers to enhance and maintain the performance of supply chain processes."

- **Dr. Arif Mehmood**, Department of Transport, Abu Dhabi, United Arab Emirates.

H. Qudrat-Ullah (ed.), *Innovative Solutions for Sustainable Supply Chains*,
Understanding Complex Systems, https://doi.org/10.1007/978-3-319-94322-0

Index

H. Qudrat-Ullah (ed.), *Innovative Solutions for Sustainable Supply Chains*, Understanding Complex Systems, https://doi.org/10.1007/978-3-319-94322-0

O

P

Q

R

S

Kolibri 360 GmbH
Ferdinand-Jühlke-Straße 7
99095 Erfurt, Deutschland
produktsicherheit@kolibri360.de